D1001060

Molecular Thermodynamics

of

Fluid-Phase Equilibria

PRENTICE-HALL INTERNATIONAL SERIES
IN THE PHYSICAL AND CHEMICAL ENGINEERING SCIENCES

NEAL R. AMUNDSON, EDITOR, *University of Minnesota*

ADVISORY EDITORS

ANDREAS ACRIVOS, *Stanford University*
JOHN DAHLER, *University of Minnesota*
THOMAS J. HANRATTY, *University of Illinois*
JOHN M. PRAUSNITZ, *University of California*
L. E. SCRIVEN, *University of Minnesota*

AMUNDSON   *Mathematical Methods in Chemical Engineering*
ARIS   *Elementary Chemical Reactor Analysis*
ARIS   *Introduction to the Analysis of Chemical Reactors*
ARIS   *Vectors, Tensors, and the Basic Equations of Fluid Mechanics*
BOUDART   *Kinetics of Chemical Processes*
FREDRICKSON   *Principles and Applications of Rheology*
HAPPEL AND BRENNER   *Low Reynolds Number Hydrodynamics*
HIMMELBLAU   *Basic Principles and Calculations in Chemical Engineering, 2nd ed.*
HOLLAND   *Multicomponent Distillation*
HOLLAND   *Unsteady State Processes with Applications in Multicomponent Distillation*
KOPPEL   *Introduction to Control Theory with Applications to Process Control*
LEVICH   *Physicochemical Hydrodynamics*
PETERSEN   *Chemical Reaction Analysis*
PRAUSNITZ   *Molecular Thermodynamics of Fluid-Phase Equilibria*
PRAUSNITZ AND CHUEH   *Computer Calculations for High-Pressure Vapor-Liquid Equilibria*
PRAUSNITZ, ECKERT, ORYE, O'CONNELL   *Computer Calculations for Multicomponent Vapor-Liquid Equilibria*
WHITAKER   *Introduction to Fluid Mechanics*
WILDE   *Optimum Seeking Methods*

PRENTICE-HALL, INC.
PRENTICE-HALL INTERNATIONAL, INC., UNITED KINGDOM AND EIRE
PRENTICE-HALL OF CANADA, LTD., CANADA

Molecular Thermodynamics
of
Fluid-Phase Equilibria

J. M. Prausnitz

Department of Chemical Engineering
University of California at Berkeley

Prentice-Hall, Inc.

Englewood Cliffs, New Jersey

PRENTICE-HALL INTERNATIONAL, INC., *London*
PRENTICE-HALL OF AUSTRALIA, PTY. LTD., *Sydney*
PRENTICE-HALL OF CANADA, LTD., *Toronto*
PRENTICE-HALL OF INDIA PRIVATE LTD., *New Delhi*
PRENTICE-HALL OF JAPAN, INC., *Tokyo*

© 1969 by Prentice-Hall, Inc.
Englewood Cliffs, N. J.

All rights reserved. No part of this book may be
reproduced in any form or by any means without
permission in writing from the publisher.

Current printing (last digit):

10   9   8   7   6   5   4

13-599639-2

Library of Congress Catalog Card No. 69-16866

Printed in the United States of America

Preface

Since the generality of thermodynamics makes it independent of molecular considerations, the expression "molecular thermodynamics" requires explanation.

Classical thermodynamics presents broad relationships between macroscopic properties, but it is not concerned with quantitative prediction of these properties. Statistical thermodynamics, on the other hand, seeks to establish relationships between macroscopic properties and intermolecular forces through partition functions; it is very much concerned with quantitative prediction of bulk properties. However, useful configurational partition functions have been constructed only for nearly ideal situations and, therefore, statistical thermodynamics is at present insufficient for many practical purposes.

Molecular thermodynamics seeks to overcome some of the limitations of both classical and statistical thermodynamics. Molecular phase-equilibrium thermodynamics is concerned with application of molecular physics and chemistry to the interpretation, correlation, and prediction of the thermodynamic properties used in phase-equilibrium calculations. It is an engineering science, based on classical thermodynamics but relying on molecular physics and statistical thermodynamics to supply insight into the behavior of matter. In application, therefore, molecular thermodynamics is rarely exact; it must necessarily have an empirical flavor.

In the present work I have given primary attention to gaseous and liquid mixtures. I have been concerned with the fundamental problem of how best to calculate fugacities of components in such mixtures; the analysis should therefore be useful to engineers engaged in design of equipment for separation operations. Chapters 1, 2, and 3 deal with basic thermodynamics, and, in order to facilitate molecular interpretation of thermodynamic properties, Chapter 4 presents a brief discussion of intermolecular forces. Chapter 5 is devoted to calculation of fugacities in gaseous mixtures and Chapter 6 is concerned with excess functions of liquid mixtures. Chapter 7 serves as an introduction to the theory of liquid solutions with attention to both "physical" and "chemical" theories. Fugacities of gases dissolved in liquids are discussed in Chapter 8 and those of solids dissolved in liquids in Chapter 9. Finally, Chapter 10 considers fluid-phase equilibria at high pressures.

While it is intended mainly for chemical engineers, others interested in fluid-phase equilibria may also find the book useful. It should be of value to university seniors or first-year graduate students in chemistry or chemical engineering who have completed a standard one-year course in physical chemistry and who have had some previous experience with classical thermodynamics.

The subjects discussed follow quite naturally from my own professional activities. Phase-equilibrium thermodynamics is a vast subject, and no attempt has been made to be exhaustive. I have arbitrarily selected those topics with which I am familiar and have omitted others which I am not qualified to discuss; for example, I do not consider solutions of metals or electrolytes. In essence, I have written about those topics which interest me, which I have taught in the classroom, and which have comprised much of my research. As a result, emphasis is given to results from my own research publications, not because they are in any sense superior, but because they encompass material with which I am most closely acquainted.

In the preparation of this book I have been ably assisted by many friends and former students; I am deeply grateful to all. Helpful comments were given by J. C. Berg, R. F. Blanks, P. L. Chueh, C. A. Eckert, M. L. McGlashan, A. L. Myers, J. P. O'Connell, Otto Redlich, Henri Renon, F. B. Sprow, and H. C. Van Ness. Generous assistance towards improvement of the manuscript was given by R. W. Missen and by C. Tsonopoulos who also prepared the index. Many drafts of the manuscript were cheerfully typed by Mrs. Irene Blowers and Miss Mary Ann Williams and especially by my faithful assistant for over twelve years, Mrs. Edith Taylor, whose friendship and conscientious service deserve special thanks.

Much that is here presented is a reflection of what I have learned from my teachers of thermodynamics and phase equilibria: G. J. Su, R. K.

Toner, R. L. Von Berg, and the late R. H. Wilhelm; and from my colleagues at Berkeley: B. J. Alder, Leo Brewer, K. S. Pitzer (now at Stanford University), and especially J. H. Hildebrand, whose strong influence on my thought is evident on many pages.

I hope that I have been able to communicate to the reader some of the fascination I have experienced in working on and writing about phase-equilibrium thermodynamics. To think about and to describe natural phenomena, to work in science and engineering—all these are not only useful but they are enjoyable to do. In writing this book I have become aware that for me phase-equilibrium thermodynamics is a pleasure as well as a profession; I shall consider it a success if a similar awareness can be awakened in those students and colleagues for whom this book is intended. *Felix qui potuit rerum cognoscere causas.*

Finally, I must recognize what is all too often forgotten—that no man lives or works alone, but that he is molded by those who share his life, who make him what he truly is. Therefore I dedicate this book to Susie, who made it possible, and to Susi and Toni, who prepared the way.

J. M. PRAUSNITZ

Berkeley, California

Contents

4. Intermolecular Forces and the Theory of Corresponding States   52

5. Fugacities in Gas Mixtures   89

6. Fugacities in Liquid Mixtures: Excess Functions

7. Fugacities in Liquid Mixtures: Theories of Solutions

8. Solubilities of Gases in Liquids   351

9. Solubility of Solids in Liquids   385

10. High-Pressure Equilibria   407

Appendix I. Outline of a Proof of the Uniformity of Intensive Potentials as a Criterion of Phase Equilibrium   456

Molecular Thermodynamics

of

Fluid-Phase Equilibria

The Phase-Equilibrium Problem

1

We live in a world of mixtures—the air we breathe, the food we eat, the gasoline in our automobiles. Wherever we turn, we find that our lives are linked with materials which consist of a variety of chemical substances. Many of the things we do are concerned with the transfer of substances from one mixture to another; for example, in our lungs, we take oxygen from the air and dissolve it in our blood, while carbon dioxide leaves the blood and enters the air; in our coffee maker, water-soluble ingredients are leached from the coffee grain into the water; and when someone stains his tie with gravy he relies on some cleaning fluid to dissolve and thereby remove the greasy spot. In each of these common daily experiences, as well as in many others in physiology, home life, industry, and so on, there is a transfer of a substance from one phase to another. This occurs because when two phases are brought into contact, they tend to exchange their constituents until the composition of each phase attains a constant value; when that state is reached we say that the phases are in equilibrium. The equilibrium compositions of two phases are often very much different from one another and it is precisely this difference which enables us to separate mixtures by distillation, extraction, and other phase-contacting operations.

The final, or equilibrium, phase compositions depend on many variables, such as the temperature and pressure, and on the chemical nature and concentration of the various substances involved. Phase-equilibrium thermodynamics seeks to establish the relations between the various properties (in particular, temperature, pressure, and composition) which ultimately prevail when two or more phases reach a state of equilibrium wherein all tendency for further change has ceased.

Since so much of life is concerned with the interaction between different phases, it is evident that phase-equilibrium thermodynamics is a subject of fundamental importance in many sciences, physical as well as biological. It is of special interest in chemistry and chemical engineering since so many operations in the manufacture of chemical products consist of phase-contacting: Extraction, adsorption, distillation, leaching, and absorption are essential unit operations in chemical industry and an understanding of any one of them is based, at least in part, on the science of phase equilibrium.

1.1   Essence of the Problem

We want to relate quantitatively the variables which describe the state of equilibrium of two or more homogeneous phases which are free to interchange energy and matter. By a homogeneous phase at equilibrium we mean any region in space where the intensive properties are everywhere the same.† Intensive properties are those which are independent of the mass, size, or shape of the phase; we are concerned primarily with the intensive properties temperature, density, pressure, and composition (usually expressed in terms of mole fractions). We want to describe the state of two or more phases which are free to interact and which have reached a state of equilibrium. Then, given some of the equilibrium properties of the two phases, our task is to predict the remaining ones.

The type of problem which phase-equilibrium thermodynamics seeks to solve is illustrated schematically in Fig. 1-1. We suppose that two multicomponent phases, α and β, have reached an equilibrium state and we are given the temperature T of the two phases and the mole fractions $x_1^\alpha, x_2^\alpha, \ldots$, of phase α. Our task then, is to find the mole fractions, $x_1^\beta, x_2^\beta, \ldots$, of phase β and the pressure P of the system. Alternatively, we might know $x_1^\alpha, x_2^\alpha, \ldots$, and P and be asked to find $x_1^\beta, x_2^\beta, \ldots$, and T, or our problem might involve still other combinations of known and unknown variables. The number of intensive properties which must be specified to fix unambiguously the state of equilibrium is given by the Gibbs phase rule:

$$\begin{matrix} \text{Number of independent} \\ \text{intensive properties} \end{matrix} = \begin{matrix} \text{Number of components} \\ - \text{ Number of phases} + 2. \end{matrix} \quad (1.1\text{-}1)$$

†We are here neglecting all body forces such as those due to gravity, electric or magnetic fields, surface forces, etc.

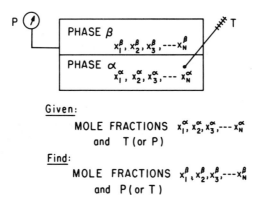

Fig. 1-1   Statement of problem.

For example, for a two-component, two-phase system, the number of independent intensive properties is two.   In such a system the intensive properties of interest usually are x_1^α, x_1^β, T, and P.†   Two of these, any two, must be specified before the remaining two can be found.

How shall we go about solving the problem illustrated in Fig. 1-1? What theoretical framework is available which might give us a basis for finding a solution?   It is when this question is raised that we turn to thermodynamics.

1.2   Application of Thermodynamics to Phase-Equilibrium Problems

One of the characteristics of modern science is abstraction.   By describing a difficult, real problem in abstract, mathematical terms, it is sometimes possible to obtain a simple solution to the problem not in terms of immediate physical reality, but in terms of mathematical quantities which are suggested by an abstract description of the real problem.   Thermodynamics provides the mathematical language in which an abstract solution of the phase-equilibrium problem is readily obtained.

Application of thermodynamics to phase equilibria in multicomponent systems is shown schematically in Fig. 1-2.   The real world and the real problem are represented by the lower horizontal line while the upper horizontal line represents the world of abstraction.   The three-step application of thermodynamics to a real problem consists of an indirect mental process; instead of attempting to solve the real problem within the world of physically realistic variables, the indirect process first projects the problem into the abstract world, then seeks a solution within that world, and

†Since $\sum_i x_i = 1$ for each phase, x_2^α and x_2^β are not additional variables in this case.

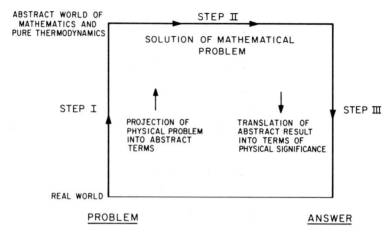

Fig. 1-2   Three-step application of thermodynamics to phase-equilibrium problems.

finally projects this solution back into physical reality. The solution of a phase-equilibrium problem using thermodynamics involves three steps: In Step I, the real problem is translated into an abstract, mathematical problem; in Step II a solution is found to the mathematical problem, and in Step III the mathematical solution is translated back into physically meaningful terms.

The essential feature of Step I is to define appropriate and useful mathematical functions in order to facilitate Step II. The profound insight of Gibbs, who in 1875 defined such .a function—the chemical potential—made it possible to achieve the goal of Step II; the mathematical solution to the phase-equilibrium problem is given by the remarkably simple result that at equilibrium, the chemical potential of each component must be the same in every phase.

The really difficult step is the last one, Step III. Thanks to Gibbs, Steps I and II present no further problems and essentially all work in this field, after Gibbs, has been concerned with Step III. From the viewpoint of a formal theoretical physicist, the phase-equilibrium problem has been solved completely by Gibbs' relation for the chemical potentials. A pure theoretician may require nothing further, but someone who is concerned with obtaining useful numerical answers to real problems must face the task of translating the abstract results of Step II into the language of physical reality.

Our concern in this book is concentrated almost exclusively on Step III. In Chap. 2 we briefly review some of the important concepts which lead to Gibbs' equation, viz., that the chemical potential of each component must be the same in all equilibrated phases. In a sense, we may call Chap. 2 historical since it reproduces, in perhaps more modern terminology, work that was completed many years ago. However, in all the remaining chapters, we address ourselves to the contemporary problem of how quantitatively to relate the chemical potential to the primary variables temperature, pressure, and composition. We should point out at once that this problem, designated by Step III, is mostly outside the realm of classical thermodynamics and much of the material in later chapters cannot be called thermodynamics in the strict sense. Classical thermodynamics by itself gives us important but also severely limited information on the relation between the abstract chemical potential and the real, experimentally accessible quantities temperature, pressure, and composition. For quantitative, numerical results, Step III must also utilize concepts from statistical thermodynamics and from molecular physics.

To solve problems of the type illustrated in Fig. 1-1, we must make the transition from what we have, viz., the abstract thermodynamic equation of equilibrium, toward what we want, viz., quantitative information about temperature, pressure and phase compositions. Thanks to Gibbs, the thermodynamic equation of equilibrium is now well known and we need not concern ourselves with it except as a place to start. In any problem concerning the equilibrium distribution of some component i between two phases α and β, we must always begin with the relation

$$\mu_i^\alpha = \mu_i^\beta \qquad (1.2\text{-}1)$$

where μ is the chemical potential. It is then that our problem begins; we must now ask how μ_i^α is related to T, P, and x_1^α, x_2^α, ..., and similarly how μ_i^β is related to T, P, and x_1^β, x_2^β, ... To establish these relations it is convenient to introduce certain auxiliary functions such as fugacity and activity. These functions do not solve the problem for us, but they facilitate our efforts to find a solution since they make the problem somewhat easier to visualize; fugacity and activity are quantities which are much closer to our physical senses than the abstract concept of a chemical potential. Suppose, for example, that phase α is a vapor and phase β is a liquid. Then, as discussed further in subsequent chapters, Eq. (1.2-1) can be rewritten

$$\varphi_i y_i P = \gamma_i x_i f_i^0 \qquad (1.2\text{-}2)$$

where, in the vapor phase, y_i is the mole fraction and φ_i is the fugacity coefficient, and, in the liquid phase, x_i is the mole fraction, γ_i is the activity coefficient, and f_i^0 is the fugacity of component i at some fixed condition known as the standard state. The details of Eq. (1.2-2) are not important just now; they are covered later on. What is important to note is the procedure whereby the highly abstract equation (1.2-1) has been transformed into the not quite so abstract equation (1.2-2). Equation

(1.2-2), unlike Eq. (1.2-1), at least has in it explicitly three of the variables of interest, x_i, y_i, and P. Equation (1.2-2) is no more and no less funda- mental than Eq. (1.2-1); one is merely a mathematical transformation of the other, and any claim which Eq. (1.2-2) may have to being more useful is only a consequence of a fortunate choice of useful auxiliary functions in bringing about the transformation.

Much of this utility comes from the concept of ideality. If we define mixtures with certain properties as ideal mixtures, we then find, as a result of our choice of auxiliary functions, that the equation of equilibrium can be simplified further; for example, for a mixture of ideal gases $\varphi_i = 1$, and for ideal liquid mixtures at low pressures, $\gamma_i = 1$ when f_i^0 is given by the saturation pressure of pure liquid i at the temperature of interest. We thus find that some of the auxiliary functions (such as φ_i and γ_i) are useful because they are numerical factors, frequently of the order of unity, which establish the connection between real mixtures and those which by judicious choice have been defined as ideal mixtures. From the viewpoint of formal thermodynamics, Eq. (1.2-2) is no better than Eq. (1.2-1); but from the viewpoint of experimental chemistry and chemical engineering, Eq. (1.2-2) is preferable because it provides a convenient frame of reference.

In the general case we cannot assume ideal behavior and we must then establish two relations, one for φ_i and one for γ_i:

$$\varphi_i = \mathcal{f}_\varphi(T, P, y_1, y_2, \ldots) \qquad (1.2\text{-}3)$$

$$\gamma_i = \mathcal{f}_\gamma(T, P, x_1, x_2, \ldots). \qquad (1.2\text{-}4)$$

In Chaps. 3 and 5 we discuss in detail what we can say about the func- tion $\mathcal{f}_\varphi$ in Eq. (1.2-3). In Chap. 4 we digress with a brief discussion of the nature of intermolecular forces, since the functional relationships of both Eqs. (1.2-3) and (1.2-4) are determined by forces which operate between molecules. In all of the chapters thereafter we are concerned, in one way or another, with the function $\mathcal{f}_\gamma$ in Eq. (1.2-4). However, before discuss- ing in detail various procedures for calculating fugacities and other useful auxiliary functions, we first give in Chap. 2 a brief survey of Steps I and II indicated in Fig. 1-2.

Classical Thermodynamics of Phase Equilibria

2

Thermodynamics as we know it today originated during the middle of the nineteenth century, and while the original thermodynamic formulas were applied to only a limited class of phenomena (such as heat engines), they have, as a result of suitable extensions, become applicable to a large number of problems in both physical and biological sciences. From its Greek root (*thermē*, heat; *dynamis*, force), one might well wonder what "thermodynamics" has to do with the *distribution* of various components between various phases. Indeed the early workers in thermodynamics were concerned only with systems of one component, and it was not until the monumental work of J. Willard Gibbs that thermodynamic methods were shown to be useful in the study of multicomponent systems. It was Gibbs who first saw the generality of thermodynamics. He was able to show that a thermodynamic treatment is possible for a wide variety of applications, including the behavior of chemical systems.

This chapter briefly reviews the essential concepts of the classical thermodynamic description of phase equilibria.† It begins with a com-

---

†More complete discussions are given in references listed at the end of this chapter.

bined statement of the first and second laws as applied to a closed, homogeneous system, and proceeds toward the laws of equilibrium for an open,
heterogeneous system. For our purposes here we exclude surface and
tensile effects, acceleration or change of position in an external field, such
as a gravitational or electromagnetic field (other than along a surface of
constant potential); for simplicity, we also rule out chemical and nuclear
reactions.† We are then left with the classical problem of phase equilibrium in which we consider internal equilibrium with respect to three
processes: (1) heat transfer between any two phases within the heterogeneous system; (2) displacement of a phase boundary; and (3) mass transfer of any component in the system across a phase boundary. The governing potentials in the first two processes are temperature and pressure,
respectively, and we assume prior knowledge of their existence; the
governing potential for the third process, however, is considered not to be
known a priori, and it is one of the prime responsibilities of classical
thermodynamics of phase equilibria to "discover" and exploit the appropriate "chemical potential."‡ A heterogeneous system which is in a
state of internal equilibrium is a system at equilibrium with respect to
each of these three processes.

The chapter continues with some discussion of the nature of the chemical potential and the need for standard states, and then introduces the
auxiliary functions fugacity and activity. The chapter concludes with a
very simple example of how the thermodynamic equations of phase
equilibrium may be applied to obtain a physically useful result.

2.1 Homogeneous Closed Systems

A homogeneous system is one with uniform properties throughout; that
is, a property such as density has the same value from point to point, in a
macroscopic sense. A phase is a homogeneous system. A closed system
is one which does not exchange matter with its surroundings, although it
may exchange energy. Thus in a closed system not undergoing chemical
reaction, the number of moles of each species is constant. This constraint
can be expressed as

$$dn_i = 0, \qquad i = 1, 2, \ldots, m \qquad (2.1\text{-}1)$$

where n_i is the number of moles of the ith species, and m is the number of
species present.

For a homogeneous, closed system, with the qualifications given
previously, taking into account interactions of the system with its sur-

---

† However, see the final two paragraphs of App. I (p. 458).
‡ This was first done by Gibbs in 1875.

roundings in the form of heat transfer and work of volumetric displacement, a combined statement of the first and second laws of thermodynamics is [cf. Denbigh[1]]:

$$dU \leq T_B dS - P_E dV,$$
(2.1-2)

where dU, dS and dV are, respectively, small changes in energy, entropy and volume of the *system* resulting from the interactions; each of these properties is a state function whose value in a prescribed state is independent of the previous history of the system. For the purpose of this statement, the surroundings are considered to be two distinct bodies: a constant-volume heat bath, also at constant, uniform temperature T_B, with which the system is in thermal contact only, and another external body, at constant, uniform pressure P_E, with which the system is in "volumetric" contact only through a movable, thermally insulated piston.

Since Eq. (2.1-2) is our starting point, it is important to have a better understanding of its significance and range of validity, even though we do not attempt to develop or justify it here. However, before proceeding, we need to discuss briefly three important concepts: equilibrium state, reversible process, and state of internal equilibrium.

By an equilibrium state we mean one from which there is no tendency to depart spontaneously, having in mind certain permissible changes or processes, viz., heat transfer, work of volume displacement and, for open systems (next section), mass transfer across a phase boundary. In an equilibrium state, values of the properties are independent of time and of the previous history of the system; further, they are stable, that is, not subject to "catastrophic" changes on slight variations of external conditions. We distinguish an equilibrium state from a steady state, insisting that in an equilibrium state there are no net fluxes of the kind under consideration (heat transfer, etc.) across a plane surface placed anywhere in the system.

In thermodynamics we are normally concerned with a finite change in the equilibrium state of a system or a variation in an equilibrium state subject to specified constraints. A change in the equilibrium state of a system is called a *process*. A reversible process is one in which the system is maintained in a state of virtual equilibrium throughout the process; a reversible process is sometimes referred to as one connecting a series of equilibrium states. This requires that the potential difference (between system and surroundings) causing the process to occur, be only infinitesimal; then the direction of the process can be reversed by an infinitesimal increase or decrease, as the case may be, in the potential for the system or the surroundings. Any natural or actual process occurs irreversibly; we can think of a reversible process as a limit to be approached but never attained. The inequality in Eq. (2.1-2) refers to a natural (irreversible) process and the equality to a reversible process.

By a single-phase system in a state of internal equilibrium we mean one that is homogeneous (uniform properties) even though it may be undergoing an irreversible process as a result of an interaction with its surroundings. In practice, such a state may be impossible to achieve, but the concept is useful for a discussion of the significance of Eq. (2.1-2), to which we now return.

If the interaction of the system with its surroundings occurs reversibly (reversible heat transfer and reversible boundary displacement), the equality sign of Eq. (2.1-2) applies; in that event, $T_B = T$, the temperature of the system, and $P_E = P$, the pressure of the system. Hence, we may write

$$dU = T\,dS - P\,dV.$$
(2.1-3)

The first term on the right is the heat absorbed by the system ($T\,dS = \delta Q_{rev}$), and the second term is the work done by the system ($\delta W_{rev} = P\,dV$). The form of this equation implies that the system is characterized by two independent variables or degrees of freedom, here represented by S and V.

If the interaction between system and surroundings occurs irreversibly, the inequality of Eq. (2.1-2) applies:

$$dU < T_B\,dS - P_E\,dV.$$
(2.1-4)

In this case $\delta W = P_E\,dV$, but $\delta Q \neq T_B\,dS$. However, if the system is somehow maintained in a state of internal equilibrium during the irreversible interaction, that is, if it has uniform properties, then it is a system characterized by two independent variables and Eq. (2.1-3) applies. Hence this equation may be applicable whether the process is externally reversible or irreversible. However, in the latter situation the terms $T\,dS$ and $P\,dV$ can no longer be identified with heat transfer and work, respectively.

To obtain the finite change in a thermodynamic property occurring in an actual process (from equilibrium state 1 to equilibrium state 2) the integration of an equation such as (2.1-3) must be done over a reversible path in order to use the properties of the *system*. This results in an equation of the form

$$\Delta U = U_2 - U_1 = \int_{S_1}^{S_2} T\,dS - \int_{V_1}^{V_2} P\,dV.$$
(2.1-5)

Since U is a state function, this result is independent of the path of integration, and it is independent of whether the system is maintained in a state of internal equilibrium or not during the actual process; it requires only that the initial and final states be equilibrium states. Hence the essence of classical (reversible) thermodynamics lies in the possibility of making such a calculation by constructing a convenient, reversible path to replace the actual or irreversible path of the process, which is usually not amenable to an exact description.

Equation (2.1-3) represents a fundamental thermodynamic relation.[3] If U is considered to be a function of S and V, and if this function U is known, then all other thermodynamic properties can be obtained by purely mathematical operations on this function. For example, $T = (\partial U/\partial S)_V$ and $P = -(\partial U/\partial V)_S$. While another pair of independent variables could be used to determine U, no other pair has this simple physical significance for the function U. We therefore call the group of variables U, S, V a fundamental grouping.

An important aspect of Eq. (2.1-2) is that it presents U as a potential function. If the variation dU is constrained to occur at constant S and V, then

$$dU_{S,V} \leq 0. \tag{2.1-6}$$

Equation (2.1-6) says that at constant S and V, U tends toward a minimum in an actual or irreversible process in a closed system and remains constant in a reversible process. Since an actual process is one tending toward an equilibrium state, an approach to equilibrium at constant entropy and volume is accompanied by a decrease in internal energy. Equation (2.1-6), then, provides a criterion for equilibrium in a closed system; we shall make use of this criterion later.

Other extensive thermodynamic potentials for closed systems and other fundamental groupings can be obtained by using different pairs of the four variables P, V, T, and S as independent variables on the right of Eq. (2.1-3). Partial Legendre transformations[4] enable us to use three other pairs and still retain the important property of a fundamental equation. For example, suppose we wish to interchange the roles of P and V in Eq. (2.1-3) so as to have P as an independent variable. We then define a new function which is the original function, U, minus the product of the two quantities to be interchanged with due regard for the sign of the term in the original equation. That is, we define

$$H = U - (-PV) = U + PV \tag{2.1-7}$$

where H is called the *enthalpy* of the system, and is a state function since it is defined in terms of state functions. Differentiation of Eq. (2.1-7) and substitution for dU in Eq. (2.1-3) gives

$$\boxed{dH = T\,dS + V\,dP,}\qquad\text{(2.1-8)}$$

and the independent variables are now S and P. The role of H as a potential for a closed system at constant S and P means that

$$dH_{S,P} \le 0.\qquad\text{(2.1-9)}$$

Similarly, to interchange T and S (but not P and V) in Eq. (2.1-3), we define the *Helmholtz energy*

$$A = U - TS,\qquad\text{(2.1-10)}$$

which results in

$$\boxed{dA = -S\,dT - P\,dV}\qquad\text{(2.1-11)}$$

and

$$dA_{T,V} \le 0;\qquad\text{(2.1-12)}$$

in this case the independent variables or constraints are T and V. Finally, in order to interchange both T and S and P and V in Eq. (2.1-3) so as to use T and P as the independent variables, we define the *Gibbs energy*

$$G = U - TS - (-PV) = H - TS,\qquad\text{(2.1-13)}$$

which yields

$$\boxed{dG = -S\,dT + V\,dP,}\qquad\text{(2.1-14)}$$

and

$$dG_{T,P} \le 0.\qquad\text{(2.1-15)}$$

A summary of the four fundamental equations and the roles of U, H, A and G as thermodynamic potentials is given in Table 2-1. Also included in the table are a set of identities resulting from the fundamental equations and the set of equations known as Maxwell's relations. These relations are obtained from the fundamental equations by the application of Euler's reciprocity theorem, which takes advantage of the fact that the order of differentiation in forming second partial derivatives is immaterial for continuous functions and their derivatives.

Table 2-1

SOME IMPORTANT THERMODYNAMIC RELATIONS
FOR A HOMOGENEOUS CLOSED SYSTEM

Fundamental Equations

$$dU = T\,dS - P\,dV \qquad\qquad dA = -S\,dT - P\,dV$$
$$dH = T\,dS + V\,dP \qquad\qquad dG = -S\,dT + V\,dP$$

Extensive Functions as Thermodynamic Potentials

$$dU_{S,V} \le 0 \qquad\qquad dA_{T,V} \le 0$$
$$dH_{S,P} \le 0 \qquad\qquad dG_{T,P} \le 0$$

Identities Resulting from the Fundamental Equations

$$\left(\frac{\partial U}{\partial S}\right)_V = T = \left(\frac{\partial H}{\partial S}\right)_P \qquad\qquad \left(\frac{\partial H}{\partial P}\right)_S = V = \left(\frac{\partial G}{\partial P}\right)_T$$

$$\left(\frac{\partial U}{\partial V}\right)_S = -P = \left(\frac{\partial A}{\partial V}\right)_T \qquad\qquad \left(\frac{\partial A}{\partial T}\right)_V = -S = \left(\frac{\partial G}{\partial T}\right)_P$$

Maxwell Relations Resulting from the Fundamental Equations

$$\left(\frac{\partial T}{\partial V}\right)_S = -\left(\frac{\partial P}{\partial S}\right)_V \qquad\qquad \left(\frac{\partial S}{\partial V}\right)_T = \left(\frac{\partial P}{\partial T}\right)_V$$

$$\left(\frac{\partial T}{\partial P}\right)_S = \left(\frac{\partial V}{\partial S}\right)_P \qquad\qquad \left(\frac{\partial S}{\partial P}\right)_T = -\left(\frac{\partial V}{\partial T}\right)_P$$

2.2   Homogeneous Open Systems

An open system can exchange matter as well as energy with its surroundings. We now consider how the laws of thermodynamics for a closed system must be modified to apply to an open system.

For a closed homogeneous system we considered U to be a function only of S and V; that is,

$$U = U(S, V). \tag{2.2-1}$$

In an open system, however, there are additional independent variables for which we can use the mole numbers of the various species present. Hence we must now consider U as the function

$$U = U(S, V, n_1, n_2, \ldots, n_m), \tag{2.2-2}$$

where m is the number of species. The total differential is then

$$dU = \left(\frac{\partial U}{\partial S}\right)_{V,n_i} dS + \left(\frac{\partial U}{\partial V}\right)_{S,n_i} dV + \sum_i \left(\frac{\partial U}{\partial n_i}\right)_{S,V,n_j} dn_i, \quad (2.2\text{-}3)$$

where subscript n_i refers to all mole numbers and n_j to all mole numbers other than the ith. Since the first two derivatives in Eq. (2.2-3) refer to a closed system, we may use the identities of Table 2-1. Furthermore, we define the function μ_i as

$$\mu_i \equiv \left(\frac{\partial U}{\partial n_i}\right)_{S,V,n_j} \quad (2.2\text{-}4)$$

We may then rewrite Eq. (2.2-3) in the form

$$\boxed{dU = T\,dS - P\,dV + \sum_i \mu_i\,dn_i\,,} \quad (2.2\text{-}5a)$$

which is the fundamental equation for an open system corresponding to Eq. (2.1-3) for a closed system. The function μ_i is an intensive quantity, and we expect it to depend on the temperature, pressure, and composition of the system. However, our prime task is to show that μ_i is a mass or chemical potential, as we might suspect from its position in Eq. (2.2-5a) as a coefficient of dn_i, just as T (the coefficient of dS) is a thermal potential and P (the coefficient of dV) is a mechanical potential. Before doing this, however, we consider other definitions of μ_i and the corresponding fundamental equations for an open system in terms of H, A, and G. Using the defining equations for H, A and G [Eqs. (2.1-7), (10), and (13)], we may substitute for dU in Eq. (2.2-5a) in each case and arrive at the following further three fundamental equations for an open system:

$$\boxed{\begin{aligned} dH &= T\,dS + V\,dP + \sum_i \mu_i\,dn_i & (2.2\text{-}5b) \\ dA &= -S\,dT - P\,dV + \sum_i \mu_i\,dn_i & (2.2\text{-}5c) \\ dG &= -S\,dT + V\,dP + \sum_i \mu_i\,dn_i & (2.2\text{-}5d) \end{aligned}}$$

From the definition of μ_i given in Eq. (2.2-4) and from Eqs. (2.2-5), it follows that

$$\mu_i = \left(\frac{\partial U}{\partial n_i}\right)_{S,V,n_j} = \left(\frac{\partial H}{\partial n_i}\right)_{S,P,n_j} = \left(\frac{\partial A}{\partial n_i}\right)_{T,V,n_j} = \left(\frac{\partial G}{\partial n_i}\right)_{T,P,n_j} \quad (2.2\text{-}6)$$

There are thus four expressions for μ_i where each is a derivative of an extensive property with respect to the amount of the component under consideration, and each involves a fundamental grouping of variables: U, S, V; H, S, P; A, T, V; and G, T, P. The quantity μ_i is the partial molar Gibbs energy, but it is *not* the partial molar internal energy, enthalpy, or Helmholtz energy. This is because the independent variables T and P, which are arbitrarily chosen in defining partial molar quantities, are also the fundamental independent variables for the Gibbs energy G.

2.3   Equilibrium in a Heterogeneous Closed System

A heterogeneous, closed system is made up of two or more phases with each phase considered as an open system within the overall closed system. We now consider the conditions under which the heterogeneous system is in a state of internal equilibrium with respect to the three processes of heat transfer, boundary displacement, and mass transfer.

We already have four criteria with different sets of constraints for equilibrium in a closed system, as given by the second set of equations in Table 2-1 with the equal sign in each case. However, these are in terms of the four extensive thermodynamic potentials U, H, A, and G. We can obtain more useful criteria in terms of the intensive quantities T, P, and μ_i. We expect that in order to have thermal and mechanical equilibrium in the system, the temperature and pressure must be uniform throughout the whole heterogeneous mass. If μ_i is the intensive potential governing mass transfer we expect that μ_i will also have a uniform value throughout the whole heterogeneous system at equilibrium with respect to this process. The proof of this was first given by Gibbs in 1875.† He used the function U as a starting point rather than H, A, or G, probably because of the symmetry in the expression for dU in Eq. (2.2-5a); each differential on the right is the differential of an extensive quantity and each coefficient is an intensive quantity. This means that the uniformity of all intensive potentials at equilibrium can be proved by consideration of just the one function U. Details of this proof are given in Appendix I. The general result for a closed, heterogeneous system consisting of π phases and m species is that at equilibrium with respect to the processes described earlier,

$$T^{(1)} = T^{(2)} = \ldots = T^{(\pi)}, \tag{2.3-1}$$

$$P^{(1)} = P^{(2)} = \ldots = P^{(\pi)}, \tag{2.3-2}$$

---

† *The Scientific Papers of J. Willard Gibbs*, Vol. I (New York: Dover Publications, Inc., 1961), p. 65.

$$
\left.\begin{aligned}
\mu_1^{(1)} &= \mu_1^{(2)} = \ldots = \mu_1^{(\pi)} \\
\mu_2^{(1)} &= \mu_2^{(2)} = \ldots = \mu_2^{(\pi)} \\
\vdots \quad &\quad \vdots \qquad\qquad \vdots \\
\mu_m^{(1)} &= \mu_m^{(2)} = \ldots = \mu_m^{(\pi)}
\end{aligned}\right\}, \tag{2.3-3}
$$

where the superscript in parentheses denotes the phase, and the subscript the species. This set of equations provides the basic criteria for phase equilibrium for our purposes. In the next two sections, we consider the number of independent variables (degrees of freedom) in systems of interest to us.

2.4   The Gibbs-Duhem Equation

We may characterize the intensive state of each phase present in a heterogeneous system at internal equilibrium by its temperature and pressure, and the chemical potential of each species present—a total of $m + 2$ variables. However, these are not all independently variable, and we now derive the important relation, known as the Gibbs-Duhem equation, which shows this.

Consider a particular phase within the heterogeneous system as an open, homogeneous system. The fundamental equation in terms of U [Eq. (2.2-5a)] is:

$$
dU = T\,dS - P\,dV + \sum_i \mu_i\,dn_i. \tag{2.4-1}
$$

We may integrate this equation from a state of zero mass ($U = S = V = n_1 = \ldots = n_m = 0$) to a state of finite mass (U, S, V, $n_1, \ldots, n_m$) at constant temperature, pressure and composition; along this path of integration all coefficients, including all μ_i, in Eq. (2.4-1) are constant, and the integration results in

$$
U = TS - PV + \sum_i \mu_i n_i. \tag{2.4-2}
$$

This equation may be regarded as expressing U as a function of T, P, composition, and the size of the system. The path of integration amounts to adding together little bits of the phase, each with the same temperature, pressure, and composition, to obtain a finite amount of phase. Since U is a state function, the result expressed by Eq. (2.4-2) is independent of the path of integration. Differentiation of this equation so as to obtain a general expression for dU comparable to that in Eq. (2.4-1) results in

$$
dU = T\,dS + S\,dT - P\,dV - V\,dP + \sum_i \mu_i\,dn_i + \sum_i n_i\,d\mu_i. \tag{2.4-3}
$$

Comparing Eqs. (2.4-1) and (2.4-3), we have

$$S\,dT - V\,dP + \sum_i n_i\,d\mu_i = 0, \qquad (2.4\text{-}4)$$

which is the Gibbs-Duhem equation. This is a fundamental equation in the thermodynamics of solutions and it is used extensively in Chap. 6. For now we note that it places a restriction on the simultaneous variation of T, P, and the μ_i for a single phase. Hence, of the $m + 2$ intensive variables that may be used to characterize a phase, only $m + 1$ are independently variable; a phase has $m + 1$ degrees of freedom.

2.5   The Phase Rule

When we consider the number of degrees of freedom in a heterogeneous system, we need to take into account the results of the last two sections. If the heterogeneous system is *not* in a state of internal equilibrium, but each phase is, the number of independent variables is $\pi(m + 1)$, since for each phase there are $m + 1$ degrees of freedom; a Gibbs-Duhem equation applies to each phase. However, if we stipulate that the system *is* in a state of internal equilibrium, then among the $\pi(m + 1)$ variables there are $(\pi - 1)(m + 2)$ equilibrium relations, given by Eqs. (2.3-1) to (2.3-3). Thus the number of degrees of freedom, F, which is the number of intensive variables used to characterize the system minus the number of relations or restrictions connecting them, is

$$F = \pi(m + 1) - (\pi - 1)(m + 2)$$
$$= m + 2 - \pi. \qquad (2.5\text{-}1)$$

In the type of system we have been considering, the number of species m is the same as the number of components [i.e., the number of independently variable species in a chemical sense], since we have ruled out chemical reaction and all special "chemical" restrictions.†

2.6   The Chemical Potential

The task of phase equilibrium thermodynamics is to describe quantitatively the distribution at equilibrium of every component among all the phases present. For example, in distillation of a mixture of toluene and

---

† See final paragraph of App. I (p. 458).

hexane we want to know how, at a certain temperature and pressure, the toluene (or hexane) is distributed between the liquid and the gaseous phases; or, in the extraction of acetic acid from an aqueous solution using benzene we want to know how the acetic acid distributes itself between the two liquid phases. The thermodynamic solution to the phase-equilibrium problem was obtained many years ago by Gibbs when he introduced the abstract concept of chemical potential. The goal of present work in phase-equilibrium thermodynamics is to relate the abstract chemical potential of a substance to physically measurable quantities such as temperature, pressure, and composition.

To establish the desired relation, we must immediately face one apparent difficulty: We cannot compute an absolute value for the chemical potential but must content ourselves with computing changes in the chemical potential which accompany any arbitrary change in the independent variables temperature, pressure, and composition. This difficulty is apparent rather than fundamental; it is really no more than an inconvenience. It arises because the relations between chemical potential and physically measurable quantities are in the form of differential equations which, upon integration, give only differences. These relations are discussed in more detail in Chap. 3, but one example is warranted. For a pure substance i, the chemical potential is related to the temperature and pressure by the differential equation

$$d\mu_i = -s_i dT + v_i dP, \tag{2.6-1}$$

where s_i is the molar entropy and v_i the molar volume. Integrating and solving for μ_i at some temperature T and pressure P,

$$\mu_i(T, P) = \mu_i(T', P') - \int_{T'}^{T} s_i dT + \int_{P'}^{P} v_i dP, \tag{2.6-2}$$

where the superscript r refers to some arbitrary reference point.

In Eq. (2.6-2) the two integrals on the right side can be evaluated from thermal and volumetric data over the temperature range T' to T and the pressure range P' to P. However, the chemical potential $\mu_i(T', P')$ is unknown. Hence the chemical potential at T and P can only be expressed relative to the value at the arbitrary reference state designated by T' and P'.

Our inability to compute an absolute value for the chemical potential complicates somewhat the use of thermodynamics in practical applications. This complication follows from a need for arbitrary reference states which are commonly called standard states. Successful application of thermodynamics to real systems frequently is based on a judicious choice of standard states as shown by examples discussed in later chapters. For the present it is only necessary to recognize why standard states arise at all and to remember that they introduce a constant into our equation. This constant need not give us concern since it must always cancel out when we compute for some substance the change of chemical potential which results from a change of any, or all, of the independent variables.

2.7   Fugacity and Activity

The chemical potential does not have an immediate equivalent in the physical world and it is therefore desirable to express the chemical potential in terms of some auxiliary function which might be more easily identified with physical reality. Such an auxiliary function is supplied by the concept of fugacity.

In attempting to simplify the abstract equation of chemical equilibrium, G. N. Lewis first considered the chemical potential for a pure, ideal gas and then generalized to all systems the result he obtained for the ideal case. From Eq. (2.6-1)

$$\left(\frac{\partial \mu_i}{\partial P}\right)_T = v_i. \tag{2.7-1}$$

Substituting the ideal-gas equation,

$$v_i = \frac{RT}{P}, \tag{2.7-2}$$

and integrating at constant temperature,

$$\mu_i - \mu_i^0 = RT \ln \frac{P}{P^0}. \tag{2.7-3}$$

Equation (2.7-3) says that for an ideal gas, the change in chemical potential, in isothermally going from pressure P^0 to pressure P, is equal to the product of RT and the logarithm of the pressure ratio P/P^0. Hence at constant temperature, the change in the abstract, thermodynamic quantity μ is a simple logarithmic function of the physically real quantity, pressure. The essential value of Eq. (2.7-3) is that it simply relates a mathematical abstraction to a common, intensive property of the real world. However, Eq. (2.7-3) is valid only for pure, ideal gases; in order to generalize it, Lewis defined a function f, called the *fugacity*,† by writing for an isothermal change for any component in any system, whether solid, liquid

†From the Latin *fuga*, meaning flight or escape.

or gas, pure or mixed, ideal or not,

$$\mu_i - \mu_i^0 = RT \ln \frac{f_i}{f_i^0}.$$

(2.7-4)

While either μ_i^0 or f_i^0 is arbitrary, both may not be chosen independently; when one is chosen, the other is fixed.

For a pure, ideal gas the fugacity is equal to the pressure, and for a component i in a mixture of ideal gases it is equal to its partial pressure $y_i P$. Since all systems, pure or mixed, approach ideal-gas behavior at very low pressures, the definition of fugacity is completed by the limit

$$\frac{f_i}{y_i P} \longrightarrow 1 \quad \text{as} \quad P \longrightarrow 0,$$

(2.7-5)

where y_i is the mole fraction of i.

Lewis called the ratio f/f^0 the *activity*, which is given the symbol **a**. The activity of a substance gives an indication of how "active" a substance is relative to its standard state since it provides a measure of the difference between the substance's chemical potential at the state of interest and that at its standard state. Since Eq. (2.7-4) was obtained for an isothermal change, the temperature of the standard state must be the same as the temperature of the state of interest. The compositions and pressures of the two states, however, need not be (and indeed usually are not) the same.

The relation between fugacity and chemical potential is of conceptual aid in performing the translation from thermodynamic to physical variables. It is difficult to visualize the chemical potential, but the concept of fugacity is less so. Fugacity is a "corrected pressure" which for a component in a mixture of ideal gases is equal to the partial pressure of that component. The ideal gas is not only a limiting case for thermodynamic convenience but corresponds to a well-developed physical model based on the kinetic theory of matter. The concept of fugacity, therefore, helps to make the transition from pure thermodynamics to the theory of intermolecular forces; if the fugacity is a "corrected pressure" then these corrections are due to nonidealities which, hopefully, can be interpreted by molecular considerations.

The fugacity provides a convenient transformation of the fundamental equation of phase equilibrium, Eq. (2.3-3). For phases α and β, respectively, Eq. (2.7-4) is

$$\mu_i^\alpha - \mu_i^{0\alpha} = RT \ln \frac{f_i^\alpha}{f_i^{0\alpha}},$$

(2.7-6)

and

$$\mu_i^\beta - \mu_i^{0\beta} = RT \ln \frac{f_i^\beta}{f_i^{0\beta}}. \qquad (2.7\text{-}7)$$

Substituting Eqs. (2.7-6) and (2.7-7) into the equilibrium relation, Eq. (2.3-3),

$$\mu_i^{0\alpha} + RT \ln \frac{f_i^\alpha}{f_i^{0\alpha}} = \mu_i^{0\beta} + RT \ln \frac{f_i^\beta}{f_i^{0\beta}}. \qquad (2.7\text{-}8)$$

We now consider two cases. First, suppose that the standard states for the two phases are the same, i.e., suppose

$$\mu_i^{0\alpha} = \mu_i^{0\beta}. \qquad (2.7\text{-}9)$$

In that case, it follows that

$$f_i^{0\alpha} = f_i^{0\beta}. \qquad (2.7\text{-}10)$$

Equations (2.7-8), (2.7-9), and (2.7-10) give a new form of the fundamental equation of phase equilibrium, viz.,

$$\boxed{f_i^\alpha = f_i^\beta.} \qquad (2.7\text{-}11)$$

Second, suppose that the standard states for the two phases are at the same temperature but not at the same pressure and composition. In that case, we make use of an exact relation between the two standard states:

$$\mu_i^{0\alpha} - \mu_i^{0\beta} = RT \ln \frac{f_i^{0\alpha}}{f_i^{0\beta}}. \qquad (2.7\text{-}12)$$

Substituting Eq. (2.7-12) into Eq. (2.7-8), we again have

$$\boxed{f_i^\alpha = f_i^\beta.} \qquad (2.7\text{-}11)$$

Equation (2.7-11) gives a very useful result. It tells us that the equilibrium condition in terms of chemical potentials can be replaced without loss of generality by an equation which says that the fugacities of any species must be the same in all phases. (The condition that the activities must be equal holds only for the special case where the standard states in all phases are the same.) Equation (2.7-11) is equivalent to Eq. (2.3-3), and from a strictly thermodynamic point of view, one is not preferable to the other. However, from the viewpoint of one who wishes to apply thermodynamics to physical problems, an equation which equates fugacities is more convenient than one which equates chemical potentials. In all subsequent discussion, therefore, we regard Eqs. (2.3-1), (2.3-2), and (2.7-11) as the three fundamental equations of phase equilibrium.

The chapters to follow treat in detail the relations between fugacity and the independent variables temperature, pressure, and composition. However, before discussing the details of these relations, it is desirable to give a preview of where we are going, to present an illustration of how the various concepts discussed in this chapter can, in at least one very simple case, lead to a relation possessing immediate physical utility.

2.8   A Simple Application: Raoult's Law

Consider the equilibrium distribution of a component in a binary system between a liquid phase and a vapor phase. We seek a simple relation describing the distribution of the components between the phases, i.e., an equation relating x, the mole fraction in the liquid phase, to y, the mole fraction in the vapor phase. We limit ourselves to a very simple system, the behavior of which can be closely approximated by the assumption of various types of ideal behavior.

For component 1, the equilibrium equation says that

$$f_1^V = f_1^L \qquad (2.8-1)$$

where superscript V refers to the vapor and superscript L to the liquid. We now have the problem of relating the fugacities to the mole fractions. To solve this problem, we make two simplifying assumptions, one for each phase:

Assumption 1. The fugacity f_1^V, at constant temperature and pressure, is proportional to the mole fraction y_1. That is, we assume

$$f_1^V = y_1 f_{\text{pure 1}}^V, \qquad (2.8-2)$$

where $f_{\text{pure 1}}^V$ is the fugacity of pure component 1 as a vapor at the temperature and pressure of the mixture.

Assumption 2. The fugacity f_1^L, at constant temperature and pressure, is proportional to the mole fraction x_1. That is, we assume

$$f_1^L = x_1 f_{\text{pure 1}}^L, \qquad (2.8-3)$$

where $f_{\text{pure 1}}^L$ is the fugacity of pure component 1 as a liquid at the temperature and pressure of the mixture.

Assumptions 1 and 2 are equivalent to saying that both vapor-phase and liquid-phase solutions are ideal solutions, and Eqs. (2.8-2) and (2.8-3) are statements of what is known as the Lewis fugacity rule. These assumptions are valid only for very limited conditions, as discussed more fully in later chapters. For mixtures of similar components, however, they are reasonable approximations based on the naïve but attractive supposition that the fugacity of a component in a given phase increases in proportion to its mole fraction in that phase.

Upon substituting Eqs. (2.8-2) and (2.8-3) into (2.8-1), the equilibrium relation now becomes

$$y_1 f^V_{\text{pure 1}} = x_1 f^L_{\text{pure 1}}. \qquad (2.8\text{-}4)$$

Equation (2.8-4) gives an ideal-solution relation using only mole fractions and pure-component fugacities. It is the basis of the original K charts [$K = y/x = f^L/f^V$] used in the petroleum industry. Equation (2.8-4) can be simplified further by introducing several more assumptions.

Assumption 3. Pure component-1 vapor at the temperature T and pressure P is an ideal gas. It follows that

$$f^V_{\text{pure 1}} = P. \qquad (2.8\text{-}5)$$

Assumption 4. The effect of pressure on the fugacity of a condensed phase is negligible at moderate pressures. Further, the vapor in equilibrium with pure liquid 1 at temperature T is an ideal gas. It follows that

$$f^L_{\text{pure 1}} = P^s_1, \qquad (2.8\text{-}6)$$

where P^s_1 is the saturation (vapor) pressure of pure liquid 1 at temperature T.

Substituting Eqs. (2.8-5) and (2.8-6) into (2.8-4) we obtain

$$y_1 P = x_1 P^s_1, \qquad (2.8\text{-}7)$$

which is the desired, simple relation known as Raoult's law.

Equation (2.8-7) is of limited utility because of the severe simplifying assumptions on which it is based. The derivation of Raoult's law has been given here only to illustrate the general procedure whereby relations in thermodynamic variables can, with the help of physical arguments, be translated into useful, physically significant, equations. In general, this procedure is considerably more complex but the essential problem is always the same: How is the fugacity of a component related to the measurable quantities temperature, pressure and composition? It is the task of molecular thermodynamics to provide useful answers to this question. All of the chapters to follow are concerned with techniques for establishing useful relations between the fugacity of a component in a phase and physicochemical properties of that phase. To establish such relations, we rely heavily on classical thermodynamics but we also utilize, as much as possible, concepts from statistical mechanics and molecular physics.

Review Problems in General Chemical Thermodynamics†

1. The volume coefficient of expansion of mercury at $0°C$ is $18 \times 10^{-5} \, (°C)^{-1}$. The coefficient of compressibility β is $5.39 \times 10^{-6} \, atm^{-1}$. If mercury were heated from $0°C$ to $1°C$ in a constant volume system, what pressure would be developed?

$$\beta \equiv \frac{-1}{v} \left(\frac{\partial v}{\partial P} \right)_T$$

2. The residual volume α is the difference between the ideal-gas volume and the actual gas volume. It is defined by the equation

$$\alpha = \frac{RT}{P} - v.$$

For a certain gas, α has been measured at $100°C$ and at different molar volumes; the results are expressed by the empirical equation $\alpha = 2 - (3/v^2)$, where v is in liters per g-mol.
    The velocity of sound c is given by the formula

$$c^2 = -g_c k v^2 \left(\frac{\partial P}{\partial v} \right)_T.$$

Calculate the velocity of sound for this gas at $100°C$ when its molar volume is 2.3 liters, using $k = 1.4$. The molecular weight is 100.

3. Show that when the van der Waals equation of state is written in the virial form,

$$\frac{Pv}{RT} = 1 + \frac{B}{v} + \frac{C}{v^2} + \cdots,$$

the second virial coefficient is given by

$$B = b - \frac{a}{RT}.$$

† A table of gas constants is given at the end of this chapter.

4. The second virial coefficient B of a certain gas is given by

$$B = a - \frac{b}{T^2}$$

where a and b are constants.

Compute the change in internal energy for this gas in going, at temperature τ, from very low pressure to a pressure π. Use the equation

$$z = \frac{Pv}{RT} = 1 + \frac{BP}{RT} \ .$$

5. Consider the equation of state

$$\left(P + \frac{n}{v^2 T^{1/2}}\right)(v - m) = RT$$

where n and m are constants for any gas. Assume that carbon dioxide follows this equation. Calculate the compressibility factor of carbon dioxide at 100°C and at a volume of 6.948 cubic decimeters per kilogram. Use the following data in your calculation:

Critical pressure: 72.9 atm
Critical temperature: 304.2° K

Compare your answer with the experimental value.

6. Consider an aqueous mixture of sugar at 25°C and 1 atm pressure. The activity coefficient of water is found to obey a relation of the form

$$\ln \gamma_w = A(1 - x_w)^2$$

where γ_w is normalized such that $\gamma_w \to 1$ as $x_w \to 1$ and where A is an empirical constant dependent only on temperature.

Find an expression for γ_s, the activity coefficient of sugar normalized such that $\gamma_s \to 1$ as $x_w \to 1$ (or as $x_s \to 0$). The mole fractions x_w and x_s refer to water and sugar, respectively.

7. Consider a binary liquid solution of components 1 and 2. At constant temperature (and low pressure) component 1 follows Henry's law for the mole fraction range $0 \le x_1 \le a$. Show that component 2 must follow Raoult's law for the mole fraction range $(1 - a) \le x_2 \le 1$.

8. Using only data given in the steam tables, compute the fugacity of steam at 600° F and 1000 psia.

9. The inversion temperature is the temperature where the Joule-Thomson coefficient changes sign and the Boyle temperature is the temperature where the second virial coefficient changes sign. Show that for a van der Waals gas the low-pressure inversion temperature is exactly twice the Boyle temperature.

10. A gas, designated by subscript 1, is to be dissolved in a nonvolatile liquid. At a certain pressure and temperature the solubility of the gas in the liquid is x_1 (where x is mole fraction). Assume that Henry's law holds. Show that the change in solubility with temperature is given by

$$\frac{d \ln x_1}{d \, 1/T} = \frac{-\Delta \bar{h}_1}{R}$$

where

$$\Delta \bar{h}_1 = [\bar{h}_{1 \text{ (in liquid solution)}} - h_{1 \text{ (pure gas)}}]$$

at the same pressure and temperature.

On the basis of physical reasoning alone would you expect $\Delta \bar{h}_1$ to be positive or negative?

REFERENCES

1. Denbigh, K. G., *The Principles of Chemical Equilibrium* (2nd ed.). London: Cambridge University Press, 1966. See Chapters 1 and 2.

2. Zemansky, M. W., *Heat and Thermodynamics* (5th ed.). New York: McGraw-Hill Book Company, 1968. See Chapters 16 and 18.

3. *The Scientific Papers of J. Willard Gibbs*, Vol. I. New York: Dover Publications, Inc., 1961. See pp. 55–100.

4. Callen, H. B., *Thermodynamics*. New York: John Wiley & Sons, Inc., 1960. See pp. 90–100.

5. Guggenheim, E. A., *Thermodynamics* (5th ed.). Amsterdam: North Holland Publishing Co., 1967. See Chapter 1.

Table of Gas Constants

Gas Constant R in Various Units	
Atm, cc, g-mol, °K	82.06
Atm, liters, g-mol, °K	0.08206
Chu, lb-mol, °K	1.987
Btu, lb-mol, °R	1.987
Cal, g-mol, °K	1.987
Psia, cu-ft, lb-mol, °R	10.73
Psfa, cu-ft, lb-mol, °R	1544.0
Atm, cu-ft, lb-mol, °R	0.730
Kw-hr, lb-mol, °K	0.001049
Hp-hr, lb-mol, °R	0.000780
Atm, cu-ft, lb-mol, °K	1.3145
Mm Hg, liters, g-mol, °K	62.37
In. Hg, cu-ft, lb-mol, °R	21.85
Ergs, g-mol, °K	8.314×10^7

Boltzmann Constant k	
Ergs, molecule, °K	1.38044×10^{-16}
Electron volts, molecule, °K	8.6167×10^{-5}

Thermodynamic Properties From Volumetric Data

3

For any substance, regardless of whether it is pure or a mixture, all thermodynamic properties of interest in phase equilibria can be calculated from thermal and volumetric measurements. For a given phase (solid, liquid, or gas) thermal measurements (heat capacities) give information on how thermodynamic properties vary with temperature, whereas volumetric measurements give information on how thermodynamic properties vary with pressure or density. Whenever there is a change of phase (e.g., fusion or vaporization) additional thermal and volumetric measurements are required to characterize that change.

Frequently it is useful to express certain thermodynamic functions of a substance relative to those which the same substance has as an ideal gas at the same temperature and composition and at some specified pressure or density. These relative thermodynamic functions are sometimes called configurational properties. The fugacity is such a relative function because its numerical value is always relative to that of an ideal gas at unit fugacity; in other words, the standard state fugacity f_i^0 in Eq. (2.7-4) is arbitrarily set equal to some fixed value, usually taken as 1 atm.

As indicated in the previous chapter, the thermodynamic function of primary interest is the fugacity which is directly related to the chemical potential; however, the chemical potential is directly related to the Gibbs energy which, by definition, is found from the enthalpy and entropy. Therefore a proper discussion of calculation of fugacities from volumetric properties must begin with the question of how enthalpy and entropy, at constant temperature and composition, are related to pressure. On the other hand, as indicated in Chap. 2, the chemical potential may also be expressed in terms of the Helmholtz energy in which case the first question must be how entropy and energy, at constant temperature and composition, are related to volume. The answers to these questions may readily be found from Maxwell's relations, and we can then obtain exact equations for the thermodynamic functions U, H, S, A, and G; from these we can easily derive the chemical potential and finally, the fugacity.

If we consider a homogeneous mixture at some fixed composition, we must specify two additional variables. In practical phase-equilibrium problems the common additional variables are temperature and pressure, and in Sec. 3.1 we give equations for the thermodynamic properties with T and P as independent variables. However, volumetric data are most commonly expressed by an equation of state which uses temperature and volume as the independent variables, and therefore it is a matter of practical importance to have available equations for the thermodynamic properties in terms of T and V; these are given in Sec. 3.4. The equations in Sec. 3.1 and 3.4 contain no simplifying assumptions;† they are exact and are not restricted to the gas phase but, in principle, apply equally to all phases.

In Sec. 3.3 we discuss the fugacity of a pure liquid, and in Secs. 3.2 and 3.5 we give examples based on the van der Waals equation. Finally, in Sec. 3.6 we consider briefly how the exact equations for the fugacity may, in principle, be used to solve phase-equilibrium problems subject only to the (as yet unattained) condition that we have available a reliable equation of state, valid for pure substances and their mixtures over a large density range.

3.1   Thermodynamic Properties with
Independent Variables P and T

At constant temperature and composition, we can use one of Maxwell's relations to give the effect of pressure on enthalpy and entropy:

$$dH = \left[V - T \left(\frac{\partial V}{\partial T} \right)_{P, n_T} \right] dP, \qquad (3.1\text{-}1)$$

---

†The equations in Secs. 3.1 and 3.4 do, however, assume that surface effects and all body forces due to gravitational, electric or magnetic fields, etc., can be neglected.

$$dS = -\left(\frac{\partial V}{\partial T}\right)_{P,n_T} dP. \tag{3.1-2}$$

These two relations form the basis of the derivation for the desired equations. We will not present the derivation here; it requires only straightforward integrations and has been clearly given in several, readily available publications by Beattie.[1,2,3] First, expressions for the enthalpy and entropy are found. The other properties are then calculated from the definitions of enthalpy, Helmholtz energy, and Gibbs energy:

$$U = H - PV \tag{3.1-3}$$

$$A = H - PV - TS \tag{3.1-4}$$

$$G = H - TS \tag{3.1-5}$$

$$\mu_i = \left(\frac{\partial G}{\partial n_i}\right)_{T,P,n_j} \tag{3.1-6}$$

$$RT \ln \frac{f_i}{f_i^0} = \mu_i - \mu_i^0. \tag{3.1-7}$$

The results are given in Eqs. (3.1-8) to (3.1-14). It is understood that all integrations are performed at constant temperature and constant composition. The symbols have the following meanings:

h_i^0 = molar enthalpy of pure i as an ideal gas at temperature T.
s_i^0 = molar entropy of pure i as an ideal gas at temperature T and 1 atm.
μ_i^0 = $h_i^0 - Ts_i^0$, and f_i^0 = 1 atm.
n_i = number of moles of i.
n_T = total number of moles.
y_i = mole fraction of $i = n_i/n_T$.

All extensive properties denoted by capital letters (V, U, H, S, A, and G) represent the total property for n_T moles and therefore are *not* on a molar basis. Extensive properties on a molar basis are denoted by lower case letters (v, u, h, s, a, and g). In Eqs. (3.1-10) to (3.1-13), the pressure P is in atmospheres.

$$U = \int_0^P \left[V - T\left(\frac{\partial V}{\partial T}\right)_{P,n_T}\right] dP - PV + \sum_i n_i h_i^0 \tag{3.1-8}$$

$$H = \int_0^P \left[V - T\left(\frac{\partial V}{\partial T}\right)_{P,n_T}\right] dP + \sum_i n_i h_i^0 \tag{3.1-9}$$

$$S = \int_0^P \left[\frac{n_T R}{P} - \left(\frac{\partial V}{\partial T}\right)_{P,n_T}\right] dP - R\sum_i n_i \ln y_i P + \sum_i n_i s_i^0 \tag{3.1-10}$$

$$A = \int_0^P \left[V - \frac{n_T RT}{P} \right] dP + RT \sum_i n_i \ln y_i P - PV + \sum_i n_i (h_i^0 - Ts_i^0)$$

(3.1-11)

$$G = \int_0^P \left[V - \frac{n_T RT}{P} \right] dP + RT \sum_i n_i \ln y_i P + \sum_i n_i (h_i^0 - Ts_i^0)$$

(3.1-12)

$$\mu_i = \int_0^P \left[\bar{v}_i - \frac{RT}{P} \right] dP + RT \ln y_i P + h_i^0 - Ts_i^0 \qquad (3.1\text{-}13)$$

and finally

$$\boxed{ RT \ln \frac{f_i}{y_i P} = \int_0^P \left[\bar{v}_i - \frac{RT}{P} \right] dP, } \qquad (3.1\text{-}14)$$

where $\bar{v}_i \equiv \left(\dfrac{\partial V}{\partial n_i} \right)_{T, P, n_j}$ = partial molar volume of i.

The dimensionless ratio $f_i / y_i P$ is called the fugacity coefficient and is denoted by φ_i. For a mixture of ideal gases, $\varphi_i = 1$, as shown later.

Equations (3.1-8) to (3.1-14) enable us to compute all the desired thermodynamic properties for any substance relative to the ideal-gas state at 1 atm and at the same temperature and composition, provided we have information on volumetric behavior in the form

$$V = \mathcal{f}(T, P, n_1, n_2, \ldots). \qquad (3.1\text{-}15)$$

In order to evaluate the integrals in Eqs. (3.1-8) to (3.1-14) the volumetric information required must be available not just for the pressure P, where the thermodynamic properties are desired, but for the entire pressure range 0 to P.

In Eqs. (3.1-8) and (3.1-11) the quantity V appearing in the PV product is the total volume at the system pressure P and at the temperature and composition used throughout. This volume V is found from the equation of state, Eq. (3.1-15).

For a pure component, $\bar{v}_i = v_i$, and Eq. (3.1-14) simplifies to

$$RT \ln \left(\frac{f}{P} \right)_{\text{pure } i} = \int_0^P \left[v_i - \frac{RT}{P} \right] dP, \qquad (3.1\text{-}16)$$

where v_i is the molar volume of pure i. Equation (3.1-16) is frequently expressed in the equivalent form

$$\ln \left(\frac{f}{P}\right)_{\text{pure } i} = \int_0^P \frac{(z_i - 1)}{P}\, dP, \qquad (3.1\text{-}17)$$

where z, the compressibility factor, is defined by

$$z \equiv \frac{Pv}{RT}. \qquad (3.1\text{-}18)$$

The fugacity of any component i in a mixture is given by Eq. (3.1-14), which is not only general and exact but also remarkably simple.   One might well wonder then, why there are any problems at all in calculating fugacities and, subsequently, computing phase-equilibrium relationships. The problem, it turns out, is not with Eq. (3.1-14) but rather with Eq. (3.1-15), where we have written the vague symbol $\mathcal{f}$, meaning "some function." Herein lies the difficulty: What is $\mathcal{f}$? The function $\mathcal{f}$ need not be an analytical function; sometimes we have available tabulated volumetric data which may then be differentiated and integrated numerically to yield the desired thermodynamic functions.   But this is rarely the case, especially for mixtures. Usually one must estimate volumetric behavior from limited experimental data.   There is, unfortunately, no generally valid equation of state, applicable to a large number of pure substances and to their mixtures over a wide range of conditions including the condensed state. There are some good equations of state useful for only a limited class of substances and for limited conditions; however, these equations are almost always pressure-explicit rather than volume-explicit. As a result it is necessary to express the derived thermodynamic functions in terms of the less convenient independent variables V and T as shown in Sec. 3.4.   Before concluding this section, however, let us briefly discuss some of the features of Eq. (3.1-14).

First, we consider the fugacity of a component i in a mixture of ideal gases. In that case the equation of state is:

$$V = \frac{(n_1 + n_2 + \ldots)\, RT}{P}, \qquad (3.1\text{-}19)$$

and the partial molar volume of i is

$$\bar{v}_i \equiv \left(\frac{\partial V}{\partial n_i}\right)_{T, P, n_j} = \frac{RT}{P}. \qquad (3.1\text{-}20)$$

Substituting in Eq. (3.1-14) gives:

$$f_i = y_i P. \qquad (3.1\text{-}21)$$

For a mixture of ideal gases, then, the fugacity of i is equal to its partial pressure, as expected.

Next, let us assume that the gas mixture follows Amagat's law, at all pressures up to the pressure of interest. Amagat's law states that at fixed temperature and pressure, the volume of the mixture is a linear function of the mole numbers

$$V = \sum_i n_i v_i, \qquad (3.1\text{-}22)$$

where v_i is the molar volume of pure i at the same temperature and pressure.

Another way to state Amagat's law is to say that at constant temperature and pressure, the components mix isometrically, i.e., with no change in total volume. If there is no volume change, then the partial molar volume of each component must be equal to its molar volume in the pure state. It is this equality which is asserted by Amagat's law. Differentiating Eq. (3.1-22) we have

$$\bar{v}_i \equiv \left(\frac{\partial V}{\partial n_i}\right)_{T,P,n_j} = v_i. \qquad (3.1\text{-}23)$$

Substitution in Eq. (3.1-14) yields:

$$RT \ln \frac{f_i}{y_i P} = \int_0^P \left(v_i - \frac{RT}{P}\right) dP. \qquad (3.1\text{-}24)$$

Upon comparing Eq. (3.1-24) with Eq. (3.1-16), we obtain

$$\boxed{f_i = y_i f_{\text{pure } i},} \qquad (3.1\text{-}25)$$

which is commonly called the Lewis fugacity rule. In Eq. (3.1-25) the fugacity of pure i is evaluated at the temperature and pressure of the mixture.

The Lewis fugacity rule is a particularly simple equation and is therefore widely used for evaluating fugacities of components in gas mixtures. However, it is not a reliable rule because it is based on the severe simplification introduced by Amagat's law. The Lewis fugacity rule is discussed further in Chap. 5; for present purposes it is sufficient to understand clearly how Eq. (3.1-25) was obtained. The derivation assumes additivity of the volumes of all the components in the mixture at constant temperature and pressure; at high pressures, this is frequently a very good assumption because at liquid-like densities fluids tend to mix with little or no change in volume. For example, volumetric data for the nitrogen-butane system at 340°F, shown in Fig. 3-1, indicate that at 10,000 psia the molar volume of the mixture is nearly a straight-line function of the mole fraction.[4] At first glance, therefore, one might be tempted to conclude that for this system, at 10,000 psia and 340°F, the Lewis fugacity rule should give an excellent approximation for the fugacities of the components in the mixture. A second look, however, shows that this conclusion is not justified because the Lewis fugacity rule assumes additivity of volumes not only at the pressure *P* of interest, but for the entire pressure range 0 to *P*. Figure 3-1 shows that at pressures lower than about 5000 psia, the volumetric behavior deviates markedly from additivity. As indicated by Eq. (3.1-14) the partial molar volume $\bar{v}_i$ is part of an integral and, as a result, whatever assumption one wishes to make about $\bar{v}_i$ must hold not only at the upper limit but for the entire range of integration.

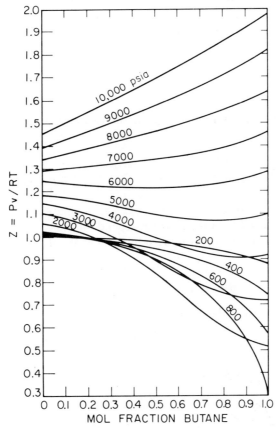

Fig. 3-1 Compressibility factors for nitrogen/butane mixtures at 340°F. Data of Evans and Watson.

3.2   Fugacity of a Component in a
Mixture at Moderate Pressures

In the previous section we first calculated the fugacity of a component in a mixture of ideal gases and then in an ideal mixture of real gases, i.e., one which obeys Amagat's law. To illustrate the use of Eq. (3.1-14) with a more realistic example, we compute now the fugacity of a component in a binary mixture at moderate pressures. In this illustrative calculation we use, for simplicity, a form of the van der Waals equation which is valid only up to moderate pressures:

$$Pv = RT + \left(b - \frac{a}{RT}\right)P + \ldots \quad \text{terms in } P^2, P^3, \text{etc.,} \quad (3.2\text{-}1)$$

where a and b are the van der Waals constants for the mixture. In order to calculate the fugacity with Eq. (3.1-14), we must first find an expression for the partial molar volume; for this purpose, Eq. (3.2-1) is rewritten on a total (rather than molar) basis by substituting $V = n_T v$:

$$V = \frac{n_T RT}{P} + n_T b - \frac{n_T a}{RT}. \quad (3.2\text{-}2)$$

We let subscripts 1 and 2 stand for the two components. Differentiating Eq. (3.2-2) with respect to n_1,

$$\bar{v}_1 = \left(\frac{\partial V}{\partial n_1}\right)_{T,P,n_2} = \frac{RT}{P} + \frac{\partial(n_T b)}{\partial n_1} - \frac{1}{RT}\frac{\partial(n_T a)}{\partial n_1}. \quad (3.2\text{-}3)$$

We must now specify a mixing rule, i.e., a relation which states how the constants a and b for the mixture depend on the composition. We use the mixing rules originally proposed by van der Waals:

$$a = y_1^2 a_1 + 2y_1 y_2 \sqrt{a_1 a_2} + y_2^2 a_2, \quad (3.2\text{-}4)$$
$$b = y_1 b_1 + y_2 b_2. \quad (3.2\text{-}5)$$

In order to utilize these mixing rules in Eq. (3.2-3), we rewrite them

$$n_T a = \frac{n_1^2 a_1 + 2n_1 n_2 \sqrt{a_1 a_2} + n_2^2 a_2}{n_T}, \quad (3.2\text{-}6)$$
$$n_T b = n_1 b_1 + n_2 b_2. \quad (3.2\text{-}7)$$

The partial molar volume for component 1 is:

$$\bar{v}_1 = \left(\frac{\partial V}{\partial n_1}\right)_{T,P,n_2} = \frac{RT}{P} + b_1 - \frac{1}{RT}\left[\frac{n_T(2n_1 a_1 + 2n_2 \sqrt{a_1 a_2}) - n_1^2 a}{n_T^2}\right].$$
$$(3.2\text{-}8)$$

In performing the differentiation it is important to remember that n_2 is held constant and that therefore n_T cannot also be constant.

Algebraic rearrangement and subsequent substitution into Eq. (3.1-14) gives the desired result:

$$\varphi_1 = \frac{f_1}{y_1 P} = \exp\left\{\left[b_1 - \frac{a_1}{RT}\right]\frac{P}{RT}\right\} \cdot \exp\left\{\frac{[a_1^{1/2} - a_2^{1/2}]^2 \, y_2^2 \, P}{(RT)^2}\right\}. \qquad (3.2\text{-}9)$$

Equation (3.2-9) contains two exponential factors to correct for non-ideality. The first correction is independent of component 2 but the second is not, since it contains a_2 and y_2. We can therefore rewrite Eq. (3.2-9) by utilizing the boundary condition

$$\text{as } y_2 \rightarrow 0, \quad f_1 \rightarrow f_{\text{pure 1}} = P \exp\left\{\left[b_1 - \frac{a_1}{RT}\right]\frac{P}{RT}\right\}. \qquad (3.2\text{-}10)$$

Upon substitution Eq. (3.2-9) becomes

$$f_1 = y_1 f_{\text{pure 1}} \exp \frac{[a_1^{1/2} - a_2^{1/2}]^2 \, y_2^2 \, P}{(RT)^2}. \qquad (3.2\text{-}11)$$

When written in this form, we see that the exponential in Eq. (3.2-11) is a correction to the Lewis fugacity rule.

Figure 3-2 presents fugacity coefficients for several hydrocarbons in binary mixtures with nitrogen. In these calculations Eq. (3.2-9) was used with $y_1 = 0.10$ and $T = 70°C$; in each case component 2 is nitrogen. For

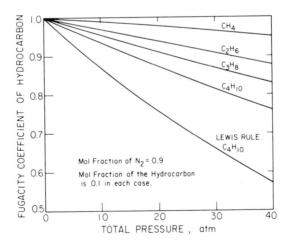

Fig. 3-2   Fugacity coefficients of light hydrocarbons in binary mixtures with nitrogen at 70°C. Calculations based on simplified form of van der Waals' equation.

comparison we also show the fugacity coefficient of butane according to the Lewis fugacity rule; in that calculation the second exponential in Eq. (3.2-9) was neglected. From Eq. (3.2-11) we see that the Lewis rule is poor for butane for two reasons: First, the mole fraction of butane is small (hence y_2^2 is near unity), and second, the difference in intermolecular forces between butane and nitrogen (as measured by $|a_1^{1/2} - a_2^{1/2}|$) is large. If the gas in excess were hydrogen or helium instead of nitrogen, the deviations from the Lewis rule for butane would be still larger.

3.3 Fugacity of a Pure Liquid or Solid

The derivation of Eq. (3.1-16) is general and not limited to the vapor phase. It may be used to calculate the fugacity of a pure liquid or that of a pure solid. Such fugacities are of importance in phase equilibrium thermodynamics because we frequently use a pure condensed phase as the standard state for activity coefficients.

To calculate the fugacity of a pure solid or liquid at given temperature T and pressure P, we separate the integral in Eq. (3.1-16) into two parts: The first part gives the fugacity of the saturated vapor at T and P^s (the saturation pressure), and the second part gives the correction due to the compression of the condensed phase to the pressure P. At the saturation pressure P^s the fugacity of saturated vapor is equal to the fugacity of the saturated liquid (or solid) because the saturated phases are in equilibrium. Let superscript s refer to saturation and superscript c to the condensed phase. Equation (3.1-16) for a pure component is now rewritten

$$RT \ln \frac{f_i^c}{P} = \int_0^{P_i^s} \left(v_i - \frac{RT}{P} \right) dP + \int_{P_i^s}^{P} \left(v_i^c - \frac{RT}{P} \right) dP. \qquad (3.3\text{-}1)$$

The first term on the right-hand side gives the fugacity of the saturated vapor which is the same as that of the saturated condensed phase. Equation (3.3-1) becomes:

$$RT \ln \frac{f_i^c}{P} = RT \ln \frac{f_i^s}{P_i^s} + \int_{P_i^s}^{P} v_i^c \, dP - RT \ln \frac{P}{P_i^s}, \qquad (3.3\text{-}2)$$

which can be rearranged to yield

$$f_i^c = P_i^s \varphi_i^s \exp \int_{P_i^s}^{P} \frac{v_i^c \, dP}{RT}, \qquad (3.3\text{-}3)$$

where $\varphi_i^s = (f/P)_i^s$.

Equation (3.3-3) gives the important result that the fugacity of a pure condensed component i at T and P is, to a first approximation, equal to P_i^s, the saturation (vapor) pressure at T. Two corrections must be applied. First, the fugacity coefficient φ_i^s corrects for deviations of the saturated vapor from ideal-gas behavior. Second, the exponential correction (often called the Poynting correction) accounts for the compression of the liquid or solid to a pressure P greater than P_i^s. In general, the volume of a liquid or a solid is a function of both temperature and pressure, but at conditions remote from critical, a condensed phase may often be regarded as incompressible and in that case the Poynting correction takes the simple form

$$\exp\left[\frac{v_i^c(P - P_i^s)}{RT}\right].$$

The two corrections just mentioned are often, but not always, small and sometimes they are completely negligible. If the temperature T is such that the saturation pressure P^s is low (say below 1 atm) then φ_i^s is very close to unity.† Figure 3-3 gives fugacity coefficients for four liquids at saturation; these coefficients were calculated with Eq. (3.3-1) using vapor-phase volumetric data. Since the liquids are at saturation conditions, no Poynting correction is required. We see that when plotted against reduced temperature the results for the four liquids are almost (but not quite) superimposable; further, we note that φ_i^s differs considerably from unity as the critical temperature is approached. The correction φ_i^s always tends to decrease the fugacity f_i^c because for all pure, saturated substances $\varphi_i^s < 1$.

The Poynting correction is an exponential function of the pressure; it is small at low pressure but may become large at extremely high pressures or at low temperatures. To illustrate, some numerical values of the Poynting correction are given in Table 3-1 for a typical, incompressible component with $v_i^c = 100$ cc/g-mol and $T = 300°K$.

Figure 3-4 shows the fugacity of compressed liquid water as a function of pressure for three temperatures; the fugacities were calculated from the thermodynamic properties reported by Keenan and Keyes.[5] The lowest temperature is much less, whereas the highest temperature is just slightly less than the critical temperature, 705°F. The saturation pressures are also indicated in Fig. 3-4, and we see that, as indicated by Eq. (3.3-3), the fugacity of compressed liquid water is more nearly equal to the saturation pressure than to the total pressure. At the highest temperature, liquid water is no longer incompressible and its compressibility cannot be neglected in the Poynting correction.

---

†There are, however, a few exceptions. Substances which have a strong tendency to polymerize (e.g., acetic acid or hydrogen fluoride) may show significant deviations from ideal gas behavior even at pressures in the vicinity of 1 atm. See Sec. 5.9.

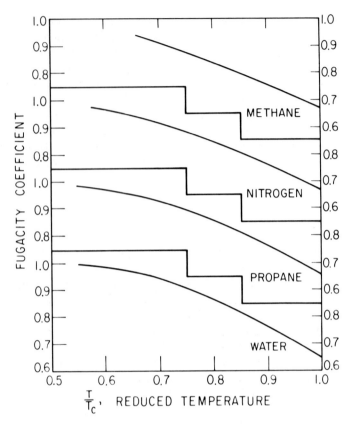

Fig. 3-3 Fugacity coefficients from vapor-phase volumetric data for four saturated liquids.

Table 3-1

The Poynting Correction

EFFECT OF PRESSURE ON FUGACITY OF A
PURE, CONDENSED AND INCOMPRESSIBLE
SUBSTANCE WHOSE MOLAR VOLUME IS
$100 \text{ cc/g-mol} (T = 300° \text{K})$

Pressure in Excess of Saturation Pressure (atm)	Poynting Correction
1	1.00405
10	1.0405
100	1.499
1000	57.0

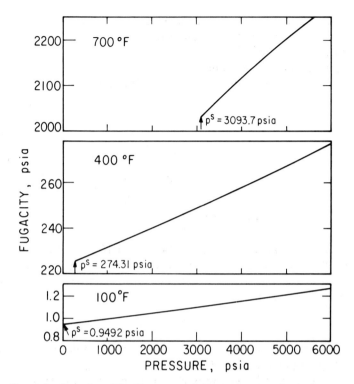

Fig. 3-4   Fugacity of liquid water at three temperatures from saturation pressure to 6000 psia.   The critical temperature of
water is 705°F.

3.4   Thermodynamic Properties with Independent Variables V and T

In Sec. 3.1 we gave expressions for the thermodynamic properties in terms of the independent variables P and T. Since volumetric properties of fluids are usually (and more simply) expressed by equations of state which are pressure-explicit, it is more convenient to calculate thermodynamic properties in terms of the independent variables V and T.

At constant temperature and composition, we can use one of Maxwell's relations to give the effect of volume on energy and entropy:

$$dU = \left[T\left(\frac{\partial P}{\partial T}\right)_{V,n_T} - P \right] dV \tag{3.4-1}$$

$$dS = \left(\frac{\partial P}{\partial T}\right)_{V,n_T} dV . \tag{3.4-2}$$

These two equations form the basis of the derivation for the desired equations. Again, we will not present the derivation but will refer to the publications of Beattie.[1,2,3]

First, expressions are found for the energy and entropy. The other properties are then calculated from their definitions:

$$H = U + PV \tag{3.4-3}$$

$$A = U - TS \tag{3.4-4}$$

$$G = U + PV - TS \tag{3.4-5}$$

$$\mu_i = \left(\frac{\partial A}{\partial n_i}\right)_{T,V,n_j} \tag{3.4-6}$$

$$RT \ln f_i / f_i^0 = \mu_i - \mu_i^0 . \tag{3.4-7}$$

The results are given in Eqs. (3.4-9) to (3.4-16). As in Sec. 3.1, it is understood that all integrations are performed at constant temperature and composition. The symbols are the same as those defined after Eq. (3.1-7) with one addition:

$$u_i^0 = h_i^0 - RT. \tag{3.4-8}$$

From the definition of h_i^0 it follows that u_i^0 is the molar energy of pure i as an ideal gas at temperature T.

The thermodynamic functions are given by the following equations. In Eqs. (3.4-11) to (3.4-14) the units of $V/n_i RT$ are atmospheres^{-1}. Since the independent variable in Eqs. (3.4-9) to (3.4-11) is the volume, no latent heat terms need to be added to these equations when they are applied to a condensed phase.

$$U = \int_V^\infty \left[P - T\left(\frac{\partial P}{\partial T}\right)_{V,n_T}\right]dV + \sum_i n_i u_i^0 \tag{3.4-9}$$

$$H = \int_V^\infty \left[P - T\left(\frac{\partial P}{\partial T}\right)_{V,n_T}\right]dV + PV + \sum_i n_i u_i^0 \tag{3.4-10}$$

$$S = \int_V^\infty \left[\frac{n_T R}{V} - \left(\frac{\partial P}{\partial T}\right)_{V,n_T}\right]dV + R\sum_i n_i \ln \frac{V}{n_i RT} + \sum_i n_i s_i^0 \tag{3.4-11}$$

$$A = \int_V^\infty \left[P - \frac{n_T RT}{V}\right]dV - RT\sum_i n_i \ln \frac{V}{n_i RT} + \sum_i n_i(u_i^0 - Ts_i^0) \tag{3.4-12}$$

$$G = \int_V^\infty \left[P - \frac{n_T RT}{V}\right]dV - RT\sum_i n_i \ln \frac{V}{n_i RT} + PV + \sum_i n_i(u_i^0 - Ts_i^0) \tag{3.4-13}$$

$$\mu_i = \int_V^\infty \left[\left(\frac{\partial P}{\partial n_i} \right)_{T,V,n_j} - \frac{RT}{V} \right] dV - RT \ln \frac{V}{n_i RT} + RT + u_i^0 - Ts_i^0$$

$$(3.4\text{-}14)$$

$$RT \ln f_i = \int_V^\infty \left[\left(\frac{\partial P}{\partial n_i} \right)_{T,V,n_j} - \frac{RT}{V} \right] dV - RT \ln \frac{V}{n_i RT}. \qquad (3.4\text{-}15)$$

Equation (3.4-15) can be rewritten in the more useful form

$$\boxed{RT \ln \frac{f_i}{y_i P} = \int_V^\infty \left[\left(\frac{\partial P}{\partial n_i} \right)_{T,V,n_j} - \frac{RT}{V} \right] dV - RT \ln z,} \qquad (3.4\text{-}16)$$

where z is the compressibility factor of the mixture.

Equation (3.4-16) gives the fugacity of component i in terms of independent variables V and T; it is similar to Eq. (3.1-14) which gives the fugacity in terms of independent variables P and T. However, in addition to the difference in the choice of independent variables there is another, less obvious difference: Whereas in Eq. (3.1-14) the key term is $\bar{v}_i$, the partial molar volume of i, in Eq. (3.4-16) the key term is $(\partial P / \partial n_i)_{T,V,n_j}$, which is not a partial molar quantity.†

For a pure component Eq. (3.4-16) becomes:

$$RT \ln \left(\frac{f}{P} \right)_{\text{pure } i} = \int_V^\infty \left[\frac{P}{n_i} - \frac{RT}{V} \right] dV - RT \ln z + RT (z - 1). \qquad (3.4\text{-}17)$$

Equation (3.4-17) is not particularly useful; for a pure component Eq. (3.1-17) is much more convenient. However, for mixtures, Eq. (3.4-16) is more useful than Eq. (3.1-14).

---

†The definition of a partial molar property is applicable only to extensive properties differentiated at constant temperature and pressure. The total volume of a mixture is related to the partial molar volumes by a summation:

$$V = \sum_i n_i (\partial V / \partial n_i)_{T,P,n_j}.$$

The analogous equation for the total pressure is not valid:

$$P \neq \sum_i n_i (\partial P / \partial n_i)_{T,V,n_j}.$$

The derivative $(\partial P / \partial n_i)_{T,V,n_j}$ should not be regarded as a partial pressure; it does not have a significance analogous to that of the partial molar volume.

Equations (3.4-9) to (3.4-16) enable us to compute all thermodynamic properties relative to the properties of an ideal gas at 1 atm and at the same temperature and composition, provided we have information on volumetric behavior in the form

$$P = \mathcal{f}(T, V, n_1, n_2, \ldots). \tag{3.4-18}$$

Almost all equations of state are pressure-explicit [Eq. (3.4-18)] rather than volume-explicit [Eq. (3.1-15)]. As a result, for phase-equilibrium problems, Eq. (3.4-16) is more useful than Eq. (3.1-14).

To use Eq. (3.4-16) for calculating the fugacity of a component in a mixture, volumetric data must be available, preferably in the form of an equation of state, at the temperature under consideration and as a function of composition and density, from zero density to the density of interest, which corresponds to the lower limit V in the integral. The molar density of the mixture, n_T/V, corresponding to the lower integration limit, must often be found from the equation of state itself since the specified conditions are usually composition, temperature, and pressure; the density is not ordinarily given. This calculation is tedious, for it is inevitably of the trial-and-error type. However, regardless of the number of components in the mixture, the calculation need only be performed once for any given composition, temperature, and pressure; since the quantity V in Eq. (3.4-16) is for the entire mixture, it can be used in the calculation of the fugacities for all the components. Only if the composition, temperature, or pressure changes must the trial-and-error calculation be repeated. It is probably because of the need for trial-and-error calculations that most authors of books on thermodynamics have paid little attention to Eq. (3.4-16). While trial-and-error calculations were highly undesirable before about 1960, in the age of computers they need cause us no concern. Given the temperature, pressure and composition, and given even a complicated equation of state for, say, half a dozen components, a modern digital electronic computer can easily calculate the molar volume of the mixture in much less than one second.

3.5 Fugacity of a Component in a Mixture
According to van der Waals' Equation

To illustrate the applicability of Eq. (3.4-16), we consider a mixture whose volumetric properties are described by van der Waals' equation:

$$P = \frac{RT}{v - b} - \frac{a}{v^2}. \tag{3.5-1}$$

In Eq. (3.5-1), v is the molar volume of the mixture and a and b are constants which depend on the composition.

In order to substitute Eq. (3.5-1) into Eq. (3.4-16), we must first transform it from a molar basis to a total basis by substituting $v = V/n_T$, where n_T is the total number of moles. Equation (3.5-1) then becomes

$$P = \frac{n_T RT}{V - n_T b} - \frac{n_T^2 a}{V^2}. \tag{3.5-2}$$

We want to calculate the fugacity of component i in the mixture at some given temperature, pressure and composition. Differentiating Eq. (3.5-2) with respect to n_i, we have

$$\left(\frac{\partial P}{\partial n_i}\right)_{T,V,n_j} = \frac{RT}{V - n_T b} + \frac{n_T RT \left(\dfrac{\partial (n_T b)}{\partial n_i}\right)}{(V - n_T b)^2} - \frac{1}{V^2} \frac{\partial (n_T^2 a)}{\partial n_i}. \tag{3.5-3}$$

Substituting into Eq. (3.4-16) and integrating, we obtain

$$RT \ln\left(\frac{f_i}{y_i P}\right) = RT \ln \frac{V - n_T b}{V} \Bigg]_V^\infty - n_T RT \left(\frac{\partial (n_T b)}{\partial n_i}\right) \frac{1}{(V - n_T b)} \Bigg]_V^\infty$$

$$+ \left(\frac{\partial (n_T^2 a)}{\partial n_i}\right) \frac{1}{V} \Bigg]_V^\infty - RT \ln z. \tag{3.5-4}$$

At the upper limit of integration, as $V \to \infty$,

$$\ln \frac{V - n_T b}{V} \to 0, \quad \frac{1}{V - n_T b} \to 0, \quad \frac{1}{V} \to 0, \tag{3.5-5}$$

and Eq. (3.5-4) becomes

$$RT \ln \frac{f_i}{y_i P} = RT \ln \frac{V}{V - n_T b} + n_T RT \left(\frac{\partial (n_T b)}{\partial n_i}\right)\left(\frac{1}{V - n_T b}\right)$$

$$- \left(\frac{\partial (n_T^2 a)}{\partial n_i}\right) \frac{1}{V} - RT \ln z. \tag{3.5-6}$$

Equation (3.5-6) gives the desired result, derived from rigorous thermodynamic relations. To proceed further, it is necessary to make assumptions concerning the composition dependence of the constants a and b. These assumptions cannot be based on thermodynamic arguments but must be obtained from molecular considerations. Suppose that we have m components in the mixture. If we interpret the constant b as a term proportional to the size of the molecules and if we assume that the molecules are spherical, then we might average the molecular diameters, giving

$$b^{1/3} = \sum_{i=1}^{m} y_i b_i^{1/3}. \tag{3.5-7}$$

On the other hand, we may choose to average the molecular volumes directly and obtain the simpler relation

$$b = \sum_{i=1}^{m} y_i b_i. \tag{3.5-8}$$

Neither Eq. (3.5-7) nor Eq. (3.5-8) is in any sense a "correct" mixing rule; both are based on arbitrary assumptions, and alternate mixing rules could easily be constructed based on different arbitrary assumptions. Equation (3.5-8) is commonly used because of its mathematical simplicity.

At moderate densities, for mixtures whose molecules are not too dissimilar in size, the particular mixing rule used for b does not significantly affect the results. However, the fugacity of a component in a mixture is sensitive to the mixing rule used for the constant a. If we interpret a as a term which reflects the strength of attraction between two molecules, then for a mixture we may want to express a by averaging over all molecular pairs. Thus

$$a = \sum_{i=1}^{m} \sum_{j=1}^{m} y_i y_j a_{ij}, \tag{3.5-9}$$

where a_{ij} is a measure of the strength of attraction between a molecule i and a molecule j. If i and j are the same chemical species, then a_{ij} is clearly the van der Waals a for that substance. If i and j are chemically not identical and if we have no experimental data for the i-j mixture, we then need to express a_{ij} in terms of a_i and a_j. This need is the cause of one of the key problems of phase-equilibrium thermodynamics. Given information on the intermolecular forces of each of two pure fluids, how can we predict the intermolecular forces in a mixture of these two fluids? There is no general answer to this question. An introduction to the study of intermolecular forces is given in the next chapter, where it is shown that only under severe limiting conditions can the forces between molecule i and molecule j be related in a simple way to the forces between two molecules i and two molecules j. Our knowledge of molecular physics is, unfortunately, not sufficient to give us generally reliable methods for predicting the properties of mixtures using only knowledge of the properties of the pure components.

For $i \neq j$, it was suggested many years ago by Berthelot, on strictly empirical grounds, that

$$a_{ij} = \sqrt{a_i a_j}. \tag{3.5-10}$$

This relation, which is often called the geometric-mean assumption, was used extensively by van der Waals and his followers in their work on mixtures. Since van der Waals' time, the geometric-mean assumption has been used for other quantities in addition to van der Waals' a; it is commonly used for those parameters which are a measure of intermolecular attraction. Long after the time of van der Waals and Berthelot, it was shown by London (see Sec. 4.4) that under certain conditions there is some theoretical justification for the geometric-mean assumption.

If we adopt the mixing rules given by Eqs. (3.5-8), (3.5-9) and (3.5-10), the fugacity for component i, given by Eq. (3.5-6), becomes

$$\ln \frac{f_i}{y_i P} = \ln \frac{v}{v-b} + \frac{b_i}{v-b} - \frac{2\sqrt{a_i} \sum_{j=1}^{m} y_j \sqrt{a_j}}{vRT} - \ln z, \tag{3.5-11}$$

where v is the molar volume and z is the compressibility factor of the mixture.

Equation (3.5-11) indicates that in order to calculate the fugacity of a component in the mixture at a given temperature, pressure, and composition, we must first compute the constants a and b for the mixture using the mixing rules given by Eqs. (3.5-8), (3.5-9), and (3.5-10). Using these constants and the equation of state, Eq. (3.5-1), we must then find the molar volume v of the mixture by trial and error; this step is the only tedious part of the calculation. Once the molar volume is known, the compressibility factor z is easily calculated and the fugacity is readily found from Eq. (3.5-11).

For a numerical example, let us consider the fugacity of hydrogen in a ternary mixture at 50°C and 300 atm containing 20 mol % hydrogen, 50 mol % methane and 30 mol % ethane. Using Eq. (3.5-11), we find the fugacity of hydrogen to be 113 atm.† From the ideal-gas law the fugacity is 60 atm, while using the Lewis fugacity rule gives 71.4 atm. These three results are significantly different from one another. No reliable experimental data are available for this mixture, but in this particular case it is

---

†The constants a and b for each component were found from critical properties. The molar volume of the mixture, as calculated from van der Waals' equation, is 1.453 ft^3/lb-mol.

probable that 113 atm is much closer to the correct value than the results of either of the two simpler calculations. However, such a conclusion cannot be generalized. The equation of van der Waals gives only an approximate description of gas-phase properties and sometimes, because of cancellation of errors, calculations based on simpler assumptions may give better results.

In deriving Eq. (3.5-11) we have shown how the rigorous expression, Eq. (3.4-16), can be used to calculate the fugacity of a component in a mixture once the equation of state for the mixture is given. In this particular calculation the relatively simple van der Waals equation was used but the same procedure can be applied to any pressure-explicit equation of state, regardless of how complex it may be.

It is frequently said that the more complicated an equation of state is, and the more constants it contains, the better representation it gives of volumetric properties. This statement is correct for a pure component if ample volumetric data are available from which to determine the constants with confidence and if the equation is used only under those conditions of temperature and pressure which were used to determine the constants. However, for predicting the properties of mixtures from pure-component data alone, the more constants one has, the more mixing rules are required, and since these rules are subject to much uncertainty, it frequently happens that a simple equation of state containing only two or three constants is better for predicting mixture properties than is a complicated equation of state containing a large number of constants.[6,7] The use of equations of state for calculating fugacities in gas-phase mixtures is discussed further in Chap. 5.

3.6   Phase Equilibria From Volumetric Properties

In Chap. 1 we indicated that the purpose of phase-equilibrium thermodynamics is to predict conditions (temperature, pressure, composition) which prevail when two or more phases are in equilibrium. In Chap. 2 we discussed thermodynamic equations which determine the state of equilibrium between phases α and β. These are:

$$\text{Equality of temperature:} \quad T^\alpha = T^\beta \qquad (3.6\text{-}1)$$

For each component i,
$$\text{equality of fugacities:} \quad f_i^\alpha = f_i^\beta \qquad (3.6\text{-}2)$$

and (for most situations),

$$\text{Equality of pressures:} \quad P^\alpha = P^\beta \qquad (3.6\text{-}3)$$

In order to find the conditions which satisfy these equations it is necessary to have a method for evaluating the fugacity of each component in phase α and in phase β. Such a method is supplied by Eqs. (3.1-14) or (3.4-16), which are valid for any component in any phase. In principle, therefore, a solution to the phase-equilibrium problem is provided completely by either one of these equations together with an equation of state and the equations of phase equilibrium.

To illustrate these ideas, let us consider vapor-liquid equilibria in a system with m components; suppose that we know the pressure P and the mole fractions $x_1, x_2, \ldots, x_m$ for the liquid phase. We want to find the temperature T and the vapor-phase mole fractions $y_1, y_2, \ldots, y_m$. We assume that we have available a pressure-explicit equation of state applicable to all components and to their mixtures over the entire density range from zero density to the density of the liquid phase. We can then compute the fugacity of each component, in either phase, by Eq. (3.4-16).

The number of unknowns is:

$y_1, y_2, \ldots, y_{m-1}$	$(m - 1)$ mole fractions†
T	temperature
v^V, v^L	molar volumes (or densities) of the equilibrium vapor and liquid phases

Total: $(m + 2)$ unknowns

The number of independent equations is:

$f_i^V = f_i^L$	m equations, where f_i^V and f_i^L are found for each component i by Eq. (3.4-16)
$P = \mathcal{f}(v^V, y_1, \ldots, T)$	
and	equation of state, applied once to liquid phase and once to vapor phase
$P = \mathcal{f}(v^L, x_1, \ldots, T)$	

Total: $(m + 2)$ independent equations

Since the number of unknowns is equal to the number of independent equations, the unknown quantities can be found by simultaneous solution

†Since $\displaystyle\sum_i^m y_i = 1$, the mth mole fraction is fixed once $(m - 1)$ mole fractions are determined.

of all the equations.† It is apparent, however, that the computational effort to do so is large, especially if the equation of state is not simple and if the number of components is high.

Calculation of vapor-liquid equilibria, along the lines just outlined, was discussed many years ago by van der Waals, who made use of the equation bearing his name, and in the period 1940–1952 extensive calculations for hydrocarbon mixtures were reported by Benedict, Webb, and Rubin who used an eight-constant equation of state.[8] The results of their calculations are given in 276 large charts; these have been published in a huge book[9] which by virtue of its size alone is likely to become a monument to the history of applied thermodynamics. With the aid of some simplifying assumptions, De Priester[10] and also Edmister and Ruby[11] condensed the charts into a more manageable form; these results are useful for estimating vapor-liquid equilibria for mixtures of light paraffinic and olefinic hydrocarbons. Calculations based on these and similar charts are always of the iterative, trial-and-error type, but they can be carried out very efficiently by an electronic computer.

Calculation of phase equilibria from volumetric data alone requires a large computational effort but, thanks to modern electronic computers, the effort does not by itself present an insuperable difficulty. The major disadvantages of this type of calculation are not computational; rather they are due to the fact that we do not have a satisfactory equation of state applicable to mixtures over a density range from zero density to liquid densities. It is this crucial deficiency which suggests that phase-equilibrium calculations based on volumetric data alone are often doubtful. In order to determine the required volumetric data with the necessary degree of accuracy, a large amount of experimental work is required; rather than make all these volumetric measurements, it is usually more economical to measure the desired phase equilibria directly. For those mixtures where the components are chemically similar (e.g. mixtures of paraffinic and olefinic hydrocarbons) an equation-of-state calculation for vapor-liquid equilibria provides a reasonable possibility because many simplifying assumptions can be made concerning the effect of composition on volumetric behavior. But even in this relatively simple situation there is much ambiguity when one attempts to predict properties of mixtures using equation-of-state constants determined from pure-component data. Benedict, et al., used eight empirical constants to describe the volumetric behavior of each pure hydrocarbon; in order to fix *uniquely* eight constants, a very large amount of experimental data is required at the outset. Even with the generous amount of data which Benedict had at his disposal, his constants could not be determined without some ambiguity.

---

†We have considered here the case where P and x are known and T and y are unknown, but similar reasoning applies to other combinations of known and unknown quantities; the number of intensive variables which must be specified is given by the phase rule.

But the essence of the difficulty arises when one must decide on mixing rules, i.e., on how these constants are to be combined for a mixture. Phase-equilibrium calculations are sensitive to the mixing rules used and in some cases it has been shown that good results can be obtained only when the mixing rules depend both on the nature of the components and on the temperature and density. In that event, there is little advantage in calculating phase equilibria from volumetric data alone.

In summary, then, the empirical-equation-of-state approach to a complete determination of phase equilibria is usually not promising because we usually do not have a sufficiently accurate knowledge of the volumetric properties of mixtures at high densities. Calculation of fugacities by Eq. (3.4-16) is practical for vapor mixtures but, certain special cases excepted, is not practical for condensed mixtures. Even in vapor mixtures the calculations are often not accurate because of our inadequate knowledge of volumetric properties. The accuracy of the fugacity calculated with Eq. (3.4-16) depends directly on the validity of the equation of state used in the calculation and, in order to determine values for the constants in even a good equation of state, one must have either a large amount of reliable experimental data, or else some sound theoretical basis for predicting volumetric properties. In a typical case we have little of either data or theory. Reliable volumetric data exist for the more common substances but these constitute only a small fraction of the number of listings in a chemical handbook. Reliable volumetric data are scarce even for binary mixtures and they are rare for mixtures containing more than two components. Considering the practically infinite number of ternary (and higher) mixtures possible, it is clear that there will never be sufficient experimental data to give an adequate empirical description of volumetric properties of mixed fluids. A strictly empirical approach to the phase-equilibrium problem is, therefore, subject to severe limitations. Progress can be achieved only by generalizing from limited, but reliable, experimental results, utilizing as much as possible the techniques based on our theoretical knowledge of molecular behavior. Significant progress in phase-equilibrium thermodynamics can be achieved only by increased use of the concepts of molecular physics. Therefore, before continuing our discussion of fugacities in Chap. 5, we turn in Chap. 4 to a brief survey of the nature of intermolecular forces.

PROBLEMS

1. Consider a mixture of N gases and assume that the Lewis fugacity rule is valid for this mixture. For this case show that the fugacity of the mixture f_M is given by

$$f_M = \prod_{i=1}^{N} f_{\text{pure } i}^{y_i}$$

where y_i is the mole fraction of component i, and $f_{\text{pure } i}$ is the fugacity of pure component i at the temperature and total pressure of the mixture.

2. A binary gas mixture contains 25 mol % A and 75 mol % B. At 50 atm total pressure and 100°C the fugacity coefficients of A and B in this mixture are, respectively, 0.65 and 0.90. What is the fugacity of the gaseous mixture?

3. At 25°C and 1 atm partial pressure, the solubility of ethane in water is very small; the equilibrium mole fraction $x_{C_2H_6} = 0.33 \times 10^{-4}$. What is the solubility of ethane at 25°C when the partial pressure is 35 atm? At 25°C, the compressibility factor of ethane is given by the empirical relation

$$z = 1 - 7.53 \times 10^{-3} P - 7.04 \times 10^{-5} P^2$$

where P is in atmospheres.

At 25°C, the saturation pressure of ethane is 41.52 atm and that of water is 0.0312 atm.

4. Consider a binary mixture of components 1 and 2. The molar Helmholtz energy change Δa is given by

$$\frac{\Delta a}{RT} = \ln \frac{v}{v - b} - y_1 \ln \frac{v}{y_1 RT} - y_2 \ln \frac{v}{y_2 RT}$$

where v is the molar volume of the mixture, b is the constant for the mixture, depending only on composition, and Δa is the molar Helmholtz energy change in going isothermally from the standard state (pure, unmixed, ideal gases at 1 atm) to the molar volume v.

The composition dependence of b is given by

$$b = y_1 b_1 + y_2 b_2$$

Find an expression for the fugacity of component 1 in the mixture.

5. Derive Eq. (3.4-17). [Hint: Start with Eq. (3.4-13)].

6. Oil reservoirs below ground frequently are in contact with underground water and, in connection with an oil drilling operation, you are asked to compute the solubility of water in a heavy oil at the underground conditions. These conditions are estimated at 280°F and 6000 psia. Experiments at 280°F and 1 atm indicate that the solubility of steam in the oil is $x_1 = 35 \times 10^{-4}$ (x_1 = mole fraction steam). Assume Henry's law in the form

$$f_1 = H(T) x_1$$

where $H(T)$ is a constant, dependent only on the temperature and f_1 is the fugacity of H_2O. Also assume that the vapor pressure of the oil is negligible at 280°F. Data for H_2O are given in the Keenan and Keyes steam tables.

REFERENCES

1. Beattie, J. A., *Chem. Rev.*, **44**, 141 (1949).

2. Beattie, J. A., and W. H. Stockmayer, *Treatise on Physical Chemistry* (H. S. Taylor and S. Glasstone, eds.), Chapter 2. Princeton, N.J.: D. Van Nostrand Co., Inc., 1942.

3. Beattie, J. A., *Thermodynamics and Physics of Matter* (F. D. Rossini, ed.), Chapter 3, Part C. Princeton, N.J.: Princeton University Press, 1955.

4. Evans, R. B., and G. M. Watson, *Chem. Eng. Data Series*, **1**, 67 (1956).

5. Keenan, J. H., and F. G. Keyes, *Thermodynamic Properties of Steam.* New York: John Wiley & Sons, Inc., 1936.

6. Ackerman, F. J., and O. Redlich, *J. Chem. Phys.*, **38**, 2740 (1963).

7. Shah, K. K., and G. Thodos, *Ind. Eng. Chem.*, **57**, No. 3, 30 (1965).

8. Benedict, M., G. B. Webb, and L. C. Rubin, *J. Chem. Phys.*, **8**, 334 (1940); **10**, 747 (1942); *Chem. Eng. Progr.* **47**, 419 (1951).

9. M. W. Kellogg Co., *Liquid-Vapor Equilibrium in Mixtures of Light Hydrocarbons*, 1950.

10. De Priester, C. L., *Chem. Eng. Progr. Symp. Ser.*, **49**, No. 7, 1 (1953).

11. Edmister, W. C., and C. L. Ruby, *Chem. Eng. Progr.*, **51**, 95 (1955).

Intermolecular Forces and the Theory of Corresponding States

4

Thermodynamic properties of any pure substance are determined by intermolecular forces which operate between the molecules of that substance. Similarly, thermodynamic properties of a mixture depend on intermolecular forces which operate between the molecules of the mixture. The case of a mixture, however, is necessarily more complicated because consideration must be given not only to interaction between molecules belonging to the same component, but also to interaction between dissimilar molecules. In order to interpret and correlate thermodynamic properties of solutions, it is therefore necessary to have some understanding of the nature of intermolecular forces. The purpose of this chapter is to give a brief introduction to the nature and variety of forces acting between molecules.

We must recognize at the outset that our understanding of intermolecular forces is far from complete and that quantitative results have been obtained for only simple and idealized models of real matter. Further, we must point out that the quantitative relations which link intermolecular forces to macroscopic properties (viz., statistical mechanics) are also at present limited to simple and idealized cases. It follows, therefore, that we can use our knowledge of intermolecular forces in only an approximate manner to interpret and generalize phase-equilibrium data. Molecular physics is always concerned with models and we must beware whenever we are tempted to substitute models for nature. Frequently, the theory of intermolecular forces gives us no more than a qualitative, or at best semiquantitative, basis for understanding phase behavior, but even such a basis can be useful for understanding and correlating experimental results.

While the separation between molecular physics and practical problems in phase behavior is still large, every year new results tend to reduce that separation. There can be no doubt that future developments in applied thermodynamics will increasingly utilize and rely upon statistical mechanics and the theory of intermolecular forces.

When a molecule is in the proximity of another, forces of attraction and repulsion strongly influence its behavior. If there were no forces of attraction, gases would not condense to form liquids and solids, and in the absence of repulsive forces, condensed matter would not show resistance to compression. The configurational properties of matter can be considered as a compromise between those forces which pull molecules together and those which push them apart; by *configurational properties* we mean those properties which depend on interaction between molecules rather than on the characteristics of molecules which are isolated. For example, the energy of vaporization for a liquid is a configurational property, but the specific heat of a gas at low pressure is not.

There are many different types of intermolecular forces, but for our limited purposes here only a few important ones are considered. These forces may be classified under the following arbitrary but convenient headings:

1. Electrostatic forces between charged particles (ions) and between permanent dipoles, quadrupoles and higher multipoles.
2. Induction forces between a permanent dipole (or quadrupole) and an induced dipole.
3. Forces of attraction (dispersion forces) and repulsion between non-polar molecules.
4. Specific (chemical) forces leading to association and complex formation, i.e., to the formation of loose chemical bonds of which hydrogen bonds are perhaps the best example.

An introductory discussion of these forces forms the substance of this chapter, with special attention given to those acting between nonpolar molecules and to the molecular theory of corresponding states.

4.1 Potential-Energy Functions

Molecules have kinetic energy as a result of their velocities relative to some fixed frame of reference; they also have potential energy as a result of their positions relative to one another. Consider two simple, spherically symmetric molecules separated by the distance r. The potential energy Γ shared by these two molecules is a function of r; the force F between the molecules is related to the potential energy by

$$F = -\frac{d\Gamma}{dr}. \qquad (4.1\text{-}1)$$

The negative of the potential energy, i.e., $-\Gamma(r)$, is the work which must be done to separate two molecules from the intermolecular distance r to infinite separation. Intermolecular forces are almost always expressed in terms of potential-energy functions. The common convention is that a force of attraction is negative and one of repulsion is positive.

In the simplified discussion above, we have assumed that the force acting between two molecules depends on their relative position as specified by only one coordinate, r. For a spherically symmetric molecule, such as an argon atom, this assumption is valid; but for more complicated molecules, other coordinates, such as the angles of orientation, may be required as additional independent variables of the potential-energy function. A more general form of Eq. (4.1-1) is

$$F(r, \theta, \phi, \ldots) = -\nabla \Gamma(r, \theta, \phi, \ldots) \qquad (4.1\text{-}2)$$

where ∇ is the gradient and θ, ϕ, ... designate whatever additional coordinates may be needed to specify the potential energy.

4.2   Electrostatic Forces

Of all intermolecular forces, those due to point charges are the easiest to understand and the simplest to treat quantitatively. If we regard two point electric charges of magnitudes e_i and e_j respectively, separated from one another *in vacuo* by the distance r, then the force between them is given by Coulomb's relation, sometimes called the *inverse-square law*:

$$F = \frac{e_i e_j}{r^2}. \qquad (4.2\text{-}1)$$

Upon integrating, the potential energy is

$$\Gamma_{ij} = \frac{e_i e_j}{r} + \text{Constant of integration.} \qquad (4.2\text{-}2)$$

The commonly adopted convention is that the potential energy is zero at infinite separation. Substituting $\Gamma = 0$ when $r = \infty$, the constant of integration in Eq. (4.2-2) vanishes.

For charged molecules (i.e., ions) e_i and e_j are integral multiples of the unit charge ϵ; therefore, the potential energy between two ions can be written

$$\Gamma_{ij} = \frac{z_i z_j \epsilon^2}{Dr},\tag{4.2-3}$$

where z_i and z_j are the ionic valences and D is the dielectric constant of the medium; D is unity in vacuo but greater than unity otherwise. When ϵ is taken as 4.802×10^{-10} esu and r is in centimeters, the potential energy is in ergs.

Electrostatic forces between ions are inversely proportional to the square of the separation and therefore they have a much longer range than other intermolecular forces which depend on higher powers of the reciprocal distance. These electrostatic forces make the dominant contribution to the configurational energy of salt crystals and are therefore responsible for the very high melting points of salts. Also, the long-range nature of ionic forces is, at least in part, responsible for the difficulty in constructing a theory of concentrated electrolyte solutions.

Electrostatic forces can arise even for those particles which do not have a net electric charge. Consider a particle having two electric charges of the same magnitude e but of opposite sign, held a distance l apart. Such a particle has an electric couple or permanent dipole moment μ defined by

$$\mu = el.\tag{4.2-4}$$

Asymmetric molecules possess permanent dipoles resulting from an uneven spatial distribution of electronic charges about the positively charged nuclei. Symmetric molecules, like argon or methane, have zero dipole moment and those molecules having very little asymmetry generally have small dipole moments. A representative selection of molecules and their dipole moments is given in Table 4-1. The common unit for dipole moments is the Debye (one Debye $= 10^{-18}$ esu-cm).

The potential energy of two permanent dipoles i and j is obtained by considering the coulombic forces between the four charges. The energy of interaction depends on the distance between dipole centers and on the relative orientations of the dipoles. If the distance r between the dipoles

Table 4-1

PERMANENT DIPOLE MOMENTS†

Molecule	μ (Debyes)	Molecule	μ (Debyes)
CO	0.10	CH_3I	1.64
C_3H_6	0.35	CH_3COOCH_3	1.67
$C_6H_5CH_3$	0.37	C_2H_5OH	1.70
PH_3	0.55	H_2O	1.84
HBr	0.80	HF	1.91
$CHCl_3$	1.05	C_2H_5F	1.92
$(C_2H_5)_2O$	1.18	$(CH_3)_2CO$	2.87
NH_3	1.47	$C_6H_5COCH_3$	3.00
$C_6H_5NH_2$	1.48	$C_2H_5NO_2$	3.70
C_6H_5Cl	1.55	CH_3CN	3.94
C_2H_5SH	1.56	$CO(NH_2)_2$	4.60
SO_2	1.61	KBr	9.07

†Taken from a more complete list of dipole moments given in Landolt-Börnstein, *Zahlenwerte und Funktionen*, 6th ed., Vol. 1, Part 3, p. 388 (Berlin: Springer, 1951); A. A. Maryott and F. Buckley, NBS Circular 537, 1953; and in A. McClellan, *Tables of Experimental Dipole Moments* (San Francisco: W. H. Freeman & Co., 1963).

is large compared to l_i and l_j, the potential energy is

$$\Gamma_{ij} = -\frac{\mu_i \mu_j}{r^3} \left[2 \cos \theta_i \cos \theta_j - \sin \theta_i \sin \theta_j \cos(\phi_i - \phi_j) \right], \quad (4.2\text{-}5)$$

where the angles θ and ϕ give the orientations of the dipole axes. The orientation making the potential energy a maximum is that corresponding to the dipoles being in the same straight line, the positive end of one facing the positive end of the other; the energy is a minimum when the dipoles are in a straight line, the positive end of one facing the negative end of the other.

In an assembly of polar molecules, the relative orientations of these molecules depend on the interplay of two factors: The presence of an electric field set up by the polar molecules tends to line up the dipoles, whereas the kinetic (thermal) energy of the molecules tends to toss them about in a random manner. We expect, therefore, that as the temperature rises, the orientations become more and more random until in the limit of very high temperature, the average potential energy due to polarity becomes vanishingly small. This expectation is confirmed by experimental evidence; whereas at low and moderate temperatures the behavior of polar gases is markedly different from that of nonpolar gases, this difference tends to disappear as the temperature increases. It was shown by Keesom[1] that at moderate and high temperatures, orientations leading to negative potential energies are preferred statistically. The average potential energy $\overline{\Gamma}_{ij}$ between two dipoles i and j at a fixed separation r is found by averaging over all orientations with each orientation weighted according to its Boltzmann factor.[2] When the Boltzmann factors are expanded in powers of $1/kT$, $\overline{\Gamma}_{ij}$ becomes

$$\overline{\Gamma}_{ij} = -\frac{2}{3}\frac{\mu_i^2\mu_j^2}{r^6 kT} + \dots \qquad (4.2\text{-}6)\dagger$$

Equation (4.2-6) indicates that for a pure polar substance ($i = j$) the potential energy varies as the fourth power of the dipole moment. A small increase in the dipole moment can therefore produce a large change in the potential energy due to permanent dipole forces. Whereas the contribution of polar forces to the total potential energy is small for molecules having dipole moments of one Debye or less, this contribution becomes increasingly significant for small molecules having larger dipole moments. A quantitative comparison of the importance of polar forces relative to some other intermolecular forces is given in Sec. 4.4.

In addition to dipole moments, it is possible for molecules to have quadrupole moments which are due to the concentration of electric charge at four separate points in the molecule. The difference between a molecule having a dipole moment and one having a linear quadrupole moment is

† This equation is based on a Boltzmann average

$$\overline{\Gamma} = \frac{\int \Gamma [\exp(-\Gamma/kT)]\,d\Omega}{\int [\exp(-\Gamma/kT)]\,d\Omega}$$

where $d\Omega$ is the element of the solid angle:

$$d\Omega = \sin\theta_1 \sin\theta_2\, d\theta_1\, d\theta_2\, d\phi.$$

However, it has been argued by Rowlinson [*Mol. Phys.*, **1**, 414 (1958)] that the proper average is

$$\overline{\Gamma} = -kT \ln \frac{\int [\exp(-\Gamma/kT)]\,d\Omega}{\int d\Omega}$$

When that average is used the numerical coefficient in Eq. (4.2-6) is $-\frac{1}{3}$ instead of $-\frac{2}{3}$.

shown schematically below:

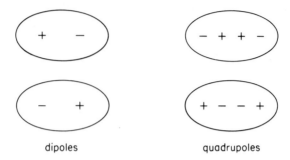

dipoles   quadrupoles

For example, carbon dioxide, a linear molecule, has no dipole moment but its quadrupole moment is sufficiently strong to affect its thermodynamic properties, which are different from those of other nonpolar molecules of similar size and molecular weight. For the simplest case, a linear molecule, the quadrupole moment Q is defined by the sum of the second moments of the charges:

$$Q = \sum_i e_i d_i^2,$$
(4.2-7)

where the charge e_i is located at a point a distance d_i away from some arbitrary origin, all charges being on the same straight line. The quadrupole moment Q is independent of the position of the origin provided the molecule has no net charge and no dipole moment. For nonlinear quadrupoles or for molecules having permanent dipoles, the definition of the quadrupole moment is more complicated.

Experimental determination of quadrupole moments is difficult and only a few measurements have been made.[3,4] Some quadrupole moments are given in Table 4-2.

Table 4-2

QUADRUPOLE MOMENTS, $(10^{26} \text{ esu-cm}^2)$†

H_2	0.63
C_2H_4	1
NH_3	1
N_2	1.5
H_2O	2
C_2H_2	3
CO_2	4.1

†A. D. Buckingham, *Quart. Rev.* (London), **13**, 183 (1959); A. D. Buckingham and R. L. Disch, *Proc. Roy. Soc.* (London), **A263**, 275 (1963).

The potential energy between a quadrupole and a dipole, or between a quadrupole and another quadrupole, is a function of the distance of separation and the angles of mutual orientation. The average potential energy is found by averaging over all orientations; each orientation is weighted according to its Boltzmann factor.[2] Upon expanding in powers of $1/kT$, we obtain:

For dipole i-quadrupole j,

$$\overline{\Gamma}_{ij} = -\frac{\mu_i^2 Q_j^2}{r^8 kT} + \ldots;$$

(4.2-8)

for quadrupole i-quadrupole j,

$$\overline{\Gamma}_{ij} = -\frac{7 Q_i^2 Q_j^2}{40 r^{10} kT} + \ldots.$$

(4.2-9)

Whereas the scientific literature dealing with dipole moments is extensive, considerably less is known about quadrupole moments and little work has been done on still higher multipoles such as octapoles and hexadecapoles. The effect of quadrupole moments on thermodynamic properties is already much less than that of dipole moments and the effect of higher multipoles is usually negligible.[5] This relative rank of importance follows from the fact that intermolecular forces due to multipoles higher than dipoles are extremely short range; for dipoles the average potential energy is proportional to the sixth power of the inverse distance of separation and for quadrupoles the average potential energy depends on the tenth power of the reciprocal distance. For still higher multipoles, the exponent is even larger.

4.3   Polarizability and Induced Dipoles

A nonpolar molecule such as argon or methane has no permanent dipole moment but when such a molecule is subjected to an electric field, the electrons are displaced from their ordinary positions and a dipole is induced. In fields of moderate strength the induced dipole moment μ^i is proportional to the field strength E; that is,

$$\mu^i = \alpha E,$$

(4.3-1)

where the proportionality factor α is a fundamental property of the substance. It is called the polarizability, which means the ease with which the molecule's electrons can be displaced by an electric field. The polarizability can be calculated in several ways, most notably from dielectric properties and from index-of-refraction data. For asymmetric molecules the polarizability is not a constant but is a function of the molecule's orientation relative to the direction of the field. Average polarizabilities of some representative molecules are given in Table 4-3.

Table 4-3

Average Polarizabilities[†]

Molecule	$\alpha(\text{cm}^3 \times 10^{25})$	Molecule	$\alpha(\text{cm}^3 \times 10^{25})$
H_2	7.9	HBr	36.1
H_2O[‡]	15.9	SO_2	37.2
Ar	16.3	Cl_2	46.1
N_2	17.6	HI	54.4
CO	19.5	$(CH_3)_2CO$	63.3
NH_3	22.6	$CHCl_3$	82.3
HF	24.6	$(C_2H_5)_2O$	87.3
CH_4	26	CCl_4	105
HCl	26.3	cyclo-C_6H_{12}	109
CO_2	26.5	$C_6H_5CH_3$	123
CH_3OH	32.3	$C_6H_5NO_2$	129
C_2H_2	33.3	$n-C_7H_{16}$	136

[†] Landolt-Börnstein, *Zahlenwerte und Funktionen,* 6th ed., Vol. 1, Part 3, p. 510 (Berlin: Springer, 1951).

[‡] N. E. Dorsey, *Properties of Ordinary Water-Substance* (New York: Reinhold Publishing Corp., 1940).

When a nonpolar molecule i is situated in an electric field set up by the presence of a nearby polar molecule j, the resultant force between the permanent dipole and the induced dipole is always attractive. The mean potential energy was first calculated by Debye and is usually associated with his name. It is given by

$$\overline{\Gamma}_{ij} = -\frac{\alpha_i \mu_j^2}{r^6}. \tag{4.3-2}$$

Polar molecules as well as nonpolar ones can have dipoles induced in an electric field. The general Debye formula, therefore, for the mean potential energy due to induction by permanent dipoles, is

$$\overline{\Gamma}_{ij} = -\frac{(\alpha_i \mu_j^2 + \alpha_j \mu_i^2)}{r^6}. \tag{4.3-3}$$

An electric field may also be caused by a permanent quadrupole moment. In that case the average potential energy of induction between a quadrupole j and a nonpolar molecule i is again attractive; if both molecules i and j have permanent quadrupole moments,

$$\bar{\Gamma}_{ij} = - \frac{3(\alpha_i Q_j^2 + \alpha_j Q_i^2)}{2r^8}.$$

(4.3-4)

For molecules with a permanent dipole moment, the potential energy due to induction is usually small when compared to the potential energy due to permanent dipoles and similarly, for molecules with a permanent quadrupole moment, the induction energy is usually less than that due to quadrupole-quadrupole interactions.

4.4   Intermolecular Forces Between Nonpolar Molecules

The concept of polarity has been known for a long time but until about 1930 there was no adequate explanation for the forces acting between nonpolar molecules. It was very puzzling, for example, why such an obviously nonpolar molecule as argon should nevertheless show serious deviations from the ideal-gas laws even at moderate pressure. In 1930 it was shown by London that so-called nonpolar molecules are, in fact, nonpolar only when viewed over a period of time; if an instantaneous photograph of such a molecule were taken, it would show that at a given instant the oscillations of the electrons about the nucleus had resulted in distortion of the electron arrangement sufficient to cause a temporary dipole moment. This dipole moment is rapidly changing its magnitude and direction and therefore averages out to zero over a short period of time; however, these quickly varying dipoles produce an electric field which then induces dipoles in the surrounding molecules. The result of this induction is an attractive force which is sometimes called the induced dipole-induced dipole force. Using quantum mechanics, London showed that, subject to certain simplifying assumptions, the potential energy between two simple, spherically symmetric molecules i and j at large distances is given by

$$\Gamma_{ij} = - \frac{3}{2} \frac{\alpha_i \alpha_j}{r^6} \left(\frac{h\nu_{0i} h\nu_{0j}}{h\nu_{0i} + h\nu_{0j}} \right),$$

(4.4-1)

where h is Planck's constant and ν_0 is a characteristic electronic frequency for each molecule in its unexcited state. This frequency is related to the variation of the index of refraction n with light frequency ν by

$$n - 1 = \frac{c}{\nu_0^2 - \nu^2},\tag{4-4-2}$$

where c is a constant. It is this relationship between index of refraction and characteristic frequency which is responsible for the name *dispersion* for the attractive force between nonpolar molecules.

For a molecule i, the product $h\nu_{0i}$ is very nearly equal to its first ionization potential I_i.† Equation (4.4-1) is therefore usually written in the form

$$\boxed{\Gamma_{ij} = -\frac{3}{2}\frac{\alpha_i\alpha_j}{r^6}\left(\frac{I_iI_j}{I_i + I_j}\right).}\tag{4.4-3}$$

If molecules i and j are of the same species, Eq. (4.4-3) reduces to

$$\Gamma_{ii} = -\frac{3}{4}\frac{\alpha_i^2 I_i}{r^6}.\tag{4.4-4}$$

Equations (4.4-3) and (4.4-4) give the important result that the potential energy between nonpolar molecules is independent of temperature and varies inversely as the sixth power of the distance between them. The attractive force therefore varies as the reciprocal seventh power. This sharp decline in attractive force as distance increases explains why it is much easier to melt or vaporize a nonpolar substance than an ionic one where the dominant attractive force varies as the reciprocal second power of the distance of separation.‡

---

†The first ionization potential is the work which must be done to remove one electron from an uncharged molecule M:

$$M \longrightarrow e^- + M^+.$$

The second ionization potential is the work needed to remove the second electron according to

$$M^+ \longrightarrow e^- + M^{++}.$$

The second ionization potential is considerably larger than the first.

In a similar manner it is possible to define third (and higher) ionization potentials.

‡Various workers, in addition to London, have derived expressions for the attractive portion of the potential function of two, spherically symmetric, nonpolar molecules. These expressions all agree on a distance dependence of r^{-6} but the coefficients differ considerably. A good review of this subject is presented by K. S. Pitzer in *Advances in Chemical Physics*, edited by I. Prigogine (New York: Interscience Publishers, Inc., 1959), Vol. 2.

London's formula is more sensitive to the polarizability than it is to the ionization potential. Furthermore, the ionization potentials of most substances do not differ very much from one another; a representative list of ionization potentials (in electron volts) is given in Table 4-4. Since the

Table 4-4

FIRST IONIZATION POTENTIALS†

Molecule	I (EV)	Molecule	I (EV)
$C_6H_5CH_3$	8.9	C_3H_8	11.2
C_6H_6	9.2	CH_3Cl	11.2
$n\text{-}C_7H_{14}$	9.5	C_2H_2	11.4
C_2H_5SH	9.7	$CHCl_3$	11.5
C_5H_5N	9.8	NH_3	11.5
$(CH_3)_2CO$	10.1	HBr	12.0
$(C_2H_5)_2O$	10.2	H_2O	12.6
$n\text{-}C_7H_{16}$	10.4	HCl	12.8
C_2H_4	10.7	CH_4	13.0
C_2H_5OH	10.7	Cl_2	13.2
HI	10.7	CO_2	13.7
C_2H_5Cl	10.8	CO	14.1
CH_3OH	10.8	H_2	15.4
CCl_4	11.0	CF_4	17.8
cyclo-C_6H_{12}	11.0	He	24.5

†Taken from a more complete list given in Landolt-Börnstein, *Zahlenwerte und Funktionen*, 6th ed., vol. 1, Part 3, p. 359 (Berlin: Springer, 1951).

polarizabilities dominate, it can be shown that the attractive potential between two dissimilar molecules is approximately given by the geometric mean of the potentials between the like molecules at the same separation. We can rewrite Eqs. (4.4-3) and (4.4-4):

$$\Gamma_{ij} = k \frac{\alpha_i \alpha_j}{r^6}, \quad \Gamma_{ii} = k \frac{\alpha_i^2}{r^6}, \quad \text{and} \quad \Gamma_{jj} = k \frac{\alpha_j^2}{r^6}, \tag{4.4-5}$$

where k is a constant which is approximately the same for the three types of interaction: i-i, j-j, and i-j. It then follows that

$$\Gamma_{ij} = \sqrt{\Gamma_{ii} \Gamma_{jj}}. \tag{4.4-6}$$

Equation (4.4-6) gives some theoretical basis for the frequently applied "geometric-mean rule" which is so often used in equations of state for gas mixtures and in theories of liquid solutions.

To show the relative magnitude of dipole, induction, and dispersion forces in some representative cases, London[6] has presented calculated potential energies for a few simple molecules. His results are given in the form

$$\Gamma_{ii} = -\frac{B}{r^6} \tag{4.4-7}$$

where B is calculated separately for the contribution due to dipole, induction and dispersion effects. In these calculations Eqs. (4.2-6), (4.3-3), and (4.4-4) were used; some results similar to those first given by London are shown in Table 4-5.

Table 4-5

RELATIVE MAGNITUDES OF INTERMOLECULAR FORCES
BETWEEN TWO IDENTICAL MOLECULES AT $0°C$

Molecule	Dipole moment (Debyes)	B (erg-cm$^6 \times 10^{60}$)		
		Dipole	Induction	Dispersion
CCl_4	0	0	0	1460
cyclo-C_6H_{12}	0	0	0	1560
CO	0.10	0.0018	0.0390	64.3
HI	0.42	0.550	1.92	380
HBr	0.80	7.24	4.62	188
HCl	1.08	24.1	6.14	107
NH_3	1.47	82.6	9.77	70.5
H_2O	1.84	203	10.8	38.1
$(CH_3)_2CO$	2.87	1200	104	486

The computed values of B indicate that the contribution of induction forces is small and that even for strongly polar substances, like ammonia, water or acetone, the contribution of dispersion forces is far from negligible.

Table 4-6 gives some calculated results for intermolecular forces between two molecules which are not alike. In these calculations Eqs. (4.2-6), (4.3-3), and (4.4-3) were used. Again we notice that polar forces are not important when the dipole moment is less than about 1 Debye and induction forces always tend to be much smaller than dispersion forces.

London's formula does not hold at very small separations where the electron clouds overlap and the forces between molecules are repulsive rather than attractive. Repulsive forces between nonpolar molecules at small distances are not understood as well as attractive forces at larger

Table 4-6

RELATIVE MAGNITUDES OF INTERMOLECULAR FORCES
BETWEEN TWO DIFFERENT MOLECULES AT $0°C$

Molecules		Dipole moment (Debyes)		B (ergs-cm^6 $\times$ 10^{60})		
(1)	(2)	(1)	(2)	Dipole	Induction	Dispersion
CCl_4	cyclo-C_6H_{12}	0	0	0	0	1510
CCl_4	NH_3	0	1.47	0	22.7	320
$(CH_3)_2CO$	cyclo-C_6H_{12}	2.87	0	0	89.5	870
CO	HCl	0.10	1.08	0.206	2.30	82.7
H_2O	HCl	1.84	1.08	69.8	10.8	63.7
$(CH_3)_2CO$	NH_3	2.87	1.47	315	32.3	185
$(CH_3)_2CO$	H_2O	2.87	1.84	493	34.5	135

distances. Theoretical considerations suggest that the repulsive potential should be an exponential function of intermolecular separation, but it is more convenient[7] to represent the repulsive potential by an inverse-power law of the type

$$\Gamma = \frac{A}{r^n},$$ (4.4-8)

where A is a positive constant and n is a number usually taken to be between 8 and 16.†

In order to take both repulsive and attractive forces between nonpolar molecules into account, it is customary to assume that the total potential energy is the sum of the two separate potentials:

$$\Gamma_{total} = \Gamma_{repulsive} + \Gamma_{attractive} = \frac{A}{r^n} - \frac{B}{r^m},$$ (4.4-9)

where A, B, n, and m are positive constants and where $n > m$. This equation was first proposed by Mie[8] and was extensively investigated by Lennard-Jones. Equation (4.4-9) forms the basis of a variety of physicochemical calculations; it has been used especially to calculate thermodynamic and transport properties of dilute nonpolar gases.[2]

†For atoms and for small molecules (e.g., He, H_2, Ne, Ar), n may be less than 8 for certain ranges of r. See I. Amdur, *Second Symposium on Thermophysical Properties* (ASME), J. F. Masi and D. H. Tsai, eds. (New York: Academic Press Inc., 1962).

4.5 Mie's Potential-Energy Function
for Nonpolar Molecules

Equation (4.4-9) gives the potential energy of two molecules as a function of their separation and it is apparent that at some distance r^* this energy is a minimum; this minimum energy is designated by Γ^*. By algebraic rearrangement, Mie's potential can be rewritten

$$\Gamma = \frac{\epsilon \, (n^n/m^m)^{1/(n-m)}}{n-m} \left[\left(\frac{\sigma}{r}\right)^n - \left(\frac{\sigma}{r}\right)^m\right], \tag{4.5-1}$$

where $\epsilon = -\Gamma^*$ and where σ is the intermolecular distance when $\Gamma = 0$.

London has shown from the theory of dispersion forces that $m = 6$ but we do not have a theoretical value for n. It is frequently convenient to let $n = 12$ in which case Eq. (4.5-1) becomes

$$\boxed{\Gamma = 4\epsilon \left[\left(\frac{\sigma}{r}\right)^{12} - \left(\frac{\sigma}{r}\right)^6\right].} \tag{4.5-2}$$

Equation (4.5-2) is called the Lennard-Jones potential.† It relates the potential energy of two molecules to their distance of separation in terms of two parameters: an energy parameter ϵ which, when multiplied by minus one, gives the minimum energy corresponding to the equilibrium separation; and a distance parameter σ which is equal to the intermolecular separation when the potential energy is zero. An illustration of Eqs. (4.5-2) and (4.5-2a) is given in Fig. 4-1.

Because of the steepness of the repulsion potential, the numerical values of r^* and σ are not far apart. For a Lennard-Jones (12-6) potential we obtain

$$\sigma = \sqrt[6]{\tfrac{1}{2}} \, r^* = 0.8909 \, r^*. \tag{4.5-3}$$

The constants A, B and n can be estimated from a variety of physical properties. Subject to several simplifying assumptions, they can be computed, for example, from the compressibility of solids at low temperatures or from specific heat data of solids or liquids. More commonly they are obtained from the variation of viscosity or self-diffusivity with tempera-

---

†Sometimes better agreement with experiment is obtained by letting $n = 9$. In that case Eq. (4.5-1) becomes

$$\Gamma = \frac{27}{4} \, \epsilon \left[\left(\frac{\sigma}{r}\right)^9 - \left(\frac{\sigma}{r}\right)^6\right]. \tag{4.5-2a}$$

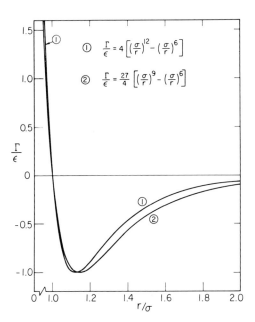

Fig. 4-1   Two forms of Mie's potential for simple, nonpolar molecules.

ture, and, most commonly, from gas-phase volumetric properties as expressed by second virial coefficients.[9]

Mie's potential applies to two nonpolar, spherically symmetric molecules which are completely isolated. In nondilute systems, and especially in condensed phases, two molecules are not isolated but have many other molecules in their vicinity. By introducing appropriate simplifying assumptions, it is possible to construct a simple theory of dense media using a form of Mie's two-body potential such as that of Lennard-Jones.[9]

Consider a condensed system at conditions not far removed from those prevailing at the triple point. We assume that the total potential energy is primarily due to interactions between nearest neighbors. Let the number of nearest neighbors in a molecular arrangement be designated by z. The total potential energy Γ_t in a system containing N molecules is then approximately given by

$$\Gamma_t = \tfrac{1}{2} N z \Gamma, \qquad\qquad (4.5\text{-}4)$$

where Γ is the potential energy of an isolated pair.† The factor $\tfrac{1}{2}$ is needed

† Even if we neglect the effect of nonnearest neighbors, Eq. (4.5-4) is not exact because it assumes additivity of two-body potentials to give the potential energy of a multibody system.

to avoid counting each pair twice. Substituting Mie's equation into Eq. (4.5-4) we have

$$\Gamma_t = \frac{1}{2} N z \left(\frac{A}{r^n} - \frac{B}{r^m} \right),$$ (4.5-5)

where r is the distance between two adjacent molecules.

Equation (4.5-5) considers only interaction between nearest-neighbor molecules. To account for additional potential energy resulting from interaction of a molecule with all of those outside its nearest-neighbor shell, numerical constants s_n and s_m (which are near unity) are introduced by rewriting the total potential energy:

$$\Gamma_t = \frac{1}{2} N z \left(\frac{s_n A}{r^n} - \frac{s_m B}{r^m} \right).$$ (4.5-6)

When the condensed system is considered as a lattice such as that existing in a regularly spaced crystal, the constants s_n and s_m can be accurately determined from the lattice geometry. For example, a molecule in a crystal of the simple-cubic type has 6 nearest neighbors at a distance r, 12 at a distance $r\sqrt{2}$, 8 at a distance $r\sqrt{3}$, and so on. The attractive energy of one molecule with respect to all of the others is then given by

$$\begin{aligned}
\Gamma_{\text{attractive}} &= A \left[\frac{6}{r^m} + \frac{12}{(\sqrt{2}\, r)^m} + \frac{8}{(\sqrt{3}\, r)^m} \right] + \cdots \\
&= \frac{6A}{r^m} \left(1 + \frac{2}{(\sqrt{2})^m} + \frac{4}{(\sqrt{3})^{m+2}} \right) + \cdots \\
&= \frac{z A s_m}{r^m}
\end{aligned}$$ (4.5-7)

and

$$s_m = 1 + \frac{2}{(\sqrt{2})^m} + \frac{4}{(\sqrt{3})^{m+2}} + \cdots$$ (4.5-8)

Similarly, s_n can be calculated for the repulsive potential. Summation constants for several geometrical arrangements are shown in Table 4-7.

Having obtained numerical values for the constants s_m and s_n, it is now possible to obtain a relation between the equilibrium distance of separation r^* of an isolated pair of molecules and the equilibrium distance of separation r_i^* between a molecule and its nearest neighbors in a condensed system. At equilibrium, the potential energy of the condensed system is a minimum; therefore

$$\left(\frac{d\Gamma_t}{dr} \right)_{r=r_i^*} = 0.$$ (4.5-9)

Table 4-7

SUMMATION CONSTANTS s_n AND s_m FOR CUBIC LATTICES†

n or m	Simple Cubic $z = 6$	Body-Centered Cubic $z = 8$	Face-Centered Cubic $z = 12$
6	1.4003	1.5317	1.2045
9	1.1048	1.2368	1.0410
12	1.0337	1.1394	1.0110
15	1.0115	1.0854	1.0033

†E. A. Moelwyn-Hughes, *Physical Chemistry* 2nd ed. (Oxford: Pergamon Press, 1961).

Substituting Eq. (4.5-6) we obtain

$$(r_i^*)^{n-m} = \frac{s_n n A}{s_m m B},$$ (4.5-10)

and comparing this result with that obtained for an isolated pair of molecules, we have

$$\left(\frac{r^*}{r_i^*}\right)^{n-m} = \frac{s_m}{s_n}.$$ (4.5-11)

Since m is equal to 6, and assuming that n is between 8 and 16, the values in Table 4-7 show that the equilibrium distance in an isolated pair is always a few percent larger than that in a condensed system. This leads to the interesting result that for a pair of adjacent molecules in a condensed system, the absolute value of the average potential energy is roughly fifty percent smaller than that corresponding to the equilibrium separation between a pair of isolated molecules.

Mie's potential, as well as other similar potentials for nonpolar, spherically symmetrical molecules, contains one independent variable r and one dependent variable Γ. When these variables are nondimensionalized with characteristic molecular constants, the resulting potential function leads to a useful generalization known as the molecular theory of corresponding states.

4.6   Molecular Theory of Corresponding States

Classical (or macroscopic) theory of corresponding states was derived by van der Waals on the basis of his well-known equation of state. It can be shown, however, that van der Waals' derivation is not strictly tied to any particular equation but can be applied to any equation of state containing two arbitrary constants in addition to the gas constant R. From the principle of continuity of the gaseous and liquid phases, van der Waals showed that at the critical point

$$\left(\frac{\partial P}{\partial v}\right)_T = \left(\frac{\partial^2 P}{\partial v^2}\right)_T = 0 \quad \text{(at critical point).} \qquad (4.6\text{-}1)$$

These relations led van der Waals to the general result that for variables v (volume), T (temperature) and P (pressure) there exists a universal function $\mathcal{f}$ such that

$$\mathcal{f}\left(\frac{v}{v_c}, \frac{T}{T_c}, \frac{P}{P_c}\right) = 0 \qquad (4.6\text{-}2)$$

is valid for all substances; the subscript c refers to the critical point. Another way of stating this result is to say that if the equation of state for any one fluid is written in reduced coordinates (i.e., v/v_c, T/T_c, P/P_c), then that equation is also valid for any other fluid.

Classical theory of corresponding states is based on mathematical properties of the macroscopic equation of state. Molecular (or microscopic) theory of corresponding states, however, is based on mathematical properties of the potential-energy function.

Intermolecular forces of a number of substances are closely approximated by the inverse-power potential function given by Eq. (4.4-9). The independent variable in this potential function is the distance between molecules. When this variable is made dimensionless, the potential function can be rewritten in a general way such that the dimensionless potential is a universal function $\mathcal{f}$ of the dimensionless distance of separation between molecules:

$$\frac{\Gamma_{ii}}{\epsilon_i} = \mathcal{f}\left(\frac{r}{\sigma_i}\right), \qquad (4.6\text{-}3)$$

where ϵ_i is an energy parameter and σ_i is a distance parameter characteristic of the interaction between two molecules of species i. For example, if the function $\mathcal{f}$ is given by the Lennard-Jones potential, then ϵ_i is the energy (times minus one) at the potential-energy minimum, and σ_i is the distance corresponding to zero potential energy. However, Eq. (4.6-3) is not restricted to the Lennard-Jones potential, nor, in fact, is it restricted to an inverse-power function as given by Eq. (4.4-9). Equation (4.6-3) merely states that the reduced potential energy (Γ_{ii}/ϵ_i) is some universal function of the reduced distance (r/σ_i).

Equation (4.6-3) expresses the microscopic theory of corresponding states; it is analogous to the macroscopic theory of corresponding states expressed by Eq. (4.6-2). Equation (4.6-3) implies that if the potential-energy function for any one fluid is written in reduced coordinates (Γ_{ii}/ϵ_i) and (r/σ_i), then that same function is also valid for any other fluid.

Once the potential-energy function of a substance is known, it is possible, at least in principle, to compute the macroscopic configurational properties of that substance by the techniques of statistical mechanics. Hence, a universal potential-energy function such as given by Eq. (4.6-3) leads to a universal equation of state and to universal values for all reduced configurational properties which can be derived from the equation of state.

The four specific assumptions which form the basis of the microscopic (or molecular) theory of corresponding states were clearly stated by Pitzer[10] and by Guggenheim.[11] They are:

1. The potential energy between two molecules depends only on the distance between them and not on their mutual orientation; that is, the force fields of the molecules are spherically symmetric.
2. The potential energy, reduced by a characteristic energy, can be written as a universal function of the intermolecular distance, reduced by a characteristic length, viz., Eq. (4.6-3).
3. Classical (rather than quantum) statistical mechanics is applicable.
4. The potential energy of all the molecules is the sum of the potential energies of all possible pairs of molecules.

In order to relate numerically the macroscopic and the microscopic theories of corresponding states, it is desirable to establish a connection between the parameters of one theory and those of the other. In the microscopic theory there are two independent parameters—an energy parameter and a distance parameter. In the macroscopic theory there appear to be three—v_c, T_c, and P_c—but only two of these are independent because, according to the theory, the compressibility factor at the critical point is the same for all fluids.

Since the critical temperature is a measure of the kinetic energy of the fluid at a characteristic state (where the liquid and gaseous states become identical), we can expect that there exists a simple proportionality between the energy parameter ϵ_i and the critical temperature T_{c_i}. Similarly, the critical volume reflects the size of the molecules; hence it is reasonable to expect a proportionality between the distance parameter σ_i and the cube root of the critical volume v_{c_i}.† For simple nonpolar molecules, i.e., those having a small number of atoms per molecule, these relationships have been found empirically for the case where the generalized function f is

---

†More rigorously, the proportionality between ϵ and T_c and between σ and $v_c^{1/3}$ can be derived theoretically from Eq. (4.6-1) and from the appropriate statistical mechanical equations.

replaced by the Lennard-Jones (12-6) potential.[2] For that particular case, we have, approximately,

$$\frac{\epsilon}{k} = 0.77 T_c,$$ (4.6-4)

and

$$\tfrac{2}{3} \pi N_A \sigma^3 = 0.75 v_c,$$ (4.6-5)

where N_A is Avogadro's constant, σ is in centimeters, and v_c is in cc/g-mol.

The parameters ϵ_i and σ_i are directly related to the macroscopic properties of the substance i but the macroscopic property used to establish this relationship need not be a critical property. For example, ϵ_i can also be evaluated from the Boyle temperature and σ_i from the molar volume at the normal boiling point. Nor is it even necessary, in principle, that the molecular parameters be determined from thermodynamic data since these parameters are also related to transport properties like viscosity and diffusivity.[2] In fact, however, potential parameters obtained from different properties of the same fluid tend to be different because the assumed potential function (e.g., Lennard-Jones) is not the "true" potential function but only an approximation. Further, the four assumptions cited above are not strictly true for most substances. If a particular potential-energy function is to be used for calculating equilibrium properties, it is best to evaluate the parameters from a property similar to the one being investigated.

An important advantage of the molecular theory, relative to the classical theory of corresponding states, is that the former permits calculation of other macroscopic properties (e.g., transport properties) in addition to those which may be calculated by classical thermodynamics from an equation of state. For the purposes of phase-equilibrium thermodynamics, however, the main advantage of the molecular theory is that it can be meaningfully extended to mixtures, thereby providing much aid in typical phase-equilibrium problems. In view of the physical significance of the parameters ϵ_i and σ_i, it is possible to make reasonable predictions of what these parameters are for the interaction between dissimilar molecules. Thus, as a first approximation, the London theory suggests that for the interaction of two unlike molecules having nearly the same size and ionization potential,

$$\epsilon_{ij} = (\epsilon_i \epsilon_j)^{1/2},$$ (4.6-6)

and, on the basis of a hard-sphere model for molecular interaction,

$$\sigma_{ij} = \tfrac{1}{2} (\sigma_i + \sigma_j).$$ (4.6-7)

Equations (4.6-6) and (4.6-7) do, in fact, provide a basis for obtaining good results for the properties of a variety of mixtures. Some applications of the molecular theory of corresponding states to phase-equilibrium calculations are discussed in later chapters.

4.7 Extension of Corresponding-States Theory to More Complicated Molecules

The theory of corresponding states as expressed by the generalized potential function [Eq. (4.6-3)] is a two-parameter theory and is therefore limited to those molecules whose energies of interaction can be adequately described in terms of a function using only two parameters. Such molecules are called simple molecules; strictly speaking, only the heavier noble gases (argon, krypton, xenon) are "simple" but the properties of several others are closely approximated by Eq. (4.6-3). A simple molecule is one whose force field has a high degree of symmetry, which is equivalent to saying that the potential energy is determined only by the distance of separation and not by the relative orientation between two molecules. Nonpolar (or slightly polar) molecules like methane, oxygen, nitrogen and carbon monoxide are therefore nearly simple molecules. For more complex molecules, however, it is necessary to introduce at least one additional parameter in the potential function and thereby to construct a three-parameter theory of corresponding states. This can be done in several ways but the most convenient for practical purposes is to divide molecules into different classes, each class corresponding to a particular extent of deviation from simple-molecule behavior.[12] This extension of the theory of corresponding states relaxes the second assumption of the four listed in the previous section. Assumptions 3 and 4 are unaffected. Assumption 1 is relaxed somewhat in the sense that we now assume that it is sufficient to use an average potential function wherein we have averaged out all effects of asymmetry in the intermolecular forces. Extension of corresponding-states theory is mostly concerned with Assumption 2.

In the three-parameter theory of corresponding states, Eq. (4.6-3) still applies but the generalized function $\mathcal{L}$ is now different for each class. The class must be designated by some third parameter; for practical considerations a convenient parameter is one which is easily calculated from readily available data. Different parameters have been proposed by Riedel, Meissner and Sefarian, Rowlinson, and Pitzer; Pitzer's proposal is, perhaps, the most useful because his third parameter is calculated from experimental data which tend to be accurate as well as accessible. Pitzer defines an *acentric factor* ω which is a measure of the acentricity, i.e., the noncentral nature of intermolecular forces.

The definition of the acentric factor is arbitrary and chosen for convenience. According to two-parameter (simple-fluid) corresponding-states behavior, the reduced saturation pressures of all liquids should be a universal function of their reduced temperatures; in fact, however, they are not, and Pitzer uses this empirical result as a measure of deviation from simple-fluid behavior. For simple fluids, it has been observed that at a temperature equal to $\frac{7}{10}$ of the critical, the saturation pressure P^s divided by the critical pressure P_c is given by

$$\frac{P^s}{P_c} = \frac{1}{10} \qquad \left(\frac{T}{T_c} = 0.7\right). \tag{4.7-1}$$

Pitzer therefore defines the acentric factor by

$$\omega \equiv - \log_{10}\left(\frac{P^s}{P_c}\right)_{T/T_c = 0.7} - 1.000. \tag{4.7-2}$$

For simple fluids $\omega \approx 0$ and for more complex fluids, $\omega > 0$. Some acentric factors for typical fluids are given in Table 4-8. The acentric factor is easily determined from a minimum of experimental information; the data required are the critical temperature and pressure and the vapor pressure at a reduced temperature of 0.7. This reduced temperature is almost always close to the normal boiling point where vapor-pressure data are most likely to be available.

Table 4-8

ACENTRIC FACTORS

CH_4	0.013	C_6H_6	0.211
O_2	0.021	CO_2	0.225
N_2	0.040	$CH_3C_6H_{11}$	0.235
CO	0.049	$CH_3C_6H_5$	0.241
C_2H_4	0.085	NH_3	0.25
C_2H_6	0.105	C_4H_3O-CHO†	0.292
CF_4	0.174	$CFCl_3$	0.295
C_2H_5SH	0.186	n-C_6H_{14}	0.298
CCl_4	0.193	iso-C_8H_{18}	0.303
n-C_4H_{10}	0.200	H_2O	0.344
cyclo-C_6H_{12}	0.209	n-C_8H_{18}	0.398

† Furfural

The three-parameter corresponding-states theory asserts that all fluids having the same acentric factor must have the same reduced configurational properties at the same reduced temperature and pressure. Pitzer, et al., have tabulated these reduced configurational properties as determined from experimental data for representative fluids.[13]

Inclusion of a third parameter very much improves the accuracy of corresponding-states correlations. For most fluids (other than those which are highly polar or those which associate strongly by hydrogen bonding) the accuracy of the gas-phase compressibility factors given in Pitzer's tables is 2 percent or better, and for many gases it is much better.

4.8   Specific (Chemical) Forces

In addition to physical intermolecular forces which have been briefly described in the previous sections, there are specific forces of attraction which lead to the formation of new molecular species; such forces are called chemical forces. A good example of such a force is that which exists between ammonia and hydrogen chloride; in this case, a new species, ammonium chloride, is formed. Such forces, in effect, constitute the basis of the entire science of chemistry and it is impossible to discuss them adequately in a few pages. However, it is important to recognize that chemical forces can, in many cases, be of major importance in determining thermodynamic properties of solutions. Whereas in the previous sections we were able to write some simple formulas for the potential energy of physically interacting molecules, we cannot give simple quantitative relations which describe on a microscopic level interaction between chemically reactive molecules. Instead, we briefly discuss in a more qualitative manner some relations between chemical forces and properties of solutions.

There are numerous types of specific chemical effects which are of importance in the thermodynamics of solutions. For example, the solubility of silver chloride in water is very small; however, if some ammonia is added to the solution, the solubility rises by several orders of magnitude due to the formation of a silver-ammonia complex. Acetone is a much better solvent for acetylene than for ethylene because acetylene, unlike ethylene, can form a hydrogen bond with the carbonyl oxygen in the solvent. As a result of an electron donor-electron acceptor interaction, iodine is more soluble in aromatic solvents like toluene and xylene than in paraffinic solvents like heptane or octane. Finally, an example is provided by a well-known industrial process for the absorption of carbon dioxide in ethanol amine; carbon dioxide is readily soluble in this solvent because of specific chemical interaction between (acidic) solute and (basic) solvent, leading to complex formation.

The main difference between a physical and a chemical force lies in the criterion of saturation: Chemical forces are saturated but physical forces are not. The saturated nature of chemical forces is intimately connected with the theory of the covalent bond and also with the law of multiple proportions which says that the ratio of atoms in a molecule is a small, integral number. If two hydrogen atoms meet, they have a strong tendency to form a hydrogen molecule H_2, but once having done so they have no appreciable further tendency to form a molecule H_3. Hence the attractive force between hydrogen atoms is "satisfied" (or saturated) once the stable H_2 molecule is formed. On the other hand, the purely physical force between, say, two argon atoms knows no such "satisfaction." Two argon atoms which are attracted to form a doublet still have a tendency to attract a third argon atom, and a triplet has a further tendency to attract a fourth. It is true that in the gaseous state doublets are much more frequent than triplets but this results from the fact that in the dilute state a two-body collision is much more probable than a three-body collision. In the condensed or highly concentrated state there are aggregates of many argon atoms.

Chemical effects in solution are generally classified in terms of *association* or *solvation*. By the former we mean the tendency of some molecules to form polymers; for example, acetic acid consists primarily of dimers due to hydrogen bonding. By solvation we mean the tendency of molecules of different species to form complexes; for example, a solution of sulfur trioxide in water shows strong solvation by the formation of sulfuric acid. This particular example illustrates a severe degree of solvation but there are many cases where the solvation is much weaker; for example, there is a tendency for chloroform to solvate with acetone due to hydrogen bonding between the primary hydrogen of the chloroform and the carbonyl oxygen of the acetone, and this tendency has a profound effect on the properties of chloroform/acetone solutions. Chloroform also forms hydrogen bonds with di-isobutyl ketone but in this case the extent of complexing is much smaller because of steric hindrance, and as a result mixtures of chloroform with di-isobutyl ketone behave more ideally than do mixtures of chloroform with acetone. Solvation effects in solution are very common. These effects tend to produce negative deviations from Raoult's law since they necessarily decrease the volatility of the original components.

It is easy to see that whenever solvation occurs in solution it has a marked effect on the thermodynamic properties of that solution. It is, perhaps, not quite so obvious that association effects, when they occur, are also of major importance. The reason for this is that the extent of association is a strong function of the composition, especially in the range which is dilute with respect to the associating component. Pure methanol, for example, exists primarily as dimer, trimer and tetramer, but when methanol is dissolved in a large excess of hexane, it exists primarily as a monomer. As the methanol concentration rises, more polymers are formed; the fraction of methanol molecules which exists in the associated form is strongly dependent on the number of methanol molecules present per unit volume of solution, and as a result the fugacity of methanol is a highly nonlinear function of its mole fraction.

The ability of a molecule to solvate or associate is closely related to its electronic structure. For example, if we want to compare the properties of aluminum trichloride and antimony trichloride we note immediately an important difference in their electronic structures:

In the trichloride, antimony has its complete octet of electrons, and therefore its chemical forces are saturated. Aluminum, however, has only six electrons and has a strong tendency to add two more. As a result, aluminum chloride solvates easily with any molecule which can act as an electron donor whereas antimony chloride does not. This difference explains, at least in part, why aluminum chloride, unlike antimony chloride, is an excellent catalyst for some organic reactions such as, for example, the Friedel-Crafts reaction.

4.9   Hydrogen Bonds

The most common chemical effect encountered in the thermodynamics of solutions is that due to the hydrogen bond. While the "normal" valence of hydrogen is unity, many hydrogen-containing compounds behave as if hydrogen were bivalent. Extensive studies of hydrogen fluoride vapor, for example, show that the correct formula is $(HF)_n$, where n depends on temperature, and may be as much as six. The only reasonable way to explain this is to write the structure of hydrogen fluoride in this manner:

$$H—F---H—F---H—F$$

where the solid line indicates the "normal" bond and the dashed line an "auxiliary" bond. Similarly, studies on the crystal structure of ice show that each hydrogen is "normally" bonded to one oxygen atom and additionally attached to another oxygen atom:

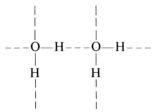

It appears that two sufficiently negative atoms X and Y (which may be identical) may, in suitable circumstances, be united with hydrogen according to X—H---Y. Consequently, molecules containing hydrogen linked to an electronegative atom (as in acids, alcohols and amines) show strong tendencies to associate with each other and to solvate with other molecules possessing accessible electronegative atoms.

The major difference between a hydrogen bond and a normal covalent bond is the former's relative weakness. The bond strength of most hydrogen bonds lies in the neighborhood of 2–10 kilocalories per gram-mole whereas the usual covalent bond strength is in the region 50–100 kilocalories per gram-mole. As a result, the hydrogen bond is broken rather easily and it is for this reason that hydrogen-bonding effects usually decrease at higher temperatures where the kinetic energy of the molecules is sufficient to break these loose bonds.

The strong effect of hydrogen bonding on physical properties is best illustrated by comparing some thermodynamic properties of two isomers such as dimethyl ether and ethyl alcohol. These molecules both have the formula C_2H_6O but strong hydrogen bonding occurs only in the alcohol. Some of the properties are shown in Table 4-9. Due to the additional

Table 4-9

SOME PROPERTIES OF THE ISOMERS ETHANOL
AND DIMETHYL ETHER

	Ethanol	Dimethyl Ether
Normal boiling point, °C	78	−25
Latent heat of evaporation at normal boiling point, kcal/g-mol	10.19	4.45
Trouton's constant,† cal/g-mol, °K	29.0	17.9
Solubility in water at 18°C and 1 atm, g/100 g	∞	7.12

†Trouton's constant is the entropy of vaporization at the normal boiling point.

cohesive forces in the hydrogen-bonded alcohol, the boiling point, heat of vaporization, and Trouton's constant are appreciably larger than those of the ether. Also, since ethanol can readily solvate with water, it is infinitely soluble in water whereas dimethyl ether is only partially soluble.

Hydrogen bonding between molecules of the same component can frequently be detected by studying the thermodynamic properties of a solution wherein the hydrogen-bonded substance is dissolved in a non-

polar, relatively inert solvent. When a strongly hydrogen-bonded substance such as ethanol is dissolved in an excess of a nonpolar solvent (such
as hexane or cyclohexane), hydrogen bonds are broken until, in the limit
of infinite dilution, all the alcohol molecules exist as monomers rather
than as dimers, trimers or higher aggregates. This follows simply from
the law of mass action: In the equilibrium $n\text{A} \rightarrow \text{A}_n$ (where n is any inte-
ger greater than one) the fraction of A molecules which are monomeric
(i.e., not polymerized) increases with falling total concentration of all A
molecules, polymerized or not. As the total concentration of A molecules
in the solvent approaches zero, the fraction of all A molecules which are
monomers approaches unity. The strong dependence of the extent of
polymerization on solute concentration results in characteristic thermodynamic behavior as shown in Figs. 4-2 and 4-3. Figure 4-2 gives the
enthalpy of mixing per mole of solute as a function of solute mole fraction

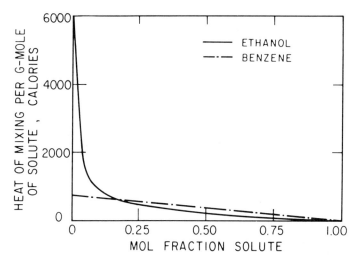

Fig. 4-2 Hydrogen bonding in solution. Heat effects for two
chemically different solutes in cyclohexane at 20°C.

at constant temperature and pressure. The behavior of strongly hydrogen-bonded ethanol is contrasted with that of nonpolar benzene and the
qualitative difference makes it clear that the effect of the solvent on the
ethanol molecules is markedly different from that on the benzene molecules. When benzene is mixed isothermally with a paraffinic or naph-
thenic solvent, there is a small absorption of heat and a small expansion
due to physical (essentially dispersion) forces. However, when ethanol is
dissolved in an "inert" solvent, hydrogen bonds are broken and, since
such breaking requires energy, much heat is absorbed; further, since a

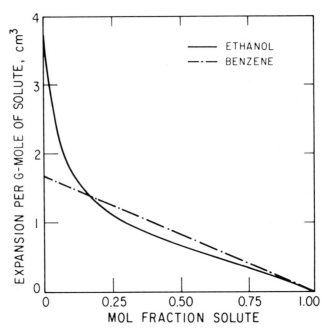

Fig. 4-3   Hydrogen bonding in solution. Volumetric effects for
two chemically different solutes in n-hexane at 6°C.

hydrogen-bonded network of molecules tends to occupy somewhat less
space than the sum of the individual nonbonded molecules, there is appreciable expansion in the volume of the mixture, as shown in Fig. 4-3.

Evidence for hydrogen-bond formation between dissimilar molecules
can be obtained in a variety of ways; a good survey is given by Pimentel
and McClellan.[14] We shall not discuss all of them here but merely discuss briefly two types of thermodynamic evidence which are best illustrated by an example, viz., hydrogen bonding between acetone and
chloroform.

Dolezalek[15] observed in 1908 that the partial pressures for liquid mixtures of acetone and chloroform were lower than those calculated from
Raoult's law and he interpreted this negative deviation as a consequence
of complex formation between the two dissimilar species. However, negative deviations from ideality can result from causes other than complex
formation; furthermore, a binary liquid mixture which forms weak complexes between the two components may nevertheless have partial pressures slightly larger than those calculated from Raoult's law.[16]   Thus
Dolezalek's evidence, while pertinent, is not completely convincing.

More direct evidence for hydrogen bond formation has been obtained
by Campbell and Kartzmark[17] who measured freezing points and heats of
mixing for mixtures of acetone with chloroform and with carbon tetra-

chloride. By comparing the properties of these two systems, it becomes evident that there is a large difference between the interaction of acetone with chloroform and the interaction of acetone with carbon tetrachloride, a molecule similar to chloroform except that the latter possesses an electron-accepting hydrogen atom which is capable of interacting with the electron-donating oxygen atom in acetone.

Figure 4-4 gives freezing point data for the two systems. The mixtures

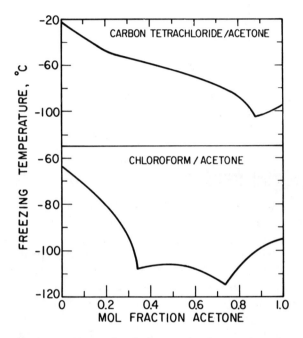

Fig. 4-4   Evidence for hydrogen bonds: Freezing-point data.

containing carbon tetrachloride exhibit simple behavior with a eutectic being formed at $-105°C$ and 87.5 mol % acetone. However, the mixtures containing chloroform show more complicated behavior: Two eutectics are formed, one at $-106°C$ and 31 mol % acetone and the other at $-115°C$ and 74 mol % acetone, and there is a convex central section whose maximum is just at 50 mol %. This indicates that the compound $(CH_3)_2CO---HCCl_3$ exists in the solid state although it is readily dissociated in the liquid state. The existence of such a compound is excellent evidence for strong interaction between the two dissimilar molecules. Since the maximum in the diagram occurs at the midpoint on the composition axis, we conclude that the complex has a 1:1 stoichiometric ratio as we would expect from the structure of the molecules.

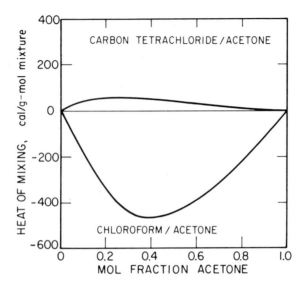

Fig. 4-5   Evidence for hydrogen bonds: Heat-of-mixing data.

The enthalpies of mixing for the two binary systems are shown in Fig. 4-5. Unfortunately, the original reference does not mention the temperature for these data but it is probably 25°C or some other temperature nearby. The enthalpy of mixing of acetone with carbon tetrachloride is positive (heat is absorbed), whereas the enthalpy of mixing of acetone with chloroform is negative (heat is evolved), and it is almost one order of magnitude larger. These data provide strong support for the contention that a hydrogen bond is formed between acetone and chloroform. The effect of physical intermolecular forces (dipole, induction, dispersion) causes a small amount of heat to be absorbed as shown by the data for the carbon tetrachloride mixtures; in the chloroform mixtures, however, there is a chemical heat effect which not only cancels the physical contribution to the observed heat of mixing but, since it is so much larger, causes heat to be evolved. Since energy is needed to break hydrogen bonds it necessarily follows that heat is liberated when hydrogen bonds are formed.

The freezing-point data and the calorimetric results show that in the liquid phase an equilibrium exists, having the form

$$(CH_3)_2C{=}O + HCCl_3 \;\rightleftarrows\; (CH_3)_2C{=}O{-}{-}{-}HCCl_3$$

It can be shown from the law of mass action that when acetone is very dilute in chloroform (i.e., when the molar ratio of chloroform to acetone is very much larger than unity) all of the acetone in solution is complexed with chloroform. For this reason the enthalpy data shown in Fig. 4-5 were replotted by Campbell and Kartzmark on coordinates of enthalpy of mixing per mole of acetone versus mole fraction of acetone. From the intercepts of this plot (zero mol % acetone) $\Delta h = -2090$ cal/g-mol for the system containing chloroform and $\Delta h = +625$ calories for the system containing carbon tetrachloride. From these results we can calculate that the enthalpy of complex formation, i.e., the enthalpy of the hydrogen bond, is $-2090 - 625 = -2715$ cal/g-mol. In view of the experimental uncertainties which are magnified by this particular method of data reduction, Campbell and Kartzmark give $\Delta h = -2.7 \pm 0.1$ kcal/g-mol as the enthalpy of hydrogen bond formation. This result is in fair agreement with enthalpy data for similar hydrogen bonds determined by different measurements.

4.10 Electron-Donor-Acceptor Complexes

While the consequences of hydrogen bonding are probably the most common chemical effect in solution thermodynamics, chemical effects may also result from other kinds of bonding forces leading to loose complex formation between electron donors and electron acceptors[18,19] which sometimes are called "charge-transfer complexes." The existence of donor-acceptor complexes can frequently be established by ultraviolet spectroscopy and, subject to certain simplifying assumptions, spectroscopic data may be used to give a quantitative measure of their stability.[20,21]

The general basis of such measurements is given schematically in Fig. 4-6 which shows two optical cells; in the upper cell we have in series two separate, dilute solutions of components A and B in some "inert" solvent. In the lower cell there is a single solution of A and B dissolved in the same "inert" solvent. We put equal numbers of A molecules in

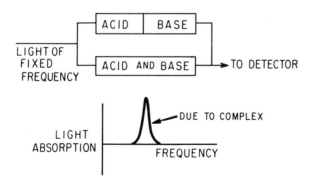

Fig. 4-6 Spectroscopic measurement of acid-base complex.

the top and bottom cells and we do the same with B. Monochromatic light of equal intensity is passed through both cells. If a complex is formed between A and B, and if the light frequency is in the absorption range of the complex, then light absorption is larger in the lower cell. On the other hand, if no complex is formed, light absorption is the same in both cells. The quantitative difference in light absorption provides a basis for subsequent calculations of complex stability and when such spectroscopic measurements are performed at different temperatures, it is possible to calculate the enthalpy and entropy of complex formation.

Table 4-10 gives some results for complex formation between s-trinitrobenzene (an electron acceptor) and aromatic hydrocarbons (electron

Table 4-10

SPECTROSCOPIC EQUILIBRIUM CONSTANTS AND ENTHALPIES FOR FORMATION OF s-TRINITROBENZENE/AROMATIC COMPLEXES DISSOLVED IN CYCLOHEXANE AT 20°C†

Aromatic	Equilibrium Constant (l/g-mol)	Δh(kcal/g-mol)
Benzene	0.88	−1.47
Mesitylene	3.51	−2.30
Durene	6.02	−2.72
Pentamethyl benzene	10.45	−3.55
Hexamethyl benzene	17.50	−4.37

†C. C. Thompson, Jr., and P. A. D. de Maine, *J. Phys. Chem.*, **69**, 2766 (1965).

donors). The results show that complex stability rises with the number of methyl groups on the benzene ring in agreement with various other measurements which have shown that π-electrons on the aromatic ring become more easily displaced when methyl groups are added; for example, the ionization potentials of methyl-substituted benzenes fall with increasing methyl substitution.† For s-trinitrobenzene with aromatics, complex stability is strong. For complexes of aromatics with other, more common polar organic molecules, complex stability is much weaker but not negligible as indicated by the experimental results of Weimer[21] shown in Table 4-11. For the polar solvents listed, Weimer found no complex formation with saturated hydrocarbons and as a result we may expect the thermodynamic properties of solutions of these polar solvents with aromatics to be significantly different from those of solutions of the same solvents with paraffins and naphthenes as has been observed.[22,23] The tendency of polar solvents to form complexes with unsaturated hydrocarbons, but not with saturated hydrocarbons, supplies a basis for various

†A lower ionization potential means that an electron can be removed more easily.

Table 4-11

SPECTROSCOPIC EQUILIBRIUM CONSTANTS FOR FORMATION
OF POLAR SOLVENT/p-XYLENE COMPLEXES DISSOLVED IN
n-HEXANE AT 25°C†

Polar Solvent	Equilibrium Constant (l/g-mol)
Acetone	0.25
Cyclohexanone	0.15
Triethylphosphate	0.14
Methoxyacetone	0.12
Cyclopentanone	0.12
γ-Butyrolactone	0.09
N-methyl pyrrolidone	0.09
Propionitrile	0.07
Nitromethane	0.05
Nitroethane	0.05
2-Nitropropane	0.05
Citraconic Anhydride	0.04
2-Nitro-2-methyl propane	0.03

†R. F. Weimer and J. M. Prausnitz, *Spectrochim. Acta*, **22**, 77 (1966).

commercial separation processes in the petroleum industry, including the well-known Edeleanu and Udex processes.

Both physical and chemical forces play an important role in determining properties of solutions. In many cases chemical forces can be neglected, thereby leading to purely "physical" solutions, but in other cases chemical forces predominate. The dissolved condition therefore, represents a state of wide versatility where in one extreme the solvent is merely a diluent with respect to the solute, while in the other extreme it is a chemical reactant.

The aim in applying thermodynamic methods to phase-equilibrium problems is to order, interpret, correlate, and finally, to predict properties of solutions. The extent to which this aim can be fulfilled depends in large measure on the degree of our understanding of intermolecular forces which are responsible for the molecules' behavior. Classical and statistical thermodynamics can define useful functions and derive relationships between them, but the numerical values of these functions cannot be determined by thermodynamics alone. Determination of numerical values, either by theory or experiment, is, strictly speaking, outside the realm of thermodynamics; such values depend directly on the microscopic physics and chemistry of the molecules involved. Future progress, therefore, in applications of phase-equilibrium thermodynamics, is possible only with increased knowledge of intermolecular forces.

PROBLEMS

1. (a) Consider one molecule of nitrogen and one molecule of ammonia 25 Å apart and at a temperature well above room temperature. Compute the force acting between these molecules. Assume that the potential energies due to various causes are additive and use the simple, spherically symmetric formulas for these potentials. Use c.g.s. units.

(b) Consider a gaseous mixture of N_2 and CO and compare it with a gaseous mixture of N_2 and Ar. Which mixture is more likely to follow Amagat's law (approximately)? Why?

2. Consider a diatomic molecule such as carbon monoxide. What is meant by the force constant of the C–O bond? How is it measured? Why is the heat capacity of CO significantly larger than that of argon?

3. Consider a spherical nonpolarizable molecule of radius 3 Å having at its center a dipole moment of 2 Debye. This molecule is dissolved in a nonpolar liquid having a dielectric constant equal to 3.5. Calculate the energy which is required to remove this molecule from the solution; express your answer in calories.

Suppose that you wished to evaporate hydrogen cyanide from a solution in carbon tetrachloride or in octane. Which case is likely to require more heat?

4. (a) For a nonpolar molecule how can the polarizability be computed from data on index of refraction?

(b) What would happen to the heat of vaporization of a liquid if it were subjected to a strong electric field?

5. What is meant by the terms "electron affinity" and "ionization potential"? How are these concepts related to the Lewis definition of acids and bases? In the separation of aromatics from paraffins by extraction, why is liquid SO_2 a better solvent than liquid NH_3?

6. Suppose that you want to measure the dipole moment of chlorobenzene in an inert solvent such as *n*-heptane. What experimental measurements would you make and how would you use them to compute the dipole moment?

7. Consider a binary solution containing components 1 and 2. The Lennard-Jones parameters ϵ_{11}, ϵ_{22} and σ_{11} and σ_{22} are known. Assuming that $\sigma_{12} = \frac{1}{2}(\sigma_{11} + \sigma_{22})$, express ϵ_{12} in terms of pure-component parameters. Also find the conditions under which $\epsilon_{12} = (\epsilon_{11}\epsilon_{22})^{1/2}$.

8. What is a hydrogen bond? Cite all the experimental evidence you can which supports the conclusion that phenol is a hydrogen-bonded substance.

9. Qualitatively compare the activity coefficient of acetone when dissolved in carbon tetrachloride with that when dissolved in chloroform.

10. (a) Vapor-phase spectroscopic data clearly show that sulfur dioxide and normal butene-2 form a complex. However, thermodynamic data at 0°C show that liquid mixtures of these components exhibit slight positive deviations from Raoult's law. Is this possible? Or do you suspect that there may be experimental error?

(b) Qualitatively compare the excess Gibbs energies of mixtures containing sulfur dioxide and normal butene-2 with those containing sulfur dioxide and isobutene. Explain.

11. Consider a polymer like cellulose nitrate. Explain why (as observed) a mixture of two polar solvents is frequently more effective in dissolving this polymer than either polar solvent by itself.

12. (a) What are the assumptions of the molecular theory of corresponding states?

(b) Does the Kihara potential necessarily violate any of these assumptions? Explain.

(c) Which of these assumptions, if any, are *not* obeyed by hydrogen?

(d) Does the molecular theory of corresponding states have anything at all to say about C_p (specific heat at constant pressure)? If so, what?

13. On the basis of the Stockmayer potential devise a corresponding-states theorem for polar substances. How many reducing parameters are there and what are they?

14. (a) What is a Gouy balance? What is a Bohr magneton?

(b) Some of the solution properties of oxygen are distinctly different from those of nitrogen or argon. What is a possible explanation of such differences?

15. Sketch *qualitatively* a plot of compressibility factor z vs. mole fraction of amine y, for the following mixtures at 170°C and 25 atm:

I. Dimethylamine/hydrogen
II. Dimethylamine/hydrogen chloride

In your sketch, indicate on the ordinate where $z = 1$. Sketch z for the entire region $y = 0$ to $y = 1$. Briefly explain your reasoning.

Critical temperatures of the pure components (°C):

Dimethylamine	165
Hydrogen	-240
Hydrogen chloride	51.4

16. Using your knowledge of intermolecular forces, explain the following observations:

I. At 30°C, the solubilities of ethane and acetylene in *n*-octane are about the same. However, at the same temperature, the solubility of acetylene in dimethyl formamide is very much larger than that of ethane.

II. At 10°C and 600 psia total pressure the K value for benzene in the methane/benzene system is much larger than that in the hydrogen/benzene system at the same temperature and pressure ($K \equiv y/x$). However, at 10°C and 50 psia the two K values are nearly the same.

REFERENCES

1. Keesom, W. H., *Physik. Z.*, **23**, 225 (1922).

2. Hirschfelder, J. O., C. F. Curtiss and R. B. Bird, *Molecular Theory of Gases and Liquids*. New York: John Wiley & Sons, Inc., 1954.

3. Buckingham, A. D., *Quart. Rev.* (London), **13**, 183 (1959).

4. Buckingham, A. D., and R. L. Disch, *Proc. Roy. Soc.* (London), **A263**, 275 (1963).

5. Parsonage, N. G., and R. L. Scott, *J. Chem. Phys.*, **37**, 304 (1962).

6. London, F., *Trans. Faraday Soc.*, **33**, 8 (1937).

7. Amdur, I., et al., *J. Chem. Phys.*, **22**, 644, 670, 1071 (1954).

8. Mie, G., *Ann. Physik*, **11**, 657 (1903).

9. Moelwyn-Hughes, E. A., *Physical Chemistry* (2nd ed.). London: Pergamon Press, 1961.

10. Pitzer, K. S., *J. Chem. Phys.*, **7**, 583 (1939).

11. Guggenheim, E. A., *J. Chem. Phys.*, **13**, 253 (1945).

12. Pitzer, K. S., D. Z. Lippman, R. F. Curl, Jr., C. M. Huggins, and D. E. Petersen, *J. Am. Chem. Soc.*, **77**, 3427, 3433 (1955); **79**, 2369 (1957); *Ind. Eng. Chem.*, **50**, 265 (1958).

13. A complete tabulation is given in G. N. Lewis, M. Randall, K. S. Pitzer and L. Brewer, *Thermodynamics* (2nd ed.), Appendix 1. New York: McGraw-Hill Book Company, 1961.

14. Pimentel, G. C., and A. L. McClellan, *The Hydrogen Bond*. San Francisco: W. H. Freeman & Co., 1960.

15. Dolezalek, F., *Z. Phys. Chem.* **64**, 727 (1908).

16. Booth, D., F. S. Dainton and K. J. Ivin, *Trans. Faraday Soc.*, **55**, 1293 (1959).

17. Campbell, A. N., and E. M. Kartzmark, *Can. J. Chem.*, **38**, 652 (1960).

18. Briegleb, G., *Elektronen-Donator-Acceptor Komplexe*. Berlin: Springer, 1961.

19. Andrews, L. J., and R. M. Keefer, *Molecular Complexes in Organic Chemistry*. San Francisco: Holden-Day, Inc., 1964.

20. Rossotti, F. J. C., and H. Rossotti, *The Determination of Stability Constants*. New York: McGraw-Hill Book Company, 1961.

21. Weimer, R. F., and J. M. Prausnitz, *Spectrochim. Acta*, **22**, 77 (1966).

22. Orye, R. V., and J. M. Prausnitz, *Trans. Faraday Soc.*, **61**, 1338 (1965).

23. Orye, R. V., R. F. Weimer and J. M. Prausnitz, *Science*, **148**, 74 (1965).

Fugacities in Gas Mixtures

5

It was shown in Chap. 2 that the basic equation of equilibrium between two phases α and β, which are at the same temperature, is given by the equality of the fugacities for any component i in these phases:

$$f_i^\alpha = f_i^\beta.$$

In many cases of interest one of the phases is a gaseous mixture; in this chapter we discuss methods for calculating the fugacity of a component in such a mixture.

Formal thermodynamics for calculating fugacities from volumetric data was discussed in Chap. 3; the two key equations are:

$$\ln \varphi_i = \frac{1}{RT} \int_0^P \left[\left(\frac{\partial V}{\partial n_i} \right)_{T,P,n_j} - \frac{RT}{P} \right] dP, \tag{1}$$

and

$$\ln \varphi_i = \frac{1}{RT} \int_V^\infty \left[\left(\frac{\partial P}{\partial n_i} \right)_{T,V,n_j} - \frac{RT}{V} \right] dV - \ln z, \tag{2}$$

where the fugacity coefficient φ_i is defined by

$$\varphi_i \equiv \frac{f_i}{y_i P},\qquad\qquad(3)$$

and z is the compressibility factor of the mixture.

Equation (1) is used whenever the volumetric data are given in volume-explicit form; i.e., whenever

$$V = \mathcal{f}_V(T, P, n_1, \dots).\qquad\qquad(4)$$

Equation (2) is used in the more common case when the volumetric data are expressed in pressure-explicit form; i.e., whenever

$$P = \mathcal{f}_P(T, V, n_1, \dots).\qquad\qquad(5)$$

The mathematical relation between volume, pressure, temperature and composition is called the *equation of state* and most forms of the equation of state are of the pressure-explicit type. Therefore Eq. (2) is frequently more useful than Eq. (1). At low or moderate pressures, it is sometimes possible to describe the volumetric properties of a gaseous mixture in a volume-explicit form, in which case Eq. (1) can be used; more often, however, the volumetric properties are much better represented in pressure-explicit form, requiring the use of Eq. (2).

Equations (1) and (2) are exact and if the information needed to evaluate the integrals is at hand, then the fugacity coefficient can be calculated exactly. The problem of calculating fugacities in the gas phase, therefore, is equivalent to the problem of estimating volumetric properties. Techniques for estimating such properties must come not from thermodynamics, but rather from molecular physics; it is for this reason that Chap. 4, dealing with intermolecular forces, precedes Chap. 5.

5.1 The Lewis Fugacity Rule

A particularly simple and popular approximation for calculating fugacities in gas-phase mixtures is given by the Lewis rule; the thermodynamic basis of the Lewis rule has already been given (Sec. 3.1). The assumption on which the rule rests states that at constant temperature and pressure the molar volume of the mixture is a linear function of the mole fraction. This assumption (Amagat's law) must hold not only at the pressure of interest but for all pressures up to the pressure of interest.

As shown in Sec. 3.1, the fugacity of component i in a gas mixture can be related to the fugacity of pure gaseous i at the same temperature and pressure by the exact relation:

$$RT \ln \frac{f_i}{y_i f_{\text{pure } i}} = \int_0^P (\bar{v}_i - v_i) dP. \tag{5.1-1}$$

According to Amagat's law, $\bar{v}_i = v_i$, and assuming validity of this equality over the entire pressure range $0 \rightarrow P$, the Lewis fugacity rule follows directly from Eq. (5.1-1):

$$f_i = y_i f_{\text{pure } i} \quad \text{(at same } T \text{ and } P), \tag{5.1-2}$$

or, in equivalent form,

$$\varphi_i = \varphi_{\text{pure } i} \quad \text{(at same } T \text{ and } P). \tag{5.1-3}$$

In effect, the Lewis rule assumes that at constant temperature and pressure, the fugacity coefficient of i is independent of the composition of the mixture and is independent of the nature of the other components which are in the mixture. These are drastic assumptions. On the basis of our knowledge of intermolecular forces, we recognize that for component i, deviations from ideal-gas behavior (as measured by φ_i), depend not only on temperature and pressure, but also on the relative amounts of component i and other components $j, k, \ldots$; further, we recognize that φ_i must depend on the chemical nature of these other components with which component i can interact. The Lewis rule, however, precludes such dependence; according to it, φ_i is a function only of temperature and pressure but not of composition.

Nevertheless, the Lewis rule is frequently preferred because of its simplicity, and for its practical usefulness in certain calculations. We may expect that the partial molar volume of a component i is close to the molar volume of pure i at the same temperature and pressure whenever the intermolecular forces experienced by a molecule i in the mixture are similar to those which it experiences in the pure state. In more colloquial terms, a molecule i which feels "at home" while it is "with company" is going to possess properties in the mixture which are close to those it has in the pure state. It therefore follows that for a component i, the Lewis fugacity rule

1. is always a good approximation at low pressures where the gas phase is nearly ideal;
2. is always a good approximation at any pressure whenever i is present in large excess (say, $y_i > 0.9$). The Lewis rule becomes exact in the limit as $y_i \rightarrow 1$;
3. is often a fair approximation over a wide range of composition and

pressure whenever the physical properties of all the components are nearly the same (e.g., N_2—CO or O_2—Ar);

4. is almost always a poor approximation at moderate and high pressures whenever the molecular properties of the other components are significantly different from those of i and when i is not present in excess. If y_i is small and the molecular properties of i differ very much from those of the dominant component in the mixture, the error introduced by the Lewis rule is often extremely large. (See Sec. 5.14.)

One of the practical difficulties encountered in the use of the Lewis rule for vapor-liquid equilibria results from the frequent necessity of introducing an arbitrary hypothetical state. At the temperature of the mixture, it often happens that P, the total pressure, exceeds P_i^s, the saturation pressure of pure component i. In that case $\varphi_{\text{pure } i}$, the fugacity coefficient of pure gas i at the temperature and pressure of the mixture, is, strictly speaking, meaningless because pure gas i cannot physically exist at these conditions. The calculation of $\varphi_{\text{pure } i}$ under such conditions requires arbitrary assumptions about the nature of the hypothetical pure gas and consequently, further inaccuracies may result from use of the Lewis rule.

In summary, then, the Lewis fugacity rule is attractive because of convience but it has no general validity. However, when applied to certain limiting situations it frequently provides good approximations.

5.2   The Virial Equation of State

As indicated at the beginning of this chapter, the problem of calculating fugacities for components in a gaseous mixture is equivalent to the problem of establishing a reliable equation of state for such a mixture; once such an equation of state exists, the fugacities can be found by straightforward computation. Such computation involves no difficulties in principle although it may be tedious and, since a large amount of trial-and-error computation is sometimes necessary, use of an electronic computer may be advisable.

Many equations of state have been proposed and each year additional ones appear in the technical literature, but almost all of them are essentially empirical in nature. A few (e.g., the equation of van der Waals) have at least some theoretical basis, but all empirical equations of state are based on more or less arbitrary assumptions which are not generally valid. Since the constants which appear in an empirical equation of state for a pure gas have at best only approximate physical significance, it is very difficult (and frequently impossible) to justify mixing rules for expressing the constants of the mixture in terms of the constants of the pure components which comprise the mixture. As a result, such relationships introduce further arbitrary assumptions and it has been found repeatedly that for typical empirical equations of state, one set of mixing rules may work well for one or several mixtures but work poorly for others.[1]

The constants which appear in a gas-phase equation of state reflect the nonideality of the gas; the fact that there is a need for any constants at all follows from the existence of intermolecular forces. Therefore, to establish the composition dependence of the constants (i.e., mixing rules) it is important that the constants in an equation of state have a clear physical significance. For reliable results, it is desirable to have a theoretically meaningful equation of state in order that mixture properties may be related to pure-component properties with a minimum of arbitrariness.

The virial equation of state for gases has a sound theoretical foundation and is free of arbitrary assumptions. The virial equation gives the compressibility factor as a power series in the reciprocal molar volume $1/v$:

$$z = \frac{Pv}{RT} = 1 + \frac{B}{v} + \frac{C}{v^2} + \frac{D}{v^3} + \dots \qquad (5.2\text{-}1)\dagger$$

In Eq. (5.2-1), B is the second virial coefficient, C is the third virial coefficient, D is the fourth, and so on. All the virial coefficients are independent of pressure or density and for pure components they are functions only of the temperature. The unique advantage of the virial equation follows from the fact that, as shown a little later, there is a theoretical relation between the virial coefficients and the intermolecular potential. Further, in a gaseous mixture, the virial coefficients depend on the composition in an exact and particularly simple manner.

The compressibility factor is sometimes written as a power series in the pressure:

$$z = \frac{Pv}{RT} = 1 + B'P + C'P^2 + D'P^3 + \dots, \qquad (5.2\text{-}2)$$

where the coefficients B', C', D', ... are again independent of pressure or density. For mixtures, however, these coefficients depend on the composition in a more complicated way than do those appearing in Eq. (5.2-1). The relations between the coefficients in Eqs. (5.2-1) and (5.2-2) are

---

†Equation (5.2-1) is frequently written in the equivalent form

$$z = 1 + B\rho + C\rho^2 + D\rho^3 + \dots, \qquad (5.2\text{-}1a)$$

where ρ, the molar density, is equal to $1/v$.

derived in Appendix II with the results

$$B' = \frac{B}{RT},$$ (5.2-3)

$$C' = \frac{C - B^2}{(RT)^2},$$ (5.2-4)

$$D' = \frac{D - 3BC + 2B^3}{(RT)^3}.$$ (5.2-5)

Equation (5.2-1) is usually superior to Eq. (5.2-2) in the sense that when the series is truncated after the third term, the experimental data are reproduced by Eq. (5.2-1) over a wider range of densities (or pressures) than by Eq. (5.2-2), provided the virial coefficients are treated as physically significant parameters. In that case, the second virial coefficient B is properly evaluated from low-pressure P-V-T data by the definition

$$B = \lim_{\rho \to 0} \left[\frac{\partial z}{\partial \rho} \right]_T.$$ (5.2-6)

Similarly, the third virial coefficient must also be evaluated from P-V-T data at low pressures; it is defined by

$$C = \lim_{\rho \to 0} \frac{1}{2} \left[\frac{\partial^2 z}{\partial \rho^2} \right]_T.$$ (5.2-7)

Reduction of P-V-T data to yield second and third virial coefficients is illustrated in Fig. 5-1, taken from the work of Douslin[2] on methane. In addition to Douslin's data obtained at the Bureau of Mines, Fig. 5-1 also shows experimental results of several other investigators. The coordinates of Fig. 5-1 follow from rewriting the virial equation in the form

$$v \left(\frac{Pv}{RT} - 1 \right) = B + \frac{C}{v} + \dots.$$ (5.2-8)

When isothermal data are plotted as shown, the intercept on the ordinate gives B, and C is found from the limiting slope as $1/v \to 0$. For mixtures, the same procedure is used, but in addition to isothermal conditions, each straightline plot must also be at constant composition.

An illustration of the applicability of Eqs. (5.2-1) and (5.2-2) at two different temperatures is given in Figs. 5-2 and 5-3 which are based on Michels' very accurate volumetric data for argon.[3,4] Using only low-pressure data along an isotherm, B and C were calculated as indicated by Eq. (5.2-8). These coefficients were then used to predict the compressibil-

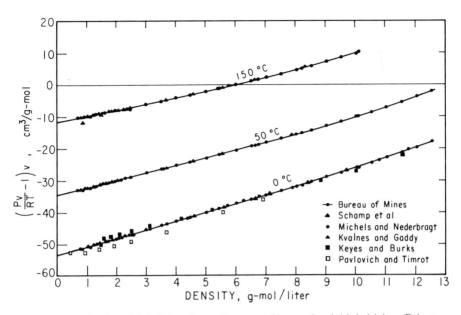

Fig. 5-1   Reduction of *P-V-T* data for methane to yield second and third virial coefficients.

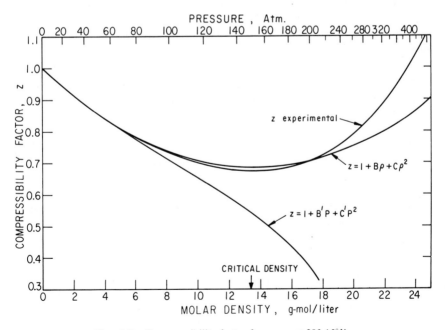

Fig. 5-2   Compressibility factor for argon at 203.15°K.

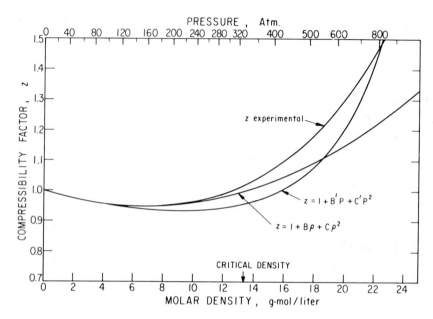

Fig. 5-3   Compressibility factor for argon at 298.15°K.

ity factors at higher pressures (or densities). One prediction is based on Eq. (5.2-1) and the other on Eq. (5.2-2) together with Eqs. (5.2-3) and (5.2-4). Experimentally determined isotherms are also shown. In both cases, Eq. (5.2-1) is more successful than Eq. (5.2-2).†

For many different gases it has been observed that Eq. (5.2-1), when truncated after the third term (that is, when D and all higher virial co-efficients are neglected), gives a very good representation of the compressibility factor up to about one half the critical density and a fair representation nearly up to the critical density.

For higher densities the virial equation is, at present, of no practical interest. Experimental as well as theoretical methods are not as yet sufficiently developed to obtain reliable quantitative results for fourth and higher virial coefficients. It is, however, applicable to moderate densities such as are commonly encountered in many typical vapor-liquid and vapor-solid equilibrium problems.

The significance of the virial coefficients lies in their direct relation to intermolecular forces. In an ideal gas the molecules exert no forces on one

†The comparisons shown in Figs. 5-2 and 5-3 are for the case where Eqs. (5.2-1) and (5.2-2) are truncated after the quadratic terms. When similar comparisons are made with these equations truncated after the linear terms, it often happens that, because of compensating errors, Eq. (5.2-2) provides a better approximation at higher densities than Eq. (5.2-1). See Chueh and Prausnitz.[38]

another. No such gas exists, but when the mean distance between molecules becomes very large (low molar density), all gases tend to behave as ideal gases. This is not surprising since intermolecular forces diminish rapidly with increasing intermolecular distance and therefore forces between molecules at low density are extremely weak. However, as the density rises, molecules come into closer proximity with one another and, as a result, interact more frequently. The purpose of the virial coefficients is to take these interactions into account. The physical significance of the second virial coefficient is that it takes into account deviations from ideal behavior which result from collisions or interactions that involve two molecules at a time. Similarly, the third virial coefficient takes into account deviations from ideal behavior which result from a collision of three molecules. The physical significance of the higher virial coefficients follows in an analogous manner.

From statistical mechanics we can derive relations between virial coefficients and intermolecular potential functions.[5] For simplicity, let us consider a gas composed of simple, spherically symmetric molecules such as methane or argon. The potential energy between two such molecules is designated by $\Gamma(r)$, where r is the distance between molecular centers. The second and third virial coefficients are given as functions of $\Gamma(r)$ and temperature by the relations

$$B = 2\pi N_A \int_0^\infty (1 - e^{-\Gamma(r)/kT})r^2\, dr \qquad (5.2\text{-}9)$$

and

$$C = \frac{-8\pi^2 N_A^2}{3} \int_0^\infty \int_0^\infty \int_{|r_{12}-r_{13}|}^{r_{12}+r_{13}} f_{12} f_{13} f_{23} r_{12} r_{13} r_{23}\, dr_{12}\, dr_{13}\, dr_{23}, \qquad (5.2\text{-}10)\dagger$$

where $f_{ij} \equiv \exp(-\Gamma_{ij}/kT) - 1$, and N_A is Avogadro's number.

Similar expressions can be written for the fourth and higher virial coefficients. While Eqs. (5.2-9) and (5.2-10) refer to simple, spherically symmetric molecules, we do not want to imply that the virial equation is applicable only to such molecules; rather, it is valid for all uncharged (electrically neutral) molecules, polar or nonpolar, regardless of molecular complexity. However, in a complex molecule the intermolecular potential depends not only on the distance between molecular centers but also on

†Equation (5.2-10) assumes, for convenience, that the potential energy of the three molecules 1, 2 and 3 is given by the sum of the three binary potential energies (additivity assumption):

$$\Gamma_{123}(r_{12}, r_{13}, r_{23}) = \Gamma_{12}(r_{12}) + \Gamma_{13}(r_{13}) + \Gamma_{23}(r_{23}).$$

This assumption is unfortunately not correct although, in many cases, it provides a good approximation.

the spatial geometry of the separate molecules and on their relative orientation. In such cases it is still possible to relate the virial coefficients to the intermolecular potential, but the mathematical expressions corresponding to Eqs. (5.2-9) and (5.2-10) are necessarily much more complicated.

5.3   Extension to Mixtures

Perhaps the most important advantage of the virial equation of state for application to phase equilibrium problems lies in its direct extension to mixtures. This extension requires no arbitrary assumptions. The composition dependence of all virial coefficients is given by a generalization of the statistical mechanical derivation used to derive the virial equation for pure gases.

First, let us consider the second virial coefficient which takes into account interactions between two molecules. In a pure gas, the chemical identity of each of the interacting molecules is always the same; in a mixture, however, there are various types of two-molecule interactions depending on the number of components which are present. In a binary mixture containing species i and j, there are three types of two-molecule interactions, which can be designated by i-i, j-j, and i-j. For each of these interactions there is a corresponding second virial coefficient which depends on the intermolecular potential between the molecules under consideration. Thus B_{ii} is the second virial coefficient of pure i which depends on Γ_{ii}; B_{jj} is the second virial coefficient of pure j which depends on Γ_{jj}; and B_{ij} is the second virial coefficient corresponding to the i-j interaction as determined by Γ_{ij}, the potential energy between molecules i and j. If i and j are spherically symmetric molecules, B_{ij} is given by the same expression as that given in Eq. (5.2-9):

$$B_{ij} = 2\pi N_A \int_0^\infty (1 - e^{-\Gamma_{ij}(r)/kT}) r^2 \, dr. \qquad (5.3-1)$$

The three second virial coefficients B_{ii}, B_{jj} and B_{ij} are functions only of the temperature; they are independent of density (or pressure) and what is most important, they are independent of composition. Since the second virial coefficient is concerned with interactions between *two* molecules, it can be rigorously shown that the second virial coefficient of a mixture is a *quadratic* function of the mole fractions y_i and y_j. For a binary mixture of components i and j,

$$B_{\text{mixture}} = y_i^2 B_{ii} + 2 y_i y_j B_{ij} + y_j^2 B_{jj}. \qquad (5.3-2)$$

For a mixture of m components the second virial coefficient is given by a rigorous generalization of Eq. (5.3-2):

$$B_{\text{mixture}} = \sum_{i=1}^{m} \sum_{j=1}^{m} y_i y_j B_{ij}.$$

(5.3-3)

The third virial coefficient of a mixture is related to the various C_{ijk} coefficients which take into account interactions between the three molecules i, j, and k; in a pure gas, the chemical identity of these three molecules is always the same but in a mixture, molecules i, j, and k may belong to different chemical species. In a binary mixture, for example, there are four C_{ijk} coefficients. Two of these correspond to the pure-component third virial coefficients and two of them are cross-coefficients. Since third virial coefficients take into account interactions between *three* molecules, it can be rigorously shown that the third virial coefficient of a mixture is a *cubic* function of the mole fractions. For a binary mixture of components i and j,

$$C_{\text{mixture}} = y_i^3 C_{iii} + 3 y_i^2 y_j C_{iij} + 3 y_i y_j^2 C_{ijj} + y_j^3 C_{jjj}.$$

(5.3-4)

Equation (5.3-4) can be rigorously generalized for a mixture of m components:

$$C_{\text{mixture}} = \sum_{i=1}^{m} \sum_{j=1}^{m} \sum_{k=1}^{m} y_i y_j y_k C_{ijk}.$$

(5.3-5)

The coefficient C_{ijk} is related to the intermolecular potentials Γ_{ij}, Γ_{ik}, and Γ_{jk} by an equation of the same form as that of Eq. (5.2-10):

$$C_{ijk} = \frac{-8\pi^2 N_A^2}{3} \int_0^{\infty} \int_0^{\infty} \int_{|r_{ij} - r_{ik}|}^{r_{ij}+r_{ik}} f_{ij} f_{ik} f_{jk} r_{ij} r_{ik} r_{jk} \, dr_{ij} \, dr_{ik} \, dr_{jk},$$

(5.3-6)

where $f_{ij} \equiv \exp(-\Gamma_{ij}/kT) - 1$, $f_{ik} \equiv \exp(-\Gamma_{ik}/kT) - 1$, and $f_{jk} \equiv \exp(-\Gamma_{jk}/kT) - 1$.

The fourth, fifth, and higher virial coefficients of a gaseous mixture are related to the composition and to the various potential functions in an analogous manner: The nth virial coefficient of a mixture is a polynomial function of the mole fractions of degree n.

Equations (5.3-3) and (5.3-5) are rigorous results from statistical mechanics and are not subject to any assumptions other than those upon which the virial equation itself is based. The proof for these equations is not simple and no attempt is made to reproduce it here. Such a proof may be found in advanced texts.† However, the physical significance of Eqs. (5.3-3) and (5.3-5) is not difficult to understand since these equations are a logical consequence of the physical significance of the individual virial coefficients; each of the individual virial coefficients describes a particular interaction and the virial coefficient of the mixture is merely a summation of the individual virial coefficients, appropriately weighted with respect to the composition of the mixture.

Extension of the virial equation to gas mixtures is based on theoretical rather than empirical grounds and it is this feature of the virial equation which makes it so useful for phase-equilibrium problems. Empirical equations of state, which contain constants having only empirical significance, are useful for pure components but cannot be extended to mixtures without the use of arbitrary mixing rules for combining the constants. The extension of the virial equation to mixtures, however, follows in a simple and rigorous way from the theoretical nature of the equation.

5.4 Fugacities from the Virial Equation

Once we have decided on a relation which gives the volumetric properties of a mixture as a function of temperature, pressure, and composition, we can readily compute the fugacities.

The virial equation for a mixture, truncated after the third term, is given by

$$z_{mix} = \frac{Pv}{RT} = 1 + \frac{B_{mix}}{v} + \frac{C_{mix}}{v^2}, \tag{5.4-1}$$

where z_{mix} is the compressibility factor of the mixture, v is the molar volume of the mixture, and B_{mix} and C_{mix} are the virial coefficients of the mixture as given by Eqs. (5.3-3) and (5.3-5). The fugacity coefficient for any component i in a mixture of m components is obtained by substitution in Eq. (3.4-16). When the indicated differentiations and integrations

---

†See, for example, Hirschfelder, et al.,[6] or T. L. Hill.[5] It is proven in these, as well as in other references, that the nth virial coefficient is a polynomial in the mole fractions of degree n. These polynomial relations do not require the assumption of pairwise additivity of potential functions. That assumption is customarily used in evaluating the individual virial coefficients from potential functions but it is not required for establishing the composition dependence of the third (and higher) virial coefficients.

are performed, we obtain

$$\ln \varphi_i = \frac{2}{v} \sum_{j=1}^{m} y_j B_{ij} + \frac{3}{2} \cdot \frac{1}{v^2} \sum_{j=1}^{m} \sum_{k=1}^{m} y_j y_k C_{ijk} - \ln z_{\text{mix}}. \qquad (5.4\text{-}2)$$

In Eq. (5.4-2) the summations are over *all* components, including component i. For example, for component 1 in a binary mixture, Eq. (5.4-2) becomes

$$\ln \varphi_1 = \frac{2}{v} (y_1 B_{11} + y_2 B_{12}) + \frac{3}{2} \cdot \frac{1}{v^2} [y_1^2 C_{111} + 2y_1 y_2 C_{112} + y_2^2 C_{122}]$$

$$- \ln z_{\text{mix}}. \qquad (5.4\text{-}3)$$

Similarly, for component 2,

$$\ln \varphi_2 = \frac{2}{v} (y_2 B_{22} + y_1 B_{12}) + \frac{3}{2} \cdot \frac{1}{v^2} [y_2^2 C_{222} + 2y_1 y_2 C_{122} + y_1^2 C_{112}]$$

$$- \ln z_{\text{mix}}. \qquad (5.4\text{-}4)$$

Equation (5.4-2) is one of the most useful equations in the thermodynamics of phase equilibria. It relates the fugacity of a component in the vapor phase to its partial pressure through the theoretically valid virial equation of state. The only practical limitation of Eq. (5.4-2) lies in its restriction to moderate densities; otherwise its application is general. It may be applied to any component in a gaseous mixture regardless of whether or not that component can exist as a pure vapor at the temperature and pressure of the mixture; no hypothetical states are introduced. Further, Eq. (5.4-2) is not limited to binaries but is applicable without further assumptions to mixtures containing any desired number of components. Finally, Eq. (5.4-2) is valid for all types of (nonionized) molecules, polar and nonpolar, regardless of molecular complexity, although it is unfortunately true that theoretical calculation of the various B and C coefficients from statistical mechanics is, at present, restricted for practical purposes to rather simple substances. However, this limitation is due not to any failure of the virial equation or of the thermodynamic equations in Sec. 3.4; rather, it is a result of our present inability to describe adequately the intermolecular potential between molecules of highly complex electronic structure.

5.5   Calculation of Virial Coefficients
from Potential Functions

In previous sections we have discussed the nature of the virial equation of
state and, in Eq. (5.4-2), we outlined the way it may be used to calculate
the fugacity of a component in a gaseous mixture. We must now consider
how to calculate the virial coefficients which appear in Eq. (5.4-2) and, in
order to do so, we make use of our discussion of intermolecular forces as
given in Chap. 4.

First, we must recognize that the first term on the right hand side of
Eq. (5.4-2) is frequently much more important than the second term; in
fact, in many cases at low or moderate densities, the second term is
sufficiently small to allow us to neglect it entirely. This is fortunate because we can estimate B's with much more accuracy than we can estimate
C's.

Equation (5.3-1) gives the relationship between the second virial coefficient B_{ij} and the intermolecular potential function $\Gamma_{ij}(r)$ for spherically
symmetrical molecules i and j, where i and j may, or may not, be

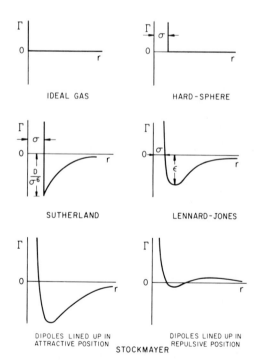

Fig. 5-4   Potential functions with zero, one, or
two adjustable parameters.

chemically identical. If the potential function $\Gamma_{ij}(r)$ is known, then B_{ij} can be calculated by integration as indicated by Eq. (5.3-1) and similarly, if the necessary potentials are known, C_{ijk} can be found from Eq. (5.3-6). Such integrations have been performed for many types of potential functions corresponding to different molecular models. A few of these models are illustrated in Figs. 5-4 and 5-5. We now give a brief discussion of each of them with reference to second virial coefficients, followed by a short section on third virial coefficients. Although engineering applications of potential functions are at present limited, there can be no

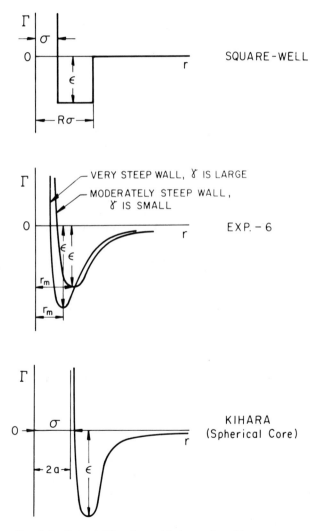

Fig. 5-5   Potential functions with three adjustable parameters.

doubt that future developments in phase-equilibrium thermodynamics will increasingly utilize these functions for solving practical problems.

Ideal-gas potential

The simplest (and trivial) case is to assume that $\Gamma = 0$ for all values of the intermolecular distance r. In that case the second, third and higher virial coefficients are zero for all temperatures and the virial equation reduces to the ideal-gas law.

Hard-sphere potential

This model takes into account the nonzero size of the molecules but neglects attractive forces. It considers molecules to be like billiard balls; for hard-sphere molecules there are no forces between the molecules when their centers are separated by a distance larger than σ, the hard-sphere diameter, but the force of repulsion becomes infinitely large when they touch, at a separation equal to σ. The potential function $\Gamma(r)$ is given by

$$\left.\begin{array}{ll} \Gamma(r) = 0 & \text{for } r > \sigma, \\ \Gamma(r) = \infty & \text{for } r \leq \sigma. \end{array}\right\} \qquad (5.5\text{-}1)$$

Substituting into Eqs. (5.3-1) we obtain for a pure component

$$B = \tfrac{2}{3} \pi N_A \sigma^3. \qquad (5.5\text{-}2)$$

For mixtures, the second virial coefficient B_{ij} ($i \neq j$) is:

$$B_{ij} = \frac{2}{3} \pi N_A \left(\frac{\sigma_i + \sigma_j}{2}\right)^3. \qquad (5.5\text{-}3)$$

The hard-sphere model gives a highly oversimplified picture of real molecules since, for a given gas, it predicts second virial coefficients which are independent of temperature. These results are in strong disagreement with experiment but give a rough approximation for the behavior of simple molecules at temperatures far above their critical temperatures. For example, helium or hydrogen have very small forces of attraction and, near room temperature, where the kinetic energy of these molecules is much larger than their potential energy, the size of the molecules is the most significant factor which contributes to any deviation from ideal-gas behavior. Therefore, at high reduced temperatures, the hard-sphere model provides a reasonable but rough approximation. Since Eq. (5.5-1) requires only one characteristic constant, the hard-sphere model is called a one-parameter model.

Sutherland potential

According to London's theory of dispersion forces, the potential energy of attraction varies inversely as the sixth power of the distance of separation. When this result is combined with the hard-sphere model, the potential function becomes:

$$\left.\begin{array}{ll} \Gamma = \infty & \text{for } r \leq \sigma \\[2mm] \Gamma = \dfrac{-D}{r^6} & \text{for } r > \sigma, \end{array}\right\} \qquad (5.5\text{-}4)$$

where D is a constant depending on the nature of the molecule. London's equation [Eq. (4.4-4)] suggests that D is proportional to the ionization potential and to the square of the polarizability. The Sutherland model provides a large improvement over the hard-sphere model and it is reasonably successful in fitting experimental second virial coefficient data with its two adjustable parameters. Like the hard-sphere model, however, it predicts that at high temperatures the second virial coefficient approaches a constant value whereas, in fact, the best available data show that it goes through a weak maximum at a temperature very much higher than the critical temperature. This limitation is not serious in typical phase-equilibrium problems where such high reduced temperatures are almost never encountered except, perhaps, for helium.

Lennard-Jones' form of Mie's potential

As discussed in Chap. 4, Lennard-Jones' form of Mie's equation is

$$\Gamma = 4\epsilon \left[\left(\frac{\sigma}{r}\right)^{12} - \left(\frac{\sigma}{r}\right)^{6} \right], \qquad (5.5\text{-}5)$$

where ϵ is the depth of the energy well (minimum potential energy) and σ is the collision diameter, i.e., the separation where $\Gamma = 0$. Equation (5.5-5) gives what is probably the best-known two-parameter potential for small, nonpolar molecules. In Lennard-Jones' formula the repulsive wall is not vertical but has a finite slope; this implies that if two molecules have very high kinetic energy, they may be able to interpenetrate to separations smaller than the collision diameter σ. Potential functions with this property are sometimes called "soft-sphere" potentials. The Lennard-Jones potential correctly predicts that at a temperature very much larger than

ϵ/k (k is Boltzmann's constant) the second virial coefficient goes through a maximum. The temperature where $B = 0$ is called the Boyle temperature.

When Lennard-Jones' potential is substituted into the statistical mechanical equation for the second virial coefficient [Eq. (5.2-9)], the required integration is not simple. However, numerical results have been obtained[6] and they are shown in Fig. 5-6 which gives the reduced (dimensionless) virial coefficient as a function of the reduced (dimensionless) temperature. The reducing parameter for the virial coefficient itself is proportional to the collision diameter σ raised to the third power and that for the temperature is proportional to the characteristic energy ϵ.

For many substances, second virial coefficients, as well as other thermodynamic and transport properties, have been interpreted and correlated successfully with the Lennard-Jones potential. Unfortunately, however, it has frequently been observed that for a given gas, one set of parameters (ϵ and σ) is obtained from data reduction of one property (e.g., the second virial coefficient) while another set is obtained from data reduction of a different property (e.g., viscosity). If the Lennard-Jones potential were the *true* potential then the parameters ϵ and σ should be the same for all properties of a given substance.

But even if attention is restricted to the second virial coefficient alone, there is good evidence that the Lennard-Jones potential is only an approximation, albeit a very good one in certain cases. It has been shown by Michels[7] that his highly accurate data for the second virial coefficient of argon over the temperature range -140 to $+150°C$ cannot be fitted with the Lennard-Jones potential within the experimental error using only one

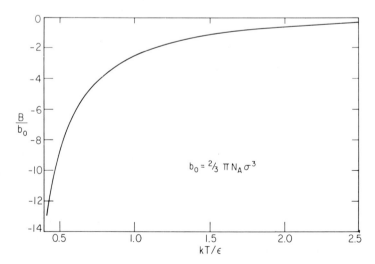

Fig. 5-6   Second virial coefficients calculated from potential of Lennard-Jones.

set of parameters. This conclusion can be supported through a very revealing series of calculations suggested by Michels.[3] We take an experimental value of B corresponding to a certain temperature and then arbitrarily assume a value for ϵ. We now calculate the corresponding value of b_0† which we must have to force agreement between the experimental B and that calculated from the Lennard-Jones function. Next, we repeat the calculation at the same temperature assuming some other value of ϵ. In this way we obtain a curve on a plot of ϵ versus b_0. We now perform the same series of calculations for another experimental value of B at a different temperature and again obtain a curve; where the two curves intersect should be the "true" value of ϵ and b_0. However, we find that when we repeat these calculations for several different temperatures, all the curves do not intersect at one point as they should if the Lennard-Jones potential were exactly correct. Such a plot is shown in Fig. 5-7; instead of a point of intersection the curves define an area which gives a region rather than a unique set of potential parameters. Therefore, we can conclude that even for a spherically symmetric, nonpolar molecule such as argon, the Lennard-Jones potential is not completely satisfactory.[4] Such a conclusion, however, was reached only because Michels' data are of unusually high accuracy and were measured over a large temperature range. For many practical calculations the Lennard-Jones potential is entirely adequate. Lennard-Jones parameters for 13 fluids are given in Table 5-1.

Table 5-1

PARAMETERS FOR THE LENNARD-
JONES POTENTIAL OBTAINED
FROM SECOND VIRIAL
COEFFICIENT DATA†

	σ (Å)	ϵ/k (°K)
Ar	3.499	118.13
Kr	3.846	162.74
Xe	4.100	222.32
CH_4	4.010	142.87
N_2	3.694	96.26
C_2H_4	4.433	202.52
C_2H_6	5.220	194.14
C_3H_8	5.711	233.28
$C(CH_3)_4$	7.420	233.66
n-C_4H_{10}	7.152	223.74
C_6H_6	8.443	247.50
CO_2	4.416	192.25
n-C_5H_{12}	8.540	217.69

†L. S. Tee, S. Gotoh, and W. E. Stewart, *I&EC Fundamentals*, **5**, 356 (1966).

†$b_0 \equiv \frac{2}{3} \pi N_A \sigma^3$.

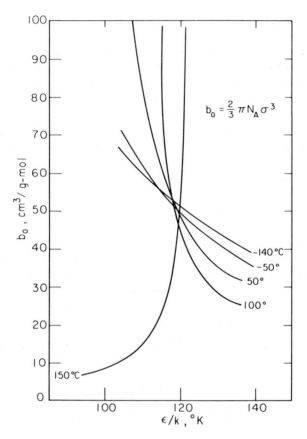

Fig. 5-7 Lennard-Jones parameters calculated from second-virial-coefficient data for argon. If perfect representation were given by Lennard-Jones' potential, all isotherms would intersect at one point.

The square-well potential

The Lennard-Jones potential is not a simple mathematical function and, in order to simplify the calculations, a crude potential was proposed having the general shape of the Lennard-Jones function. This crude potential is obviously an unrealistic simplification since it has two discontinuities, but its mathematical simplicity and flexibility make it useful for practical calculations. The flexibility arises from the fact that the square-well potential contains three adjustable parameters: the collision diameter, σ, the well depth (minimum potential energy), ϵ, and the reduced well width, R. The square-well potential function is

$$\left.\begin{array}{ll} \Gamma = \infty & \text{for } r \leq \sigma \\ \Gamma = -\epsilon & \text{for } \sigma < r \leq R\sigma \\ \Gamma = 0 & \text{for } r > R\sigma. \end{array}\right\} \qquad (5.5\text{-}6)$$

The square-well model has an infinitely steep repulsive wall and therefore, like the Sutherland model, it does not predict a maximum for the second virial coefficient. Numerical results based on the square-well potential are available[6] and, with its three adjustable parameters, good agreement can be obtained between calculated and experimental virial coefficients.[8]

The exp-6 potential

A potential function for nonpolar molecules should contain an attractive term of the London type in addition to a repulsive term about which little is known except that it must depend strongly on the intermolecular distance. For the repulsive term Mie, and later Lennard-Jones, used a term inversely proportional to r, the intermolecular distance, raised to a large power. Theoretical calculations, however, have suggested that the repulsive potential is not an inverse-power function but rather an exponential function of r. A potential function which uses an exponential form for repulsion and an inverse sixth power for attraction is called an exp-6 potential. (It is also sometimes referred to as a modified Buckingham potential.) This potential function contains three adjustable parameters and is written:

$$\Gamma = \frac{\epsilon}{1 - (6/\gamma)} \left\{ \left(\frac{6}{\gamma}\right) \exp\left[\gamma\left(1 - \frac{r}{r_m}\right)\right] - \left(\frac{r_m}{r}\right)^6 \right\} \qquad (5.5\text{-}7)\dagger$$

where $-\epsilon$ is the minimum potential energy which occurs at the intermolecular separation r_m. The third parameter, γ, determines the steepness of the repulsive wall; in the limit, when $\gamma = \infty$, the exp-6 potential becomes identical with the Sutherland potential which has a hard-sphere repulsive term.

The collision diameter σ (i.e., the intermolecular distance where $\Gamma = 0$) is only slightly less than the distance r_m but the exact relation depends on the value of γ, as shown in Table 5-2. Numerical results, based

---

†Equation (5.5-7) is valid only for $r > s$, where s (a very small distance) is that value for r where Γ goes through a (false) maximum. For completeness, therefore, it should be added that $\Gamma = \infty$ for $r < s$. The quantity s, however, is not an independent parameter and has no physical significance.

Table 5-2

THE RATIO σ/r_m FOR THE EXP-6
POTENTIAL AS A FUNCTION OF
THE REPULSIVE STEEPNESS
PARAMETER γ

γ	σ/r_m
15	0.894170
18	0.906096
20	0.912249
24	0.921911
30	0.932341
40	0.943914
100	0.970041
300	0.986692
∞	1.000000

on Eq. (5.5-7), are available[9] and good agreement can be obtained between calculated and observed virial coefficients.

The Kihara potential

According to Lennard-Jones' potential, two molecules can interpenetrate completely provided they have enough energy; according to this
model, molecules consist of point centers surrounded by "soft" (i.e.,
penetrable) electron clouds. An alternate picture of molecules is to think
of them as possessing impenetrable (hard) cores surrounded by penetrable
(soft) electron clouds. This picture leads to a model proposed by Kihara.
In crude mechanical terms, Kihara's model (for spherically symmetric
molecules) considers a molecule to be a hard billiard ball with a foam-
rubber coat; a Lennard-Jones molecule, by contrast, is a soft ball made
exclusively of foam rubber. Kihara[10, 11] writes a potential function identical to that of Lennard-Jones except that the intermolecular distance is
taken not as that between molecular centers but rather as the minimum
distance between the surfaces of the molecules cores. The Kihara potential
is

$$
\left.
\begin{aligned}
\Gamma &= \infty && \text{for } r < 2a \\
\Gamma &= 4\epsilon\left[\left(\frac{\sigma - 2a}{r - 2a}\right)^{12} - \left(\frac{\sigma - 2a}{r - 2a}\right)^{6}\right] && \text{for } r \geq 2a
\end{aligned}
\right\}, \qquad (5.5-8)
$$

where a is the radius of the spherical, molecular core, ϵ is the depth of the
energy well, and σ is the collision diameter, i.e., the distance r between
molecular centers when $\Gamma = 0$.

Equation (5.5-8) is written for the special case of a spherical core, but a more general form has been presented by Kihara for cores having other shapes such as rods, tetrahedra, triangles, prisms, etc.[11,12,13] Numerical results, based on Kihara's potential, are available for second virial coefficients for several core geometries and in particular for reduced (spherical) core sizes a^*, where $a^* \equiv 2a/(\sigma - 2a)$. When $a^* = 0$, the results are identical to those obtained from Lennard-Jones' potential. Since it is a three-parameter function, Kihara's potential is successful in fitting thermodynamic data of a large number of nonpolar molecules, including some rather complex substances whose properties are represented poorly by the two-parameter Lennard-Jones potential. Figure 5-8 shows reduced second virial coefficients calculated from Kihara's potential.

In our discussion of Lennard-Jones' potential we indicated that Michels' highly accurate second virial coefficient data for argon could not be represented over a large temperature range by the Lennard-Jones potential using only one set of potential parameters. However, these same data can be represented within the very small experimental error by the Kihara potential using only one set of parameters.[14,15,16,17] The ability of Kihara's potential to do what Lennard-Jones' potential cannot do is hardly surprising since the former potential has three adjustable parameters whereas the latter has only two. In fitting data for argon, the three Kihara parameters were determined by trial and error until the deviation between experimental and theoretical second virial coefficients

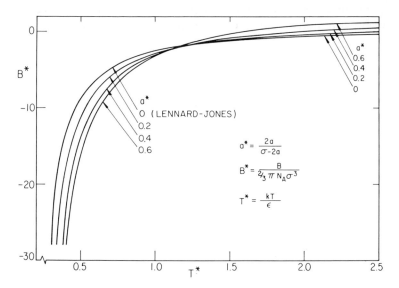

Fig. 5-8   Second virial coefficients calculated from Kihara's potential with a spherical core of radius a.

reached a minimum which was less than the experimental error. It is interesting to note, however, that the magnitude of the core diameter obtained by this procedure is physically reasonable when compared to the diameter of the "impenetrable core" of argon as calculated from its electronic structure. Figure 5-9 shows results of a theoretical calculation of the electron density as a function of distance. While the results shown appear to justify confidence in the Kihara potential, the agreement indicated must not be considered "proof" of its validity. The "true" potential between two argon atoms is undoubtedly quite different from that given by Eq. (5.5-8) especially at very small separations. However, it appears that for practical calculation of common thermodynamic properties (except those at very high temperatures) Kihara's potential is perhaps the most useful potential function now available.

One practical application of Kihara's potential is for prediction of second virial coefficients at low temperatures where experimental data are scarce and difficult to obtain. To illustrate, Fig. 5-10 shows predicted and observed second virial coefficients for krypton at low temperatures; two sets of predictions were made, one with the Kihara potential and the other with the Lennard-Jones potential. In both cases potential parameters were obtained from experimental measurements made at room temperature and above. It is evident from Fig. 5-10 that even for such a "simple" substance as krypton, the Kihara potential is significantly superior to the Lennard-Jones potential. Kihara parameters for fourteen fluids are given in Table 5-3.

For mixtures, Kihara's potential gives B_{ij} $(i \neq j)$ when the pure-component core parameters and the unlike pair potential parameters ϵ_{ij}

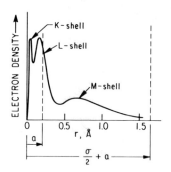

Fig. 5-9   Charge distribution in argon [quoted by C. A. Coulson, *Valence*, 2nd ed. (London: Oxford University Press, 1962)].

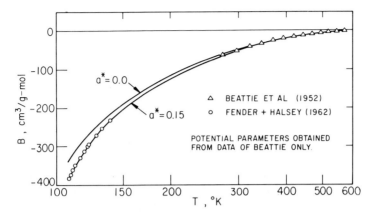

Fig. 5-10   Second virial coefficient for krypton.   Predictions at low temperature based on Lennard-Jones potential ($a^* = 0$) and on Kihara potential.

Table 5-3

PARAMETERS FOR THE KIHARA POTENTIAL
(SPHERICAL CORE) OBTAINED FROM
SECOND VIRIAL COEFFICIENT DATA[†]

	a^*	σ (Å)	ϵ/k (°K)
Ar	0.121	3.317	146.52
Kr	0.144	3.533	213.73
Xe	0.173	3.880	298.15
CH_4	0.283	3.565	227.13
N_2	0.250	3.526	139.2[‡]
O_2	0.308	3.109	194.3[§]
C_2H_6	0.359	3.504	496.69
C_3H_8	0.470	4.611	501.89
CF_4	0.500	4.319	289.7[§]
$C(CH_3)_4$	0.551	5.762	557.75
$n\text{-}C_4H_{10}$	0.661	4.717	701.15
C_6H_6	0.750	5.335	832.0[‡]
CO_2	0.615	3.760	424.16
$n\text{-}C_5H_{12}$	0.818	5.029	837.82

[†]L. S. Tee, S. Gotoh and W. E. Stewart, *I&EC Fundamentals*, **5**, 363 (1966).
[‡]A. E. Sherwood and J. M. Prausnitz, *J. Chem. Phys.*, **41**, 429 (1964).
[§]C. E. Hunt, unpublished results.

and σ_{ij} are specified. The latter two are frequently related to the pure-component parameters by empirical mixing rules. However, the core parameter for the i-j interaction can be derived exactly from the core parameters for the i-i and j-j interactions even for nonspherical cores.[10,11,13]

The difficulty of determining "true" intermolecular potentials from second virial coefficient data is illustrated in Figs. 5-11 and 5-12 which show several potential functions for argon and for neopentane. Each of these functions gives a good prediction of the second virial coefficient; the three-parameter potentials give somewhat better predictions than do the two-parameter potentials, but all of them are in fairly good agreement with the experimental data. However, the various potential functions differ very much from one another, especially for neopentane. Figures 5-11 and 5-12 give striking evidence that agreement between a particular set of experimental results and those calculated from a particular model should not be regarded as proof that the model is correct. Models are very useful in molecular thermodynamics but one must not confuse utility with truth. Figures 5-11 and 5-12 provide a powerful illustration of A. N. Whitehead's advice to scientists: "Seek simplicity but distrust it."

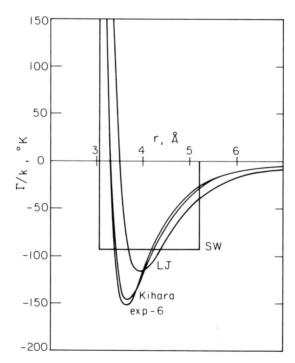

Fig. 5-11   Potential functions for argon as determined from second-virial-coefficient data.

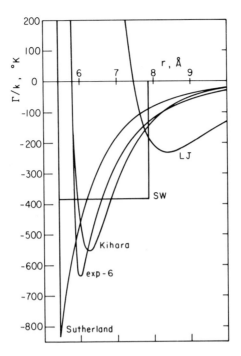

Fig. 5-12   Potential functions for neopentane as determined from second-virial-coefficient data.

The Stockmayer potential

All of the potential functions previously described are applicable only to nonpolar molecules. We now briefly consider molecules which have a permanent dipole moment; for such molecules Stockmayer proposed a potential which adds to the Lennard-Jones formula for nonpolar forces an additional term for the potential energy due to dipole-dipole inter-actions. Dipole-induced dipole interactions are not considered explicitly although it may be argued that since these forces, like London forces, are proportional to the inverse sixth power of the intermolecular separation, they are, in effect, included in the attractive term of the Lennard-Jones formula. For polar molecules, the potential energy is a function not only of the intermolecular separation but also of the relative orientation. Stockmayer's potential is:

$$\Gamma = 4\epsilon \left[\left(\frac{\sigma}{r} \right)^{12} - \left(\frac{\sigma}{r} \right)^{6} \right] + \frac{\mu^2}{r^3} f_\theta (\theta_1, \theta_2, \theta_3), \qquad (5.5\text{-}9)$$

where f_θ is a known function of the angles θ_1, θ_2 and θ_3 which determine the relative orientation of the two dipoles [see Eq. (4.2-5)]. This potential function contains only two adjustable parameters since the dipole moment μ is an independently determined physical constant.

The collision diameter σ in Eq. (5.5-9) is the intermolecular distance where the potential energy due to forces other than dipole-dipole forces becomes equal to zero.

Numerical results, based on Stockmayer's potential, are available for the second virial coefficient.[18] Figure 5-13 shows reduced second virial

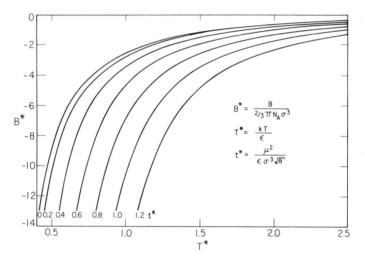

Fig. 5-13   Second virial coefficients as calculated from Stockmayer's potential for polar molecules.

coefficients calculated from Stockmayer's potential as a function of re-duced temperature and reduced dipole moment. The top curve (zero dipole moment) is for nonpolar Lennard-Jones molecules and it is evident that the effect of polarity is to lower (algebraically) the second virial coefficient due to the increased forces of attraction, especially at low temperatures, as suggested by Keesom's formula (see Sec. 4.2). Stock-mayer's potential has been used successfully to fit experimental second virial coefficient data for a variety of polar molecules; parameters for fourteen polar gases are given in Table 5-4.

The various potential models discussed above may be used to calculate B_{ij} as well as B_{ii}. The calculations for B_{ij} are exactly the same as those for B_{ii} except that the potential Γ_{ij} must be used rather than the potential Γ_{ii}.

Table 5-4

PARAMETERS FOR STOCKMAYER'S POTENTIAL
FOR POLAR FLUIDS†

	μ (Debye)	σ (Å)	ϵ/k (°K)
Acetonitrile	3.94	4.38	219
Nitromethane	3.54	4.16	290
Acetaldehyde	2.70	3.68	270
Acetone	2.88	3.67	479
Ethanol	1.70	2.45	620
Chloroform	1.05	2.98	1060
n-Butanol	1.66	2.47	1125
n-Butyl amine	0.85	1.58	1020
Methyl formate	1.77	2.90	684
n-Propyl formate	1.92	3.06	877
Methyl acetate	1.67	2.83	895
Ethyl acetate	1.76	2.99	956
Ethyl ether	1.16	3.10	935
Diethyl amine	1.01	2.99	1180

†R. F. Blanks and J. M. Prausnitz, *A. I. Ch. E. Journal*,
8, 86 (1962).

5.6   Third Virial Coefficient

In the previous section attention was directed to the second virial coefficient. We now want to consider briefly our limited knowledge concerning third virial coefficients.

In Eqs. (5.2-10) and (5.3-6) we gave expressions for the third virial coefficient in terms of three intermolecular potentials. In the derivation of these equations an important simplifying assumption was made, viz., we assumed pairwise additivity of potentials. The third virial coefficient takes into account deviations from ideal-gas behavior due to three-molecule interactions, and for a collision of three molecules i, j, and k, we really need to know Γ_{ijk}, the potential energy of the three-molecule assembly. However, in the derivation of Eqs. (5.2-10) and (5.3-6) it was assumed that

$$\Gamma_{ijk} = \Gamma_{ij} + \Gamma_{ik} + \Gamma_{jk}. \qquad (5.6\text{-}1)$$

Equation (5.6-1) says that the potential energy of the three molecules i, j, and k is equal to the sum of the potential energies of the three pairs ij, ik, and jk. This assumption of pairwise additivity of intermolecular potentials is a common one in molecular physics because little is known about three-, four- (or higher) body forces. For an m-body assembly the additivity assumption takes the form

$$\Gamma_{1,2,3,\ldots m} = \sum_{\substack{\text{all possible} \\ ij \text{ pairs}}} \Gamma_{ij}. \tag{5.6-2}$$

We also used this assumption in Sec. 4.5 where we briefly considered some properties of the condensed state. While there is no rigorous proof, it may well be that because of cancellation effects, the assumption of pairwise additivity becomes better as the number of particles increases. However, it is very likely that the assumption is somewhat in error for a three-body assembly.[19] Therefore, calculations for the third virial coefficient using Eq. (5.2-10) must be considered as approximat'ons.

For any realistic potential, the calculation of third virial coefficients is complicated and, to obtain numerical results, we require an electronic computer. Numerical computations have been carried out for several potential functions and results for pure nonpolar components are available.[9,11,20] For example, Fig. 5-14 gives reduced third virial coefficients as calculated from Kihara's potential. In these calculations, a spherical core was used and pairwise additivity was assumed. The reduced third

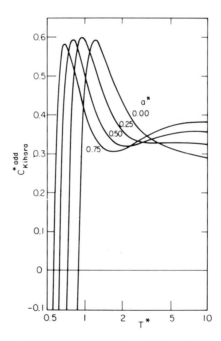

Fig. 5-14 Third virial coefficient from Kihara potential assuming additivity.

virial coefficient, reduced temperature, and reduced core are defined by

$$C^* = \frac{C}{(\frac{2}{3}\pi N_A \sigma^3)^2}, \qquad T^* = \frac{kT}{\epsilon}, \qquad a^* = \frac{2a}{\sigma - 2a}$$

where $-\epsilon$ is the minimum energy in the potential function, σ is the intermolecular distance when the potential is zero, and a is the core radius. For $a^* = 0$, the results shown are those obtained from Lennard-Jones' potential.

Some efforts have been made to include nonadditivity corrections in the calculation of third virial coefficients.[9,11] These corrections are based on a quantum-mechanical relation first derived by Axilrod and Teller[21] for the potential of three spherical, nonpolar molecules at separations where London dispersion forces dominate. The nonadditive correction is a function of the polarizability and at lower temperatures it is quite large; its overall effect is that it approximately doubles the calculated third virial coefficient at its maximum, steepens the slope near the peak value and shifts the maximum to a lower reduced temperature.

Calculated and observed third virial coefficients for argon are shown in Fig. 5-15. Calculated results are based on four potential functions; for each of these, the parameters were determined from second virial coefficient data. The solid lines include the nonadditivity correction but the dashed lines do not, and it is clear that the nonadditivity correction is appreciable. In these calculations, allowance has been made only for non-

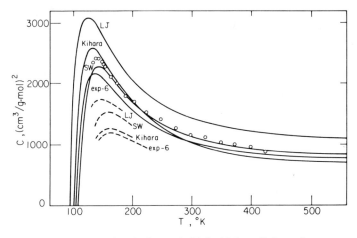

Fig. 5-15 Calculated and observed third virial coefficients for argon. Solid lines include nonadditivity corrections for attractive forces. Dashed lines show a portion of calculated results assuming additivity. Circles represent experimental data of Michels.

additivity of potentials in the attractive region. Little is known about nonadditivity in the repulsive region but when approximate corrections are made for nonadditivity of repulsive forces, they tend to cancel, in part, the nonadditivity corrections for attractive forces.[22,23]

Due to experimental difficulties, there are few reliable values of third virial coefficients. It is therefore not possible to make a truly meaningful comparison between calculated and observed third virial coefficients; not only are experimental data not plentiful, but frequently they are of low accuracy. Even when calculated from very good P-V-T data, the accuracy of third virial coefficients is about one order of magnitude lower than that for second virial coefficients. To the extent that a comparison could be made, Sherwood[8] found that the Lennard-Jones potential (with nonadditivity correction) generally predicted third virial coefficients which were too high, especially for larger molecules (such as pentane or benzene) where the predictions were very poor. The three-parameter potentials (square-well, exp-6, and Kihara) gave much better predictions; however, in view of the uncertainties in the data, and because corrections for nonadditive repulsive forces have been neglected, it is not possible to give a quantitative estimate of agreement between theory and experiment.

Little work has been done on the third virial coefficient of mixtures. The cross-coefficients, assuming additivity, can be calculated by Eq. (5.3-6) and the nonadditivity correction to these cross-coefficients, based on the formula of Axilrod and Teller, can also be computed as shown by Kihara.[11] However, the results of such calculations cannot be presented in a general manner; the coefficient C_{ijk} (for $i \neq j \neq k$) is a function of five independent variables for a two-parameter potential; for a three-parameter potential eight independent variables must be specified.

An approximate method for calculating C_{ijk} was proposed by Orentlicher[24] who showed that, subject to several simplifying assumptions, a reasonable estimate of C_{ijk} can be made based on Sherwood's numerical results for the third virial coefficient of pure gases. Component i is chosen as a reference component; let $C_{ii}(T)$ stand for the third virial coefficient of pure i at the temperature T of interest. Let the potential function between molecules i and j be characterized by the collision diameter σ_{ij} and the energy parameter ϵ_{ij}. Similarly, the potential function for the i-k pair is characterized by σ_{ik} and ϵ_{ik}, and that for the j-k pair by σ_{jk} and ϵ_{jk}. Orentlicher's approximation is:

$$\frac{C_{ijk}(T)}{C_{ii}(T)} = \left(\frac{\sigma_{ij}\,\sigma_{jk}\,\sigma_{ik}}{\sigma_{ii}^3}\right)^2 \frac{[C_{ij}^*(T_{ij}^*)\,C_{jk}^*(T_{jk}^*)\,C_{ik}^*(T_{ik}^*)]^{1/3}}{C_{ii}^*(T_{ii}^*)} \qquad (5.6\text{-}3)$$

where $T_{ij}^* = kT/\epsilon_{ij}$, etc., and where the individual reduced coefficients C_{ij}^*, etc., are obtained from available tables for pure components using any one of several popular potential functions.† Orentlicher's formula appears to give good results for mixtures in which the components do not differ very much in molecular size and characteristic energy. The accuracy of the approximation is difficult to assess, but it is probably useful for mixtures at those temperatures where the third virial coefficient of each component has already passed its maximum value. Table 5-5 gives some observed and calculated third virial cross-coefficients for binary mixtures. Since the uncertainty in the experimental results is probably at least ± 100 (cc/g-mol)^2, agreement between calculated and experimental results is good for these particular mixtures.

Third virial coefficients for polar gases were calculated by Rowlinson[67] using the Stockmayer potential and assuming additivity. Since experimental data are scarce and of low accuracy, no meaningful comparison can be made between calculated and experimental results.

Table 5-5

EXPERIMENTAL AND CALCULATED THIRD VIRIAL
CROSS-COEFFICIENTS FOR SOME BINARY MIXTURES†

| Component | | Temp. | C_{112} (cc/g-mol)^2 | |
1	2	(°K)	Experimental	Calculated
Ar	N_2	273	1349	1510
		203	1706	1770
		163	2295	2420
N_2	Ar	273	1399	1340
		203	1780	1750
		163	2397	2330
CF_4	CH_4	273	4900	5250
		373	3400	3360
		473	2600	2700
		573	2400	2400
N_2	C_2H_4	323	2300	2300

†M. Orentlicher and J. M. Prausnitz, *Can. J. Chem.*, **45**, 373 (1967).

5.7 Virial Coefficients from Corresponding-States Correlations

Since there is a direct relation between virial coefficients and intermolecular potential, it follows from the molecular theory of corresponding states (Sec. 4.6) that virial coefficients can be correlated by data reduc-

---

†C_{ij}^* is a function of kT/ϵ_{ij} and, perhaps, of some additional parameter such as a_{ij}^* for the Kihara potential. This function, however, is the same as that for C_{ii}^* which in turn depends on kT/ϵ_{ii} and, perhaps, on a_{ii}^*.

tion with characteristic parameters such as critical constants. A few such correlations are given in the following paragraphs.

The major part of this section is concerned with second virial coefficients for nonpolar gases. We cannot say much about third virial coefficients because of the scarcity of good experimental data and because of the nonadditivity problem mentioned in the previous section. Further, our understanding of polar gases is not nearly as good as that of nonpolar gases, because again, good experimental data are not plentiful for polar gases, and because theoretical models, based on ideal dipoles, often provide poor approximations to the behavior of real polar molecules.

Equation (5.2-9) relates the second virial coefficient B to the intermolecular potential Γ. Following the procedure given in Sec. 4.6, we assume that the potential Γ can be written in dimensionless form by

$$\frac{\Gamma}{\epsilon} = \ell\left(\frac{r}{\sigma}\right), \tag{5.7-1}$$

where ϵ is a characteristic energy parameter, σ is a characteristic size parameter, and ℓ is a universal function of the reduced intermolecular separation. Upon substitution, Eq. (5.2-9) can then be rewritten in dimensionless form:

$$\frac{B}{2\pi N_A \sigma^3} = \int_0^\infty \left[1 - \exp\left(-\frac{\epsilon\ell(r/\sigma)}{kT}\right)\right]\left(\frac{r}{\sigma}\right)^2 d\left(\frac{r}{\sigma}\right). \tag{5.7-2}$$

If we set σ^3 proportional to the critical volume v_c, and ϵ/k proportional to the critical temperature T_c, we obtain an equation of the form

$$\frac{B}{v_c} = \ell_B\left(\frac{T}{T_c}\right) \tag{5.7-3}$$

where ℓ_B is a universal function of the reduced temperature.

Equation (5.7-3) says that the reduced second virial coefficient is a generalized function of the reduced temperature; this function can either be determined by specifying the universal potential function Γ/ϵ and integrating, as shown by Eq. (5.7-2), or by a direct analysis of experimental data for second virial coefficients.

For example, McGlashan and Potter[25] plotted on reduced coordinates experimentally determined second virial coefficients for methane, argon, krypton, and xenon; they found that the data for these four gases were very well correlated by the empirical equation

$$\frac{B}{v_c} = 0.430 - 0.886\left(\frac{T}{T_c}\right)^{-1} - 0.694\left(\frac{T}{T_c}\right)^{-2}. \tag{5.7-4}$$

Equation (5.7-4) was established from pure-component data but, utilizing the results of statistical mechanics and the molecular theory of corresponding states, it is readily extended to mixtures. To find B for a pure component we use the critical constants v_c and T_c for that component; for a mixture of m components we first recall that the second virial coefficient is a quadratic function of the mole fraction [Eq. (5.3-3)]:

$$B_{\text{mixture}} = \sum_{i=1}^{m} \sum_{j=1}^{m} y_i y_j B_{ij} .$$ (5.7-5)

To calculate the cross-coefficient B_{ij} ($i \neq j$) we again use Eq. (5.7-4), but we must now specify the parameters $v_{c_{ij}}$ and $T_{c_{ij}}$. If we use the common, semiempirical mixing rules for the characteristic parameters of the potential function Γ_{ij}, viz.,

$$\sigma_{ij} = \tfrac{1}{2} (\sigma_i + \sigma_j)$$ (5.7-6)

and

$$\epsilon_{ij} = (\epsilon_i \epsilon_j)^{1/2},$$ (5.7-7)

it then follows from our previous assumptions that

$$v_{c_{ij}} = \tfrac{1}{8} (v_{c_i}^{1/3} + v_{c_j}^{1/3})^3$$ (5.7-8)

and

$$T_{c_{ij}} = (T_{c_i} T_{c_j})^{1/2} .$$ (5.7-9)†

To illustrate the applicability of Eq. (5.7-4) for a mixture, let us make the reasonable assumption that this equation holds also for nitrogen; we can then calculate the second virial coefficient of a mixture of argon and nitrogen using only the critical volume and critical temperature for each of these pure components. For an equimolar mixture, using Eqs. (5.7-4), (5.7-5), (5.7-8), and (5.7-9), we find, at three different temperatures:

	B_{mixture} (cm^3/g-mol)	
$T(^{\circ}\text{C})$	Calc.	Expt'l.
0	-16.5	-16.3
-70	-41.3	-40.4
-130	-89.4	-88.3

The calculated results are in excellent agreement with those found experimentally by Crain and Sonntag.[26] Equation (5.7-4) gives a good representation of the second virial coefficients of small, nonpolar molecules but for larger molecules Eq. (5.7-4) is no longer satisfactory. For example, McGlashan, Potter, and

†For an empirical modification of Eq. (5.7-9), see Eqs. (5.7-29) and (5.13-21).

Wormald[25,27] measured second virial coefficients for normal alkanes and α-olefins containing up to eight carbon atoms; in order to represent their own data, as well as those of other investigators, they used an amended form of Eq. (5.7-4), viz.,

$$\frac{B}{v_c} = 0.430 - 0.886 \left(\frac{T}{T_c}\right)^{-1} - 0.694 \left(\frac{T}{T_c}\right)^{-2} - 0.0375 (n - 1) \left(\frac{T}{T_c}\right)^{-4.5},$$

$$(5.7\text{-}10)$$

where n stands for the number of carbon atoms. Clearly, for methane ($n = 1$) Eq. (5.7-10) reduces to Eq. (5.7-4). Since experimental critical volumes for α-olefins are not highly accurate, and since critical volumes for α-olefins have not been measured for $n > 4$, McGlashan, et al., suggest that for these fluids the critical volumes be calculated from the expression

$$v_c = 25.07 + 50.38n + 0.479n^2 \quad (\text{cm}^3/\text{g-mol}). \qquad (5.7\text{-}11)$$

The equation of McGlashan, Potter, Wormwald (5.7-10) is plotted in Fig. 5-16 in order to illustrate the importance of the third parameter, n, especially at lower reduced temperatures. The experimental data for hydrocarbons of different chain length clearly show the need for a third parameter in a corresponding-states correlation. This need indicates that a potential function which contains only two characteristic constants is

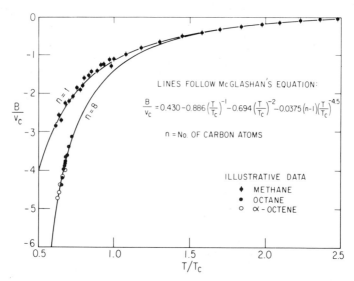

Fig. 5-16   Corresponding-states correlation of McGlashan for second virial coefficients of normal paraffins and α-olefins.

not sufficient for describing the thermodynamic properties of a series of substances which, although chemically similar, differ appreciably in molecular size.

While the addition of a third parameter significantly improves the accuracy of a corresponding-states correlation for the second virial coefficients of pure substances, it unfortunately introduces the need for an additional mixing rule when applied to the second virial coefficients of mixtures. Suppose, for example, that we want to use Eq. (5.7-10) for predicting the second virial coefficient of a mixture of propene and α-heptene. Let 1 stand for propene and 2 for heptene; how shall we compute the cross-coefficient B_{12}? For the characteristic volume and temperature we have the (at best) semitheoretical mixing rules given by Eqs. (5.7-8) and (5.7-9). We must also decide on a value of n_{12} and, since B_{12} refers to the interaction of one propene molecule with one heptene molecule, we are naturally led to the rule

$$n_{12} = \tfrac{1}{2} (n_1 + n_2) \qquad (5.7\text{-}12)$$

which, in this case, gives $n_{12} = 5$. While this procedure appears reasonable enough, we must recognize that Eq. (5.7-12) has little theoretical basis; it is strictly an ad hoc, phenomenological equation which, ultimately, can be justified only empirically. McGlashan, et al., measured volumetric properties of a nearly equimolar mixture of propene and α-heptene. The second virial coefficients of the mixture at several temperatures are shown in Fig. 5-17. The calculated results were obtained from Eq. (5.7-5); the individual coefficients B_{11}, B_{22}, and B_{12} were determined from Eq. (5.7-10) and from the mixing rules given by Eqs. (5.7-8), (5.7-9) and (5.7-12). In this system the agreement between calculated and experimental results is excellent; good agreement was also found for mixtures of propane and heptane and for mixtures of propane and octane.[25] For comparison, Fig. 5-17 also shows results calculated with the attractively simple assumption

$$B_{12} = \tfrac{1}{2} (B_{11} + B_{22}). \qquad (5.7\text{-}13)$$

Guggenheim† appropriately calls Eq. (5.7-13) the "naïve assumption." It can readily be shown that Eq. (5.7-13) follows from Amagat's law (or the Lewis rule); when Eq. (5.7-13) is substituted into the exact Eq. (5.7-5), we obtain the simple but *erroneous* result

$$B_{\text{mixture}} = y_1 B_{11} + y_2 B_{22}. \qquad (5.7\text{-}14)$$

Figure 5-17 shows that Eq. (5.7-14) is in significant disagreement with the experimental data.

†E. A. Guggenheim, *Mixtures* (Oxford: Clarendon Press, 1952).

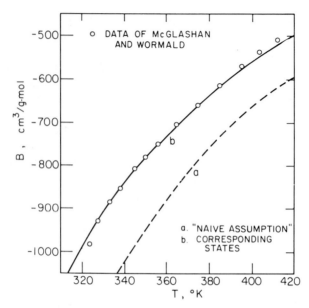

Fig. 5-17   Second virial coefficient of a mixture containing
50.05 mol % propene and 49.95 mol % α-heptene.

The correlation of Pitzer and Curl

In Secs. 4.6 and 4.7 we pointed out that the applicability of corresponding states correlations could be very much extended if we distinguish between different classes of fluids and characterize these classes with an appropriate, experimentally accessible parameter. With this extension, the theory of corresponding states is applied separately to any one class containing a limited number of fluids rather than to a very large number of fluids belonging to different classes.

The classifying parameter suggested by Pitzer is ω, the acentric factor, which is a macroscopic measure of the extent to which the force field around a molecule deviates from spherical symmetry. The acentric factor is (essentially) zero for spherical, nonpolar molecules such as the noble gases and for small, highly symmetric molecules such as methane. The operational definition of ω was given earlier [Eq. (4.7-2)]; it is:

$$\omega = -\log_{10}\left(\frac{P^s}{P_c}\right)_{T/T_c = 0.7} - 1.0000, \qquad (5.7\text{-}15)$$

where P^s is the saturation (vapor) pressure and P_c is the critical pressure. A list of some acentric factors is given in Table 4-8 on p. 74. Other classifying parameters have been proposed[28,29,30] but it appears that the acentric factor is the most practical because it is easily evaluated from experimental data which are both readily available and reasonably accurate for most common substances.

When applied to the second virial coefficient, the extended theory of corresponding states asserts that for all fluids in the same class,

$$\frac{B}{N_A \sigma^3} = \ell_\omega\left(\frac{kT}{\epsilon}\right),$$

(5.7-16)

where, as before, σ is a characteristic molecular size, and ϵ/k is a characteristic energy expressed in units of temperature. The function ℓ_ω depends on the acentric factor ω and, for any given value of ω, it applies to all substances having that value of ω.

Pitzer and Curl[31] have rewritten Eq. (5.7-16) by replacing σ and ϵ/k with macroscopic parameters. They write:

$$\frac{BP_c}{RT_c} = \ell_\omega\left(\frac{T}{T_c}\right).$$

(5.7-17)

The size parameter σ^3 has been set proportional to RT_c/P_c rather than to v_c because, as Pitzer and Curl point out, reliable experimental values are much more readily available for T_c and P_c than for v_c.

Next, Pitzer and Curl investigated the dependence of ℓ_ω on acentric factor at constant reduced temperature; they found that within the limits of experimental information then available, ℓ_ω is a linear function of ω. Thus

$$\frac{BP_c}{RT_c} = \ell^{(0)}\left(\frac{T}{T_c}\right) + \omega\ell^{(1)}\left(\frac{T}{T_c}\right).$$

(5.7-18)

The function $\ell^{(0)}$ gives the reduced second virial coefficients for simple fluids ($\omega = 0$) while $\ell^{(1)}$ is a correction function which when multiplied by ω, gives the effect of acentricity on the second virial coefficient. The two functions $\ell^{(0)}$ and $\ell^{(1)}$ were determined from experimental data for a number of nonpolar or slightly polar substances. Highly polar substances such as water, nitriles, ammonia or alcohols were not included; also the quantum gases helium, hydrogen and neon were intentionally omitted. The empirically determined functions are:

$$\ell^{(0)} = 0.1445 - 0.330\left(\frac{T}{T_c}\right)^{-1} - 0.1385\left(\frac{T}{T_c}\right)^{-2} - 0.0121\left(\frac{T}{T_c}\right)^{-3}$$

(5.7-19)

$$\ell^{(1)} = 0.073 + 0.46\left(\frac{T}{T_c}\right)^{-1} - 0.50\left(\frac{T}{T_c}\right)^{-2} - 0.097\left(\frac{T}{T_c}\right)^{-3} - 0.0073\left(\frac{T}{T_c}\right)^{-8}.$$

(5.7-20)

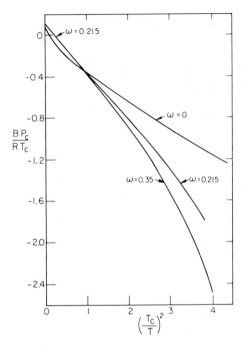

Fig. 5-18   Correlation of Pitzer and Curl for
second virial coefficients.

Figure 5-18 shows the reduced second virial coefficient as a function of reciprocal reduced temperature for three values of ω and it is clear that the introduction of a third parameter has a pronounced effect, especially at low reduced temperatures.

The equations of Pitzer and Curl provide a good correlation for the second virial coefficients of normal fluids. A normal fluid is one which is either nonpolar or else slightly polar, and which does not associate strongly (e.g., by hydrogen bonding); further, it is a fluid whose configurational properties can be evaluated to a sufficiently good approximation by classical, rather than quantum, statistical mechanics. For the three quantum gases, helium, hydrogen and neon, good experimental results are available (see App. III).

When we wish to extend the correlation of Pitzer and Curl to mixtures, we are again faced with the same problem as that mentioned in the previous section dealing with McGlashan's work on the second virial coefficients of paraffins and α-olefins. To estimate the cross-coefficient B_{12} in Eq. (5.7-5) we need three mixing rules. Following our earlier discussion, for molecules which are not very different in molecular size or type, we have

$$T_{c_{12}} = (T_{c_1} T_{c_2})^{1/2} . \qquad (5.7\text{-}21)\dagger$$

†However, see also Eqs. (5.7-29) and (5.13-21).

To obtain an expression for $P_{c_{12}}$, we first write

$$P_{c_{12}} = \frac{z_{c_{12}} R T_{c_{12}}}{v_{c_{12}}},$$ (5.7-22)

and, using Eq. (5.7-8), we obtain

$$P_{c_{12}} = \frac{z_{c_{12}} R T_{c_{12}}}{\frac{1}{8}(v_{c_1}^{1/3} + v_{c_2}^{1/3})^3}.$$ (5.7-23)

For the critical volumes v_{c_1} and v_{c_2} we can use experimental data when these are reliable; when not, we can use the equation

$$v_c = z_c \frac{R T_c}{P_c}$$ (5.7-24)

where z_c, the compressibility factor at the critical point, is given by Pitzer's correlation for normal fluids:

$$z_c = 0.291 - 0.08 \, \omega.$$ (5.7-25)

Finally, we must say something about $z_{c_{12}}$ and, because we do not know what else to do, we assume that

$$\omega_{12} = \frac{1}{2}(\omega_1 + \omega_2),$$ (5.7-26)

which immediately gives

$$z_{c_{12}} = \frac{1}{2}(z_{c_1} + z_{c_2}).$$ (5.7-27)

For those mixtures which contain helium, hydrogen or neon as one component and for the other component a normal fluid of small molecular size, good estimates of B_{12} can often be made by using effective parameters for the quantum gas in Eqs. (5.7-18), (5.7-19) and (5.7-20). The effective parameters depend somewhat on the temperature[†] but at temperatures above 80° K they are, to a good approximation:

	$T_c(°K)$	$v_c \, (cm^3/g\text{-}mol)$	z_c	ω
Helium	10.47	37.5	0.291	0
Hydrogen	43.6	51.5	0.291	0
Neon	45.5	40.3	0.291	0

It is important to recognize that the mixing rules given by Eqs. (5.7-8), (5.7-9), and (5.7-26) are only first approximations which have no general validity. For mixtures of molecules which are similar in size and chemical nature, these mixing rules are usually reliable but whenever the com-

---

[†] See also Sec. 5.13 [Eqs. (5.13-22) and (23)] and App. III.

ponents of a mixture differ appreciably in molecular size or in electronic structure they provide only rough estimates.

The geometric-mean approximation for $T_{c_{12}}$ appears to be an upper limit; for asymmetric systems (i.e., mixtures whose components differ appreciably in molecular size), $T_{c_{12}}$ is usually somewhat smaller than the geometric mean of T_{c_1} and T_{c_2}. Subject to several simplifying assumptions, it can be shown from London's dispersion formula (Sec. 4.4) that:

$$T_{c_{12}} = (T_{c_1} T_{c_2})^{1/2} \left[\frac{8 (v_{c_1} v_{c_2})^{1/2}}{(v_{c_1}^{1/3} + v_{c_2}^{1/3})^3} \right]^q \left[\frac{2 (I_1 I_2)^{1/2}}{I_1 + I_2} \right] \qquad (5.7\text{-}28)$$

where I is the ionization potential and q is a positive exponent. For spherical molecules, London's formula yields $q = 2$, but for nonspherical molecules it has been found empirically that $q = 1$ sometimes (though not always) gives better results.[32] Each of the bracketed quantities is equal to or less than unity and therefore the geometric mean for $T_{c_{12}}$ (Eq. 5.7-21) represents an upper limit.

Although Eq. (5.7-28) has some theoretical basis, it does not always provide accurate predictions for B_{12} of nonpolar gaseous mixtures. London's formula is itself subject to a considerable number of simplifying assumptions, and while Eq. (5.7-28) provides improvement over Eq. (5.7-21) in some cases, in other cases Eq. (5.7-28) has a tendency to over-correct. The crux of the problem comes from the fact that we do not as yet have sufficient understanding of intermolecular forces, even between simple molecules, to predict accurately properties of mixtures from pure-component data alone. The first bracketed quantity in Eq. (5.7-28) is usually much more important than the second one because the ionization potentials of most nonpolar materials do not differ very much.† However, even if the second term is taken into account, it is evident that with our present inadequate knowledge of intermolecular forces, corrections to the geometric-mean formula for $T_{c_{12}}$ cannot be satisfactorily correlated in terms of pure-component properties. For example, Gunn's analysis of second-virial-coefficient data for binary mixtures[33] showed that while in some cases a considerable correction to the geometric mean is required, there was no simple way to correlate the correction. Gunn writes:

$$T_{c_{12}} = (T_{c_1} T_{c_2})^{1/2} - \Delta T_{c_{12}} \qquad (5.7\text{-}29)$$

and then back-calculates $\Delta T_{c_{12}}$ from experimental B_{12} data.‡ Some of Gunn's results are shown in Fig. 5-19, and while $\Delta T_{c_{12}}$ is large for certain

---

†Helium and fluorocarbons have ionization potentials somewhat larger than those of common nonpolar fluids.

‡In reducing the data Gunn did not use Eq. (5.7-8) for $v_{c_{ij}}$, but instead used the similar and simpler rule:

$$v_{c_{ij}} = \tfrac{1}{2}(v_{c_i} + v_{c_j}). \qquad (5.7\text{-}8a)$$

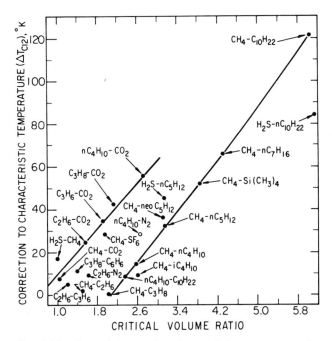

Fig. 5-19   Corrections to the characteristic temperature for second virial cross-coefficients.

mixture families (e.g., CO_2/paraffin systems), it is not a general function of the ratio of molecular sizes. For highly asymmetric systems Eqs. (5.7-8) [or (5.7-8a)] for $v_{c_{12}}$ are also not reliable. Again, Eq. (5.7-8) gives a value which appears to be an upper limit; when v_{c_1} and v_{c_2} are very different from one another, Eq. (5.7-8) gives a $v_{c_{12}}$ which is too large. This result is not surprising since a small molecule, colliding with a big one, does not "see" all of the big molecule but, in effect, collides with only part of the big molecule. Therefore the effective volume of the big molecule in a collision with a small one is less than what it is in a collision with another molecule of its own size.[34]

Deviations from the geometric-mean approximation have also been considered by Chueh;[35] his results are summarized in Sec. 5.13 and in particular in Table 5-10 (p. 158).

Polar gases

It is difficult to establish correlations for the second virial coefficients of polar gases because we do not have a simple parameter for characterizing the effect of polarity on thermodynamic properties. The dipole moment is not sufficient because in addition to its magnitude, it is necessary to know its location on the molecule; thermodynamic properties depend on whether the dipole moment is at the center of the molecule (as it is, for example, in ammonia) or at its periphery (as in p-fluorotoluene). In addition, many polar molecules form hydrogen bonds, and for such molecules the magnitude and location of the dipole moment are not sufficient to indicate their polar character. Approximate correlations have been given for the second virial coefficients of polar gases,[36,37] but unless some experimental information is available to determine the adjustable parameters, their accuracy is not high.

Third virial coefficients

Few reliable experimental data exist for third virial coefficients of pure gases, and good third virial coefficients of gaseous mixtures are very rare. In addition to the scarcity of good experimental results, it is difficult to establish an accurate corresponding-states correlation because, as briefly discussed earlier (Secs. 5.2 and 5.6), third virial coefficients do not exactly satisfy the assumption of pairwise additivity of intermolecular potentials. Fortunately, however, as shown by Sherwood,[22] nonadditivity contributions to the third virial coefficient from the repulsive part of the potential tend to cancel, at least in part, nonadditivity contributions from the attractive part of the potential; further, these contributions appear to be a function primarily of the reduced polarizability and they are important only in part of the reduced temperature range, at temperatures near or less than the critical temperature. Chueh[38] has established an approximate corresponding-states correlation for the third virial coefficient of nonpolar gases, as shown in Fig. 5-20. The reduced third virial coefficients are given by the generalized function

$$\frac{C}{v_c^2} = \mathcal{f}_c \left[\frac{T}{T_c}, d \right],$$ (5.7-30)

where

$$\mathcal{f}_c = (0.232\, T_R^{-0.25} + 0.468\, T_R^{-5})(1 - e^{(1 - 1.89\, T_R^2)})$$
$$+ d e^{-(2.49 - 2.30\, T_R + 2.70\, T_R^2)},$$ (5.7-31)

and

$$T_R = T/T_c.$$

For practical purposes, the parameter d is important only in the vicinity of $T_R \approx 1$, where deviation from simple corresponding-states behavior appears to be most noticeable; at higher temperatures, the reduced third virial coefficients correlate quite well with those of argon, for which $d = 0$. At temperatures well below the critical, the third virial coefficient is only of theoretical interest because at these temperatures condensation always occurs at a density which is sufficiently low to permit

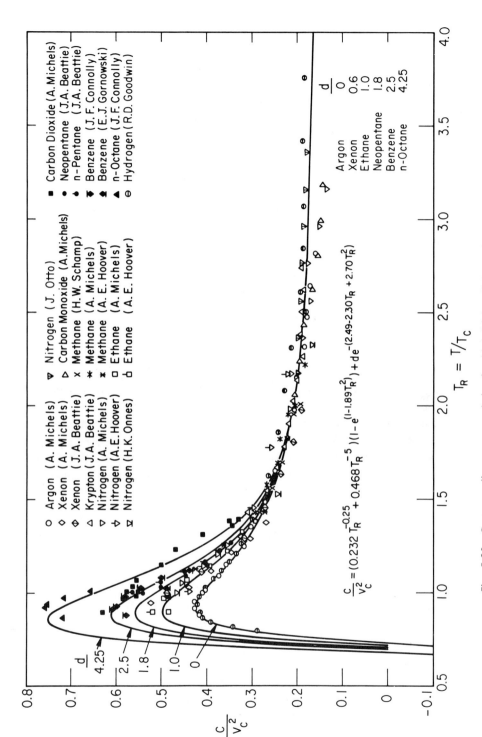

Fig. 5-20 Corresponding-states correlation for third virial coefficients of nonpolar gases.

$$\frac{C}{v_c^2} = (0.232\, T_R^{-0.25} + 0.468\, T_R^{-5})(1 - e^{(1-1.89T_R^2)}) + de^{-(2.49-2.30T_R+2.70T_R^2)}$$

133

truncation of the virial equation after the second term. Little can be said about the parameter d except that, as suggested by theory, it appears to rise with molecular size and polarizability. A quantitative correlation for d is not possible because on the one hand experimental data are neither plentiful nor precise, and on the other hand, the nature of three-body forces is not completely understood. Nevertheless, the general trend of the parameter d is sufficiently well-established to permit a reasonable estimate of d to be made for those nonpolar substances whose third virial coefficients have not been measured.

For mixtures, the approximation suggested by Orentlicher (see Sec. 5.6) is perhaps the best now available. When applied to Chueh's correlation, Orentlicher's formula for C_{ijk} takes the form

$$C_{ijk} = (C_{ij}C_{jk}C_{ik})^{1/3}, \tag{5.7-32}$$

where

$$C_{ij} = v_{c_{ij}}^2 \, f_c \left(\frac{T}{T_{c_{ij}}}, d_{ij}\right) \tag{5.7-33}$$

and $v_{c_{ij}}$ is given by Eq. (5.7-8). The generalized function f_c is that shown in Fig. 5-20 and it is given by Eq. (5.7-31). The characteristic temperature $T_{c_{ij}}$ is best determined from second-virial-coefficient data for the ij mixture. For a first estimate, $T_{c_{ij}}$ is given by the geometric-mean approximation, Eq. (5.7-21), but a better estimate can often be made by applying essentially empirical correction factors (see Table 5-10). Neither theoretical nor empirical evidence is available to suggest a good mixing rule for d_{ij}. In the absence of any pertinent information it is therefore best to use an arithmetic mean.

For quantum gases, or for mixtures containing one or more of the quantum gases, effective critical constants should be used as discussed in Sec. 5.13.

5.8 The "Chemical" Interpretation of Deviations from Gas-Phase Ideality

It was suggested many years ago that nonideal behavior of gases may be attributed to the formation of different chemical species. From the principle of Le Chatelier one can readily show that at low pressures, complex formation by association or solvation is negligible (resulting in ideal behavior); however, as the pressure rises, significant chemical conversion may occur, sometimes leading to large deviations from ideal-gas behavior. For example, in a pure gas consisting of component A, various equilibria may be postulated:

$$2 \, A \;\rightleftharpoons\; A_2, \qquad 3 \, A \;\rightleftharpoons\; A_3, \qquad \text{etc.}$$

The equilibrium constant for dimerization reflects the interaction of two molecules at a time and therefore, a relationship can be established between the dimerization equilibrium constant and the second virial coefficient. Similarly, the trimerization equilibrium constant is related to the third virial coefficient, and so on. However, regardless of the degree of association, the "chemical" viewpoint considers the forces between molecules to be of chemical, rather than physical, nature and it therefore attempts to explain nonideal behavior in terms of the formation of new chemical species. Polymerization reactions (dimerization, trimerization, etc.) result in negative deviations ($z < 1$) from ideal-gas behavior while dissociation reactions (primarily important at higher temperatures) result in positive deviations ($z > 1$).

5.9 Strong Dimerization: Carboxylic Acids

The "chemical" viewpoint is particularly justified for those systems where there are strong forces of attraction between molecules; good examples are organic acids or alcohols and other molecules capable of hydrogen bonding. To illustrate, let us consider the dimerization of acetic acid:

$$2 \, CH_3 - C\!\!\underset{OH}{\overset{O}{\diagup}} \quad \rightleftharpoons \quad CH_3 - C\!\!\underset{O-H\cdots O}{\overset{O\cdots H-O}{\diagup}}\!\!C-CH_3 \, .$$

We define an equilibrium constant in terms of the partial pressures rather than fugacities because we assume that the mixture of true species (i.e., monomers and dimers) behaves as an ideal gas.

$$K = \frac{p_{A_2}}{p_A^2} = \frac{y_{A_2}}{y_A^2 \, P}, \tag{5.9-1}$$

where P is the total pressure, y_A is the mole fraction of monomer and y_{A_2} is the mole fraction of dimer. The equilibrium constant can be calculated from P-V-T data as follows: Let v be the observed volume for one mole of acetic acid at total pressure P and temperature T. (In this connection, one mole means the formula weight of monomeric acetic acid.) Let n_A be the number of moles of monomer and n_{A_2} the number of moles of dimer; further, let α be the fraction of molecules which dimerize. Then n_T, the total number of "true" moles, is given by

$$n_T = n_A + n_{A_2} = (1 - \alpha) + \frac{\alpha}{2} = 1 - \frac{\alpha}{2}. \tag{5.9-2}$$

Since by assumption, $Pv = n_T RT$, we can compute α by substitution and obtain

$$\alpha = 2 - \frac{2 Pv}{RT}. \tag{5.9-3}$$

Since

$$y_A = \frac{n_A}{n_T}, \tag{5.9-4}$$

and

$$y_{A_2} = \frac{n_{A_2}}{n_T}, \tag{5.9-5}$$

the equilibrium constant K is related to α by

$$K = \frac{(\alpha/2)[1 - (\alpha/2)]}{(1 - \alpha)^2 P}. \tag{5.9-6}$$

Equation (5.9-6) shows that since K is a function only of temperature, α must go to zero as P approaches zero; in other words, dimerization decreases as the pressure is lowered. This pressure dependence of the degree of dimerization is a direct consequence of Le Chatelier's principle.

Figure 5-21 gives dimerization constants for acetic acid and for propionic acid as a function of temperature based on experimental P-V-T

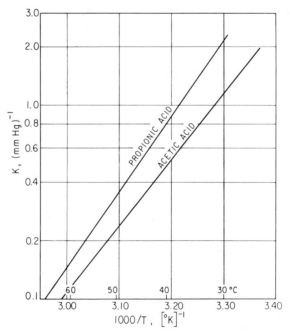

Fig. 5-21   Dimerization equilibrium constants for acetic acid and propionic acid.

measurements by MacDougall.[39,40] In these gases, dimerization is very strong even at low pressures. For example, at 40°C and $P = 12$ mm Hg, α for acetic acid is 0.8, and α for propionic acid is even larger, about 0.84. These gases show large deviations from ideal behavior even at very low pressures, well below one atmosphere.†

Since acetic and propionic acids are members of the same homologous series it appears reasonable to use the results shown in Fig. 5-21 to estimate the equilibrium constant for the solvation (or cross-dimerization) of these two acids:

$$CH_3—C\overset{\displaystyle O}{\underset{\displaystyle OH}{\Big\langle}} \;+\; CH_3—CH_2—C\overset{\displaystyle O}{\underset{\displaystyle OH}{\Big\langle}} \;\rightleftharpoons$$

$$CH_3—C\overset{\displaystyle O\text{-{}-}H—O}{\underset{\displaystyle O—H\text{-{}-}O}{\Big\langle\qquad\Big\rangle}}C—CH_2—CH_3$$

For this estimate, we assume that the enthalpy of the hydrogen bond is the same in the mixed dimer as that in either pure-component dimer. If A stands for acetic acid monomer and B stands for propionic acid monomer, we assume that there is zero enthalpy change for the reaction

$$\tfrac{1}{2}A_2 + \tfrac{1}{2}B_2 \;\rightleftharpoons\; AB.$$

However, the entropy change for this reaction is positive because distinguishability has been lost; whereas there are two different molecular species on the left side there is only one molecular species on the right and therefore the reaction proceeds toward a more probable state. The details of the entropy calculation need not concern us here;‡ the change in entropy is given by $R \ln 2$. The standard Gibbs energy of forming AB dimers from A and B monomers can now be calculated from the following

---

†Because of strong dimerization in the vapor phase, the fugacity of a pure, saturated carboxylic acid is considerably lower than its saturation (vapor) pressure even when that pressure is small. Similarly, the fugacity coefficient of a carboxylic acid in a gaseous mixture is not close to unity even at low pressures. These deviations from ideal behavior are significant in vapor-liquid equilibria and failure to take them into account can lead to serious error. (See, for example, Fig. 7-20.)

At low and moderate pressures, the fugacity of a strongly dimerized component, pure or in a mixture, is set equal to the partial pressure of that component's monomer; this partial pressure can be calculated from the dimerization equilibrium constant. Details of such calculations have been presented by various authors, notably by E. Sebastiani and L. Lacquaniti [*Chem. Eng. Sci.* **22**, 1155 (1967)] and by J. Marek [*Colln. Czech. Chem. Commun.* **19**, 1074 (1954); **20**, 1490 (1955)].

‡See R. Fowler and E. A. Guggenheim, *Statistical Thermodynamics* (London: Cambridge University Press, 1952), p. 167.

scheme:

$$2\,A \;\rightleftharpoons\; A_2 \qquad \Delta g^0_{A_2} = -RT \ln K_{A_2} \qquad\qquad (5.9\text{-}7)$$

$$2\,B \;\rightleftharpoons\; B_2 \qquad \Delta g^0_{B_2} = -RT \ln K_{B_2} \qquad\qquad (5.9\text{-}8)$$

$$A + B \;\rightleftharpoons\; AB \qquad \Delta g^0_{AB} = \tfrac{1}{2}(\Delta g^0_{A_2} + \Delta g^0_{B_2}) - RT \ln 2$$

$$= -RT \ln K_{AB}. \qquad\qquad (5.9\text{-}9)$$

where Δg^0 is the molar Gibbs energy change in the standard state. Substitution gives $K_{AB} = 2\sqrt{K_{A_2} K_{B_2}}$.

From MacDougall's data at $20°C$, $K_{A_2} = 2.44$ mm^{-1} and $K_{B_2} = 5.88$ mm^{-1}, giving $K_{AB} = 7.58$ mm^{-1}. Experimental measurements at $20°C$ by Christian[41] indicate excellent agreement between the observed properties of acetic acid/propionic acid mixtures and those calculated from the three equilibrium constants. This favorable result is a fortunate consequence of the similar chemical nature of the components which comprise the mixture; in other mixtures, where the solvating components are chemically dissimilar, it is usually not possible to predict the properties of the mixture from data for the pure components.

5.10 Weak Dimerizations and Second Virial Coefficients

The "chemical" theory of virial coefficients was proposed many years ago but more recently it has been developed by Lambert and associates[42] in connection with their studies of organic polar gases and polar gas mixtures. Following Lambert, we consider a pure polar gas A, and assume that the forces between A molecules can be divided into two classes: The first class is concerned with "normal" intermolecular forces (such as dispersion forces), which exist between nonpolar as well as polar molecules, and the second is concerned with chemical association forces leading to the formation of new chemical species. At moderate densities, the first class of forces is responsible for a "normal" or "physical" second virial coefficient whereas the second class is responsible for a dimerization equilibrium constant. The equation of state is written

$$Pv = n_T(RT + BP), \qquad\qquad (5.10\text{-}1)$$

with

$$B = B_{\text{nonpolar}} + B_{\text{polar}}.$$

Lambert has shown that B_{polar} is directly proportional to the dimerization equilibrium constant; the essential steps of the derivation follow.

Suppose that, in the absence of any dimerization, there is one mole of A. Let α equal the fraction which associates. Then, by stoichiometry, the equilibrium constant for association is given by Eq. (5.9-6). If we restrict ourselves to a small degree of association ($\alpha \ll 1$), Eq. (5.9-6), to a good approximation, becomes

$$K = \frac{\alpha}{2P}. \tag{5.10-2}$$

The gas is a mixture of monomer A and dimer A_2. Therefore, the second virial coefficient of the mixture can be written

$$n_T^2 B = n_A^2 B_{AA} + 2n_A n_{A_2} B_{AA_2} + n_{A_2}^2 B_{A_2A_2}. \tag{5.10-3}$$

Since α is very small, n_{A_2} is necessarily much smaller than n_A. Also, the total number of moles present, n_T, is given by

$$n_T = 1 - \frac{\alpha}{2} = 1 - PK. \tag{5.10-4}$$

Since $n_A \approx 1$, substitution of Eqs. (5.10-3) and (5.10-4) into (5.10-1) gives (for very small α):

$$Pv = RT + [B_{AA} - RTK]P, \tag{5.10-5}$$

where v is the volume per (stoichiometric) mole of A.

Comparing with Eq. (5.10-1), we see that

$$B_{\text{nonpolar}} = B_{AA} \quad \text{(i.e., B for \textit{pure} monomer)} \tag{5.10-6}$$

and

$$\boxed{B_{\text{polar}} = -RTK.} \tag{5.10-7}$$

Techniques for calculating B_{nonpolar} are somewhat arbitrary, but fortunately this arbitrariness is not too important for our purposes since for any case where this sort of treatment is of interest, B_{polar} is always much larger in magnitude than B_{nonpolar}. A reasonable method for estimating B_{nonpolar} is described by Carter,[43] who uses the reduced second-virial-coefficient equation of Pitzer and Curl [Eq. (5.7-18)] with the true critical constants but with a fictitious acentric factor as determined by the polar molecule's homomorph.†

Experimental data for a variety of polar gases have been reduced using Lambert's "chemical" treatment. When data are available over a range of temperature, a semilogarithmic plot of K vs. $1/T$ gives information on the enthalpy and entropy of dimer formation according to the relation

$$-\ln K = \frac{\Delta h^0}{RT} - \frac{\Delta s^0}{R}, \tag{5.10-8}$$

†The homomorph (*homo* = same, *morph* = form) of a polar molecule is a nonpolar molecule having essentially the same size and shape as those of the polar molecule.

where Δh^0 and Δs^0 are, respectively, the enthalpy and entropy of dimer formation. Some experimentally determined results are shown in Table 5-6 for typical polar gases.

Table 5-6

ENTHALPIES AND ENTROPIES OF DIMERIZATION FOR
POLAR GASES FROM SECOND
VIRIAL COEFFICIENTS†

	$-\Delta h^0$ (cal/g-mol)	$-\Delta s^0$ (cal/g-mol, °K)
Methyl chloride	3100	20.7
Water	5700	25.8
Ammonia	4400	26.8
Methyl amine	3400	22.0
Ethyl formate	4870	23.5
n-Propyl formate	5050	24.5
Methyl acetate	4540	21.3
Methyl propionate	4360	21.3
Formic acid	14,000‡	49.2‡
Acetic acid	16,000‡	54.0‡

†Taken from J. D. Lambert, *Disc. Faraday Soc.*, **15**, 226 (1953).

‡From vapor-phase compressibility measurements as reported by J. D. Lambert, *op. cit.*

Lambert's analysis of P-V-T data for the strongly associating species formic acid and acetic acid gives enthalpies and entropies of dimerization which are much larger than those for most dimers. This result is in agreement with other evidence that these gases form dimers held together by two hydrogen bonds rather than one.

Lambert's "chemical" treatment is readily extended to mixtures; at moderate densities, in a two-component mixture (say A and B) there are three equilibrium constants K_{A_2}, K_{B_2} and K_{AB}. The first and second of these can be determined from volumetric data for the pure components but the third one requires volumetric data for the mixture.

We present now two illustrations of the "chemical" theory of gas imperfection for binary mixtures. First we consider Carter's data[43] for the acetonitrile/acetaldehyde system. Carter found that the experimental second virial coefficient for the mixture was much more negative than that of either pure component, as shown in Fig. 5-22; the magnitude of the cross-virial coefficient is unexpectedly large, much larger than that predicted by Stockmayer's potential using pure-component parameters with common mixing rules uncorrected for specific interactions. When Carter's experimental data are used to calculate chemical equilibrium constants as

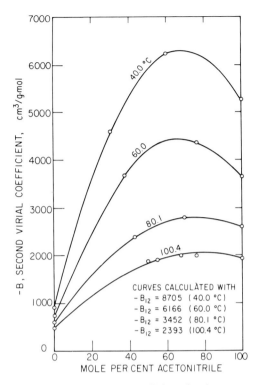

Fig. 5-22   Second virial coefficients for the aceto-
nitrile/acetaldehyde system.

described above, and when the temperature dependence of these constants is investigated using Eq. (5.10-8), it is found that the enthalpy of forma-tion of the acetonitrile/acetaldehyde complex is considerably larger in magnitude than that of either pure-component dimer.   The observed enthalpies are compared with theoretically calculated energies in Table 5-7.† The calculations are based on an electrostatic model with two struc-tures, (a) and (b).   In calculation (a) the complex (or dimer) consists of two "touching" ideal spherical dipoles such that the dipoles are lined up in a direction perpendicular to the center-to-center separation, as shown at the bottom of Table 5-7.   Structure (b) is similar except that the dipoles are now lined up in a direction parallel to the center-to-center separation. From elementary electrostatic theory it can be shown that the energy of formation of each of these structures is:

$$\text{(a)} \qquad \Delta u^0 \; = \; - \, \frac{\mu^2}{r^3} \; \left(\text{or} \; - \, \frac{\mu_i \mu_j}{r^3} \; \text{if} \; i \neq j \right) \qquad \text{(5.10-9)}$$

†For our purposes here, no distinction need be made between energy and enthalpy.

Table 5-7

ENERGIES OF COMPLEX FORMATION IN THE
ACETONITRILE/ACETALDEHYDE SYSTEM†

	$-\Delta u^0$ (energy of formation), cal/g-mol		
Complex	Calculated Parallel Structure (a)	Calculated End-to-End Structure (b)	Observed
Acetonitrile/Acetonitrile	5280	10,560	5110
Acetaldehyde/Acetaldehyde	3380	6760	3960
Acetonitrile/Acetaldehyde	4125	8250	7250

Structure (a)   Structure (b)

†J. M. Prausnitz and W. B. Carter, *A. I. Ch. E. Journal*, **6**, 611 (1960).

(b) $$\Delta u^0 = -\frac{2\mu^2}{r^3} \left(\text{or} -\frac{2\mu_i\mu_j}{r^3} \text{ if } i \neq j\right)$$ (5.10-10)

where μ is the dipole moment of the monomer and r is the distance between the two centers of the two molecules which form the complex or dimer. For acetonitrile and acetaldehyde the dipole moments are, respectively, 3.5 and 2.7 Debye. The distances r were estimated from van der Waals radii reported by Pauling.[44]

The results in Table 5-7 suggest that the structure of the complex [structure (b), perhaps] is qualitatively different from that of the dimers which seem to have the properties of structure (a). In the absence of other experimental information this interpretation of Carter's data should not be taken too seriously, but it does give a possible explanation as to why the cross-virial coefficient for this system is so different from the virial coefficients of the pure components.

For a second example we consider the system ammonia-acetylene studied by Cheh and O'Connell.[45] This system is qualitatively different from that studied by Carter because while acetaldehyde and acetonitrile are both strongly polar, in the ammonia/acetylene system only ammonia has a dipole moment. Nevertheless, in the interaction between the unlike molecules there are strong complexing forces which are not present in the interaction between like molecules; Fig. 5-23 gives B_{11}, B_{22} and B_{12} and we see that B_{12} is considerably more negative than either B_{11} or B_{22}. Cheh and

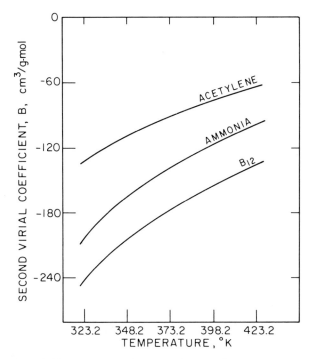

Fig. 5-23   Experimental second virial coefficients for acetylene (1) and for ammonia (2). B_{12} is the observed
cross-coefficient for binary mixtures of (1) and (2).

O'Connell, using appropriate potential functions, reduced the experimental second virial coefficients of the pure gases by taking into account
all physical dispersion, induction, dipole and quadrupole forces. Then,
using conventional mixing rules, they calculated the (physical) cross-
coefficient B_{12} and found that its magnitude was much too small, as shown
in Fig. 5-24.

The high, negative value of B_{12} can be "explained" by the formation
of a hydrogen bond between (acidic) acetylene and (basic) ammonia. This
hydrogen-bond formation affords a reasonable explanation since there is
much independent evidence for the basic properties of ammonia and there
also is some independent evidence for the acidic properties of acetylene.†

The chemical equilibrium constant K for hydrogen-bond formation is
found from

$$-RTK = \tfrac{1}{2}[B_{12\,(\text{expt})} - B_{12\,(\text{physical})}].   (5.10\text{-}11)$$

† For example, acetylene is much more soluble in oxygen-containing solvents (such as
acetone) than is ethane or ethylene.

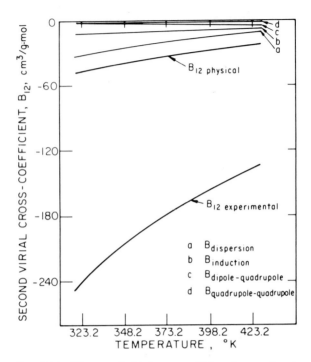

Fig. 5-24   The second virial cross-coefficient for acetylene/ ammonia mixtures.   Calculated contributions from physical forces are insufficient to account for the observed large (negative) result.

From the variation of K with temperature, the standard enthalpy and entropy of complex formation are

$$\Delta h^0 = -2210 \pm 300 \text{ cal/g-mol},$$
$$\Delta s^0 = -15.2 \pm 0.9 \text{ cal/g-mol, } {}^\circ\text{K}.$$

These values are in good agreement with what would be predicted from statistical mechanical calculations for the formation of a loosely bonded complex of ammonia and acetylene.†

These two examples illustrate that forces between molecules are often quite specific and in such cases it is unfortunately not possible to predict even approximately the properties of a mixture from those of the pure components. This fact is really not surprising if we consider a far-fetched analogy: Let us suppose that a sociologist in France has carefully studied the behavior of Frenchmen and after years of observation knows all about

---

†The factor $\frac{1}{2}$ in Eq. (5.10-11) was erroneously omitted in Reference 45. The entropy reported in Reference 45 is therefore too low by $R \ln 2$.

them. He then goes to Germany and makes a similar, thorough study of Germans. With this knowledge can he then predict the behavior of a society in which Frenchmen and Germans intermingle freely? Probably not. An analogy such as this is necessarily extreme but it serves to remind us that molecules are not inert particles floating blindly through space; instead, they are complicated "individuals" whose "personalities" are sensitive to their environment.

5.11   Fugacities at High Densities

In previous sections we have pointed out the advantages of the virial equation of state but, as indicated earlier, this equation has an important disadvantage, viz., its inapplicability to gases at high densities. The density range for practical application of the virial equation varies somewhat with the temperature but frequently the virial equation is not useful at densities larger than about 50–75% of the critical density. This disadvantage follows primarily from our limited knowledge regarding third and higher virial coefficients.

At present, gas-phase fugacities at high densities can best be calculated by using essentially empirical methods. In many cases these methods are quite reliable, although for some mixtures their reliability is often doubtful. For example, for mixtures of small, nonpolar gases not near their critical conditions very good results can usually be obtained, while for mixtures containing polar components, reliability is frequently uncertain, and for all mixtures at or near critical conditions, regardless of molecular complexity, calculated fugacities are likely to be at least somewhat in error. These limitations follow directly from our incomplete knowledge of intermolecular forces between complicated molecules and from our inability to understand properly the behavior of fluids in the vicinity of the critical state.

The general and exact equations for the calculation of fugacities from volumetric properties have been derived and discussed in Chap. 3, and the two key relations are given again by Eqs. (1) and (2) at the beginning of this chapter. We now discuss two particular methods, useful at high densities, for calculating gas-phase fugacities from volumetric properties. The two methods are similar in the sense that both are based on an ad hoc extension to mixtures of the theorem of corresponding states. In the first one, calculations are based on generalized (reduced) charts developed from pure-component properties. In the second one, calculations are performed analytically by using a generalized, empirical equation of state. We discuss here the application of these methods to gas-phase mixtures but we note, in passing, that they may also be applied to liquid-phase mixtures, as discussed briefly in Sec. 7.9.

5.12   Fugacities from Generalized Charts for
Pure Components

The theorem of corresponding states can be extended to mixtures in a straightforward way by utilizing what is often called the pseudocritical method. According to this idea, the configurational properties of a mixture may be calculated from the generalized, reduced properties of pure fluids by expressing the characteristic reducing parameters as a function of mixture composition.

According to (two-parameter) corresponding-states theory, the compressibility factor z is a universal function of reduced temperature and reduced pressure:

$$z = \mathcal{f}_z\left(\frac{T}{T_c}, \frac{P}{P_c}\right). \tag{5.12-1}$$

The pseudocritical method, introduced in 1936 by W. B. Kay,[46] assumes that the same universal function $\mathcal{f}_z$ applies to mixtures when T_c and P_c are taken, not as the (true) critical temperature and pressure of the mixture, but as pseudocritical constants, that is, characteristic parameters which depend on the composition of the mixture. Kay assumed that T_c and P_c are linear functions of the mole fractions but this particular assumption is not essential to the pseudocritical method; other functions relating T_c and P_c to the composition are consistent with the pseudocritical idea and numerous ones have been proposed.[47] Nor is the pseudocritical method limited to a two-parameter theory of corresponding states; if additional reducing parameters are used, the pseudocritical method assumes that these additional parameters also depend on the composition in some simple way. For calculations of fugacities based on the pseudocritical concept, the important assumption is always made that the functions which relate the reducing parameters to the composition are independent of density for all densities from zero to the density of interest.

We now derive an expression for φ_1, the fugacity coefficient of component 1 in a mixture. For simplicity, we confine attention to a binary mixture; it is then a simple matter to extend the result to a mixture of any number of components.[48,49]

Let φ_M be the fugacity coefficient of a binary mixture at temperature T, total pressure P and composition y_1. Then φ_1, the fugacity coefficient of component 1 in the mixture at the same temperature and pressure, is given by

$$\ln\varphi_1 = \ln\varphi_M + (1 - y_1)\left(\frac{\partial \ln \varphi_M}{\partial y_1}\right)_{T,P}. \tag{5.12-2}\dagger$$

---

†This exact thermodynamic equation follows from Euler's theorem for homogeneous functions.[50]

We now assume that the configurational properties of the mixture can be obtained from a corresponding-states correlation for pure fluids. In particular, we assume validity of the three-parameter theory of corresponding states proposed by Pitzer, where the three reducing parameters are a characteristic temperature T_c, a characteristic pressure P_c, and the acentric factor ω (see Sec. 4.7). For pure fluids, T_c and P_c are, respectively, the true critical temperature and the critical pressure. From corresponding states, therefore,

$$\ln \varphi_M = \mathcal{L}_\varphi \left(\frac{T}{T_c}, \frac{P}{P_c}, \omega \right), \tag{5.12-3}$$

where $\mathcal{L}_\varphi$ is a universal function obtained from pure-component properties. For mixtures we assume that the same function $\mathcal{L}_\varphi$ is applicable and that the parameters T_c, P_c and ω are functions of the mole fraction y_1.

We now rewrite Eq. (5.12-3) by using the chain rule:

$$\ln \varphi_1 = \ln \varphi_M + (1 - y_1) \left[\left(\frac{\partial \ln \varphi_M}{\partial T_R} \right)_{P_R, \omega} \left(\frac{\partial T_R}{\partial y_1} \right)_{T,P} \right.$$

$$\left. + \left(\frac{\partial \ln \varphi_M}{\partial P_R} \right)_{T_R, \omega} \left(\frac{\partial P_R}{\partial y_1} \right)_{T,P} + \left(\frac{\partial \ln \varphi_M}{\partial \omega} \right)_{T_R, P_R} \left(\frac{\partial \omega}{\partial y_1} \right)_{T,P} \right], \tag{5.12-4}$$

where

$$T_R = \frac{T}{T_c} \quad \text{and} \quad P_R = \frac{P}{P_c}.$$

To simplify Eq. (5.12-4), we make use of two exact thermodynamic relations,

$$\left(\frac{\partial \ln \varphi_M}{\partial T_R} \right)_{P_R, \omega} = \frac{\Delta h'_M}{R T_R T}, \tag{5.12-5}$$

and

$$\left(\frac{\partial \ln \varphi_M}{\partial P_R} \right)_{T_R, \omega} = \frac{z_M - 1}{P_R}, \tag{5.12-6}$$

where $\Delta h'_M$ is the molar enthalpy of mixture at T and zero pressure minus that at T and P, and z_M is the compressibility factor of the mixture at T and P. Further, we use the identities (at constant T and P),

$$\frac{\partial T_R}{\partial y_1} = - \frac{T}{T_c^2} \left(\frac{dT_c}{dy_1} \right) \tag{5.12-7}$$

$$\frac{\partial P_R}{\partial y_1} = -\frac{P}{P_c^2}\left(\frac{dP_c}{dy_1}\right) \tag{5.12-8}$$

and, upon substitution in Eq. (5.12-4), we obtain

$$\ln \varphi_1 = \ln \varphi_M - (1 - y_1)\left[\frac{1}{T}\left(\frac{\Delta h_M'}{RT_c}\right)\left(\frac{dT_c}{dy_1}\right) + \frac{(z_M - 1)}{P_c}\left(\frac{dP_c}{dy_1}\right)\right.$$

$$\left. - \left(\frac{\partial \ln \varphi_M}{\partial \omega}\right)_{T_R, P_R}\left(\frac{d\omega}{dy_1}\right)\right]. \tag{5.12-9}†$$

Equation (5.12-9) is the desired result but it is not yet complete. The three functions $(\Delta h_M'/RT_c)$, z_M, and $(\partial \ln \varphi_M/\partial \omega)$ are functions of T_R, P_R and ω, and they can be obtained from generalized pure-component properties only after some rule is established which expresses T_c, P_c and ω as a function of y_1. These rules are also required in order to evaluate the three derivatives dT_c/dy_1, dP_c/dy_1 and $d\omega/dy_1$. For example, the simplest assumption (Kay's rule) is to express T_c, P_c, and ω as linear functions of y_1, in which case we obtain

$$\frac{dT_c}{dy_1} = T_{c_1} - T_{c_2} \tag{5.12-10}$$

$$\frac{dP_c}{dy_1} = P_{c_1} - P_{c_2} \tag{5.12-11}$$

$$\frac{d\omega}{dy_1} = \omega_1 - \omega_2, \tag{5.12-12}$$

where T_{c_1} is the critical temperature of pure 1, etc.

†For a mixture containing m components we can utilize the same procedure to derive an expression similar to Eq. (5.12-9). Instead of starting with Eq. (5.12-2) we use for any component i

$$\ln \varphi_i = \ln \varphi_M - \sum_{\substack{j=1 \\ j \neq i}}^{m} y_j \left(\frac{\partial \ln \varphi_M}{\partial y_j}\right)_{\text{all } y_k, \, k \neq i \text{ or } j} \tag{5.12-2a}$$

and then, with the same arguments as those used before, we obtain

$$\ln \varphi_i = \ln \varphi_M - \frac{1}{T}\left(\frac{\Delta h_M'}{RT_c}\right)\sum_{\substack{j=1 \\ j \neq i}}^{m} y_j \left(\frac{dT_c}{dy_j}\right)_{y_k, \, k \neq i \text{ or } j} - \frac{(z_M - 1)}{P_c}\sum_{\substack{j=1 \\ j \neq i}}^{m} y_j \left(\frac{dP_c}{dy_j}\right)_{y_k, \, k \neq i \text{ or } j}$$

$$+ \left(\frac{\partial \ln \varphi_M}{\partial \omega}\right)_{T_R, P_R}\sum_{\substack{j=1 \\ j \neq i}}^{m} y_j \left(\frac{d\omega}{dy_j}\right)_{y_k, \, k \neq i \text{ or } j}. \tag{5.12-9a}$$

The three quantities $\Delta h'_M/RT_c$, z_M and $(\partial \ln \varphi_M/\partial \omega)$ can readily be found from Pitzer's tables.[51] To specify the first two, we require the independent variables T_R, P_R and ω as calculated from the pseudocritical hypothesis for the particular pressure, temperature and composition of interest. To specify the third quantity, only T_R and P_R are needed.†

In their study of compressibility factors for binary gaseous mixtures, Pitzer and Hultgren[52] found that while Kay's linear rule gave a fair approximation, much better results could be obtained by assuming that T_c and P_c are quadratic functions of the mole fraction. (For ω, a linear function was used.) The use of quadratic functions requires two additional parameters characteristic of the binary mixture; Pitzer and Hultgren use the relations

$$T_c = y_1 T_{c_1} + y_2 T_{c_2} + 2y_1 y_2 [2T_{c_{1/2}} - T_{c_1} - T_{c_2}] \qquad (5.12\text{-}13)$$

$$P_c = y_1 P_{c_1} + y_2 P_{c_2} + 2y_1 y_2 [2P_{c_{1/2}} - P_{c_1} - P_{c_2}] \qquad (5.12\text{-}14)$$

$$\omega = y_1 \omega_1 + y_2 \omega_2, \qquad (5.12\text{-}15)$$

where $T_{c_{1/2}}$ and $P_{c_{1/2}}$ are the pseudocritical temperature and pressure for the equimolar mixture. The parameters $T_{c_{1/2}}$ and $P_{c_{1/2}}$ depend on the intermolecular forces between unlike molecules and therefore cannot, in general, be predicted from pure-component properties.‡

For twelve binary gas mixtures, Pitzer and Hultgren[52] found significant deviations from Kay's rule; their results are given in Table 5-8. Unfortunately, even a small deviation from Kay's rule may have a significant effect on calculated thermodynamic properties, especially in the critical region. Table 5-8 shows once again that constants characteristic of a mixture cannot, with generality, be predicted from the corresponding constants for the pure components.

When Eqs. (5.12-13), (5.12-14), and (5.12-15) are substituted into Eq. (5.12-9) we obtain for the fugacity coefficients

---

†Pitzer's tables[51] for the fugacity coefficient are linear in ω; the correlation is of the form

$$\ln \varphi (T_R, P_R, \omega) = [\ln \varphi]^{(0)} (T_R, P_R) + \omega [\ln \varphi]^{(1)} (T_R, P_R).$$

Therefore

$$\left(\frac{\partial \ln \varphi}{\partial \omega} \right)_{T_R, P_R} = [\ln \varphi]^{(1)}.$$

‡To a first approximation, we might expect that

$$2T_{c_{1/2}} = T_{c_1} + T_{c_2}$$

and

$$2P_{c_{1/2}} = P_{c_1} + P_{c_2}.$$

However, in that case Eqs. (5.12-13) and (5.12-14) reduce to Kay's rule.

Table 5-8

CHARACTERISTIC PARAMETERS $T_{c_{1/2}}$ AND $P_{c_{1/2}}$ FROM
VOLUMETRIC DATA FOR TWELVE BINARY MIXTURES†

Component 1	Component 2	$\dfrac{2T_{c_{1/2}}}{T_{c_1} + T_{c_2}} - 1.0$	$\dfrac{2P_{c_{1/2}}}{P_{c_1} + P_{c_2}} - 1.0$
CH_4	C_2H_6	0.02	0.02
CH_4	C_3H_8	0.02	0.03
CH_4	$i\text{-}C_4H_{10}$	0.03	-0.01
CH_4	$n\text{-}C_4H_{10}$	0.05	$+0.01$
C_3H_8	$i\text{-}C_5H_{12}$	0.00	0.00
C_3H_8	C_6H_6	0.00	0.00
N_2	C_2H_6	-0.01	0.00
CH_4	CO_2	-0.02	-0.03
CO_2	C_2H_6	-0.04	-0.09
CO_2	C_3H_8	-0.05	-0.14
CO_2	$n\text{-}C_4H_{10}$	-0.03	-0.16
CH_4	H_2S	-0.05	-0.06

†K. S. Pitzer and G. O. Hultgren, *J. Am. Chem. Soc.*, **80**, 4793 (1958).

$$\ln \varphi_1 = \ln \varphi_M - y_2 J \tag{5.12-16}$$

$$\ln \varphi_2 = \ln \varphi_M + y_1 J, \tag{5.12-17}$$

where

$$J = \frac{1}{T} \left(\frac{\Delta h'_M}{RT_c} \right) [T_{c_1} - T_{c_2} - 2(1 - 2y_2)(2T_{c_{1/2}} - T_{c_1} - T_{c_2})]$$

$$+ \frac{(z_M - 1)}{P_c} [P_{c_1} - P_{c_2} - 2(1 - 2y_2)(2P_{c_{1/2}} - P_{c_1} - P_{c_2})]$$

$$- \left(\frac{\partial \ln \varphi_M}{\partial \omega} \right)_{T_R, P_R} (\omega_1 - \omega_2). \tag{5.12-18}$$

Equations (5.12-16) and (5.12-17), coupled with Pitzer's tables, give a convenient method for the calculation of fugacity coefficients in mixtures of nonpolar (or slightly polar) gases at high pressures, provided that it is possible to estimate the parameters $T_{c_{1/2}}$ and $P_{c_{1/2}}$. These can, in principle, be determined from volumetric data for the mixture at low pressure since, by assumption, they are independent of pressure.

Figure 5-25 gives fugacity coefficients for carbon dioxide in the carbon dioxide/butane system at 340°F and mole fraction $y_{CO_2} = 0.15$. Results obtained from the experimental data of Olds, Reamer, Sage and Lacey[53] are compared with calculations using the Lewis fugacity rule [Eq. (5.1-3)],

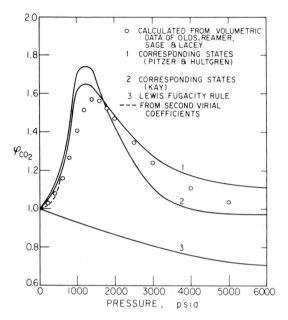

Fig. 5-25   Fugacity coefficient of carbon dioxide in the carbon dioxide/n-butane system at 340°F. The mol fraction of carbon dioxide is 0.15.

the pseudocritical concept with Kay's rule [Eqs. (5.12-10), (5.12-11), and (5.12-12)] and the pseudocritical concept with the equations of Pitzer and Hultgren [Eqs. (5.12-13), (5.12-14), and (5.12-15)]. Also shown are results for the low-pressure region based on the virial equation [Eq. (5.4-2)] neglecting third and higher virial coefficients. The Lewis rule gives very poor results and while the calculations based on Kay's rule are significantly better, they are not as good as those found by Pitzer and Hultgren's method. The comparatively good agreement with this last method is not surprising since the parameters $T_{c_{1/2}}$ and $P_{c_{1/2}}$ were found from volumetric data for the mixture.

For mixtures containing one or more of the quantum gases, effective critical constants for quantum gases should be used as discussed by Gunn and Chueh[54] [see Eqs. (5.13-22) and (5.13-23)].

5.13   Fugacities from the Redlich-Kwong
Equation of State and Its Modifications

The pseudocritical method as discussed in the previous section suffers from one disadvantage: The calculations are tedious and, if they must be performed repeatedly, it is desirable that the necessary equations be based on analytical (rather than tabular) information in order that the calculations may be more easily performed by an electronic digital computer. For this purpose it is preferable to express in the form of an equation the generalized volumetric properties given by some corresponding-states correlation, such as that of Pitzer. In other words, for convenient computer application, we would like to have an analytic, generalized equation of state giving the compressibility factor as a function of reduced temperature, reduced pressure, and whatever additional characteristic parameters we may want (such as the acentric factor). As before, we require relations which express the characteristic parameters as functions of the composition. This information, coupled with standard thermodynamic relations, is sufficient to obtain the fugacity coefficient of a component in a mixture.

To write a generalized equation which accurately reproduces the volumetric properties of gases over wide ranges of temperature and pressure is not a simple matter; such an equation is necessarily very complicated. Redlich, et al.,[55] however, have proposed such an equation which, because of its complexity, we will not reproduce here. We merely indicate that they have established an algebraic function of the form

$$z = z_{RK} + \Delta z, \qquad (5.13\text{-}1)$$

where z_{RK} is given by the Redlich-Kwong equation of state[56] and where

$$\Delta z = \pounds\left(\frac{T}{T_c}, \frac{P}{P_c}, \omega\right). \qquad (5.13\text{-}2)$$

The Redlich-Kwong equation is:

$$z_{RK} = \frac{Pv}{RT} = \frac{v}{v - b} - \frac{a}{RT^{3/2}(v + b)} \qquad (5.13\text{-}3)$$

where a and b are characteristic constants which differ from one gas to another.

Equation (5.13-3) is similar, but superior, to the equation of van der Waals; as in the van der Waals equation, the Redlich-Kwong constants a and b have an approximate physical significance: a provides a rough measure of the attractive intermolecular forces, and b gives an approximate indication of molecular size. These constants are related to the critical temperature and pressure by

$$a = \frac{\Omega_a R^2 T_c^{2.5}}{P_c} \qquad (5.13\text{-}4)$$

$$b = \frac{\Omega_b R T_c}{P_c} \tag{5.13-5}$$

where Ω_a and Ω_b are dimensionless constants.†

By using a very complicated deviation function for Δz, Eq. (5.13-1) reproduces, with good accuracy, the generalized compressibility-factor tables of Pitzer and also extends these to both lower and higher reduced temperatures. For application to mixtures, Redlich proposes the customary composition-dependences for a and b:

$$a_{\text{mixture}} = \sum_i \sum_j y_i y_j a_{ij}, \tag{5.13-6}$$

$$a_{ij} = (a_i a_j)^{1/2}, \tag{5.13-7}$$

and

$$b_{\text{mixture}} = \sum_i y_i b_i. \tag{5.13-8}$$

For the variation of T_c, P_c and ω with composition, as required in Eq. (5.13-2), Redlich, et al.,[55] propose relations which are quite complicated; they have very little theoretical justification and were established empirically. They are:

$$T_c = \left(\frac{B^2 D}{A} \right)^{2/3} \tag{5.13-9}$$

$$P_c = \left(\frac{B D^{5/4}}{A^{5/4}} \right)^{4/3} \tag{5.13-10}$$

$$\omega = \frac{C}{A} \tag{5.13-11}$$

where

$$A \equiv \sum_i \frac{y_i T_{c_i}}{P_{c_i}} \tag{5.13-12}$$

$$B \equiv \sum_i \frac{y_i T_{c_i}^{5/4}}{P_{c_i}^{1/2}} \tag{5.13-13}$$

† These constants may be found by applying to Eq. (5.13-3) the conditions at the critical point:

$$\left(\frac{\partial P}{\partial v} \right)_T = \left(\frac{\partial^2 P}{\partial v^2} \right)_T = 0.$$

This procedure yields $\Omega_a = 0.4278$ and $\Omega_b = 0.0867$ which are the values used by Redlich, et al.[55,56]

$$C \equiv \sum_i \frac{y_i \omega_i T_{c_i}}{P_{c_i}} \tag{5.13-14}$$

$$D \equiv 1 + \frac{1}{2}\left(\frac{C}{A} - \sum_i y_i \omega_i\right) . \tag{5.13-15}$$

Equation (5.13-1) not only reproduces the compressibility factors of a variety of *pure* gases; with the help of Eqs. (5.13-9) to (5.13-15) it also appears to reproduce the compressibility factors of gaseous *mixtures* with fair and sometimes excellent accuracy. In general, the equation is most reliable for mixtures of nonpolar components remote from critical conditions. To illustrate, Fig. 5-26 shows calculated and observed compressibility factors for the nitrogen/ethylene system.

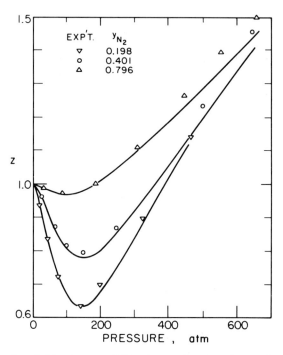

Fig. 5-26 Compressibility factors for the nitrogen/ ethylene system at 50°C calculated from modified Redlich-Kwong equation.

With Redlich's modified equation of state for the compressibility factor of a gaseous mixture, the calculation of fugacity coefficients is, in principle, straightforward. But such calculations are prohibitively tedious without an electronic computer, and fortunately, a computer program for this purpose is available.[55]

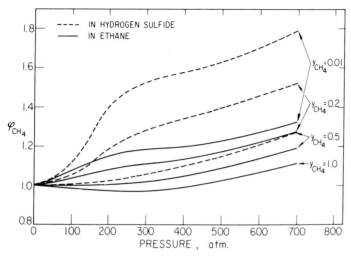

Fig. 5-27   Fugacity coefficients of methane in hydrogen sulfide and in ethane as calculated from modified Redlich-Kwong equation at 340°F.

Figure 5-27 presents calculated results for the fugacity coefficients of methane in two binary mixtures for different mole fractions of methane. In one mixture the second component is hydrogen sulfide and in the other it is ethane, and for $y_{CH_4} = 0.01$ we can see that when methane is surrounded by hydrogen sulfide molecules, its fugacity coefficient is much higher than what it is when surrounded by ethane molecules. Because of chemical dissimilarity, methane is much more "uncomfortable" with hydrogen sulfide than with ethane and therefore its fugacity (escaping tendency) is larger in the solution where it feels less "at home." A "discomfort index" is provided by the ratio of φ (methane in solution) to φ (pure methane at same temperature and pressure). Figure 5-27 shows that at $y_{CH_4} = 0.01$, $T = 340°$F and $P = 700$ atm, the "discomfort index" of methane in hydrogen sulfide is 1.61 while in ethane it is only 1.18. This "discomfort index" is, in fact, exactly the same as the activity coefficient.

For pure components the modified Redlich-Kwong equation is an improvement over the original Redlich-Kwong equation,[56] but for mixtures the improvement is usually slight, and in some cases the old equation is actually better. Failure of the modified equation to give consistently significant improvement for mixtures is probably due to the inflexibility of the mixing rules [Eqs. (5.13-6) to (5.13-15)]; it is just not reasonable to expect that an empirical equation of state should be able, with any degree of generality, to predict properties of mixtures using only properties of the pure components. For some mixtures, the mixing rules proposed by Redlich and co-workers are very good but for others they are not; it appears that whenever two components are appreciably different from one another in chemical nature and/or molecular size, the proposed mixing rules are not reliable.

The most important mixing relation is Eq. (5.13-7); even slight changes in the geometric mean assumption for a_{ij} can have a large effect on the calculated fugacities. It has therefore been proposed to retain the original Redlich-Kwong equation but to relax the assumption that a_{ij} is given by the geometric mean of a_i and a_j.[35,57,58,59]

If we retain the original Redlich-Kwong equation [Eq. (5.13-3)] with the mixing rules given by Eqs. (5.13-6) and (5.13-8), but discard Eq. (5.13-7), we obtain for the fugacity coefficient of component k in a mixture of m components:

$$
\begin{aligned}
\ln \varphi_k = {} & \ln \frac{v}{v-b} + \frac{b_k}{v-b} - \frac{2 \sum_{i=1}^{m} y_i a_{ik}}{R T^{3/2} b} \ln \frac{v+b}{v} \\[2mm]
& + \frac{ab_k}{R T^{3/2} b^2} \left[\ln \frac{v+b}{v} - \frac{b}{v+b} \right] - \ln \frac{Pv}{RT},
\end{aligned}
\qquad (5.13\text{-}16)
$$

where v is the molar volume of the mixture and a and b are the constants for the mixtures as given by Eqs. (5.13-6) and (5.13-8). Equation (5.13-16) follows from substitution of Eq. (5.13-3) into Eq. (3.4-16).

In addition to relaxing the geometric-mean assumption for a_{ij}, Chueh[35] proposed that Ω_a and Ω_b be evaluated not from critical data alone but from pure-component volumetric data for the saturated vapor. It is reasonable to base Ω_a and Ω_b on saturated vapor volume data since in typical phase-equilibrium calculations we want the fugacity coefficients in the saturated vapor mixture. Further, Chueh's proposal permits Ω_a and Ω_b to vary somewhat from one substance to another and thereby relaxes the two-parameter corresponding-states assumption implied by using universal values for Ω_a and Ω_b in Eqs. (5.13-4) and (5.13-5). Using saturated-vapor data, Chueh calculated Ω_a and Ω_b for nineteen normal fluids; these are given in Table 5-9. Also shown are the acentric factors ω for each fluid and it is evident that as the acentric factor rises, Ω_a and Ω_b deviate increasingly from the universal values 0.4278 and 0.0867.

For mixtures, it is now necessary to fix the constants a_{ij} (for $i \neq j$) and this is best accomplished by using whatever volumetric data (e.g., second virial coefficients) may be at hand for mixtures of components i and j. From Eqs. (5.13-4) and (5.13-5) we can write for a_{ij}:

$$
a_{ij} = \frac{(\Omega_{a_i} + \Omega_{a_j}) R^2 T_{c_{ij}}^{2.5}}{2 P_{c_{ij}}}
\qquad (5.13\text{-}17)
$$

Table 5-9

ACENTRIC FACTORS AND DIMENSIONLESS
CONSTANTS IN THE REDLICH-KWONG
EQUATION OF STATE FOR SATURATED VAPORS[†]

	ω	Ω_a	Ω_b
Methane	0.013	0.4278	0.0867
Nitrogen	0.040	0.4290	0.0870
Ethylene	0.085	0.4323	0.0876
Hydrogen sulfide	0.100	0.4340	0.0882
Ethane	0.105	0.4340	0.0880
Propylene	0.139	0.4370	0.0889
Propane	0.152	0.4380	0.0889
i-Butane	0.187	0.4420	0.0898
Acetylene	0.190	0.4420	0.0902
1-Butene	0.190	0.4420	0.0902
n-Butane	0.200	0.4450	0.0906
Cyclohexane	0.209	0.4440	0.0903
Benzene	0.211	0.4450	0.0904
i-Pentane	0.215	0.4450	0.0906
Carbon dioxide	0.225	0.4470	0.0911
n-Pentane	0.252	0.4510	0.0919
n-Hexane	0.298	0.4590	0.0935
n-Heptane	0.349	0.4680	0.0952
n-Octane	0.398	0.4760	0.0968

[†]P. L. Chueh and J. M. Prausnitz, *I&EC Fundamentals*, **6**, 492 (1967).

where

$$P_{c_{ij}} = \frac{z_{c_{ij}} R T_{c_{ij}}}{v_{c_{ij}}} \qquad (5.13\text{-}18)$$

$$v_{c_{ij}}^{1/3} = \frac{1}{2}(v_{c_i}^{1/3} + v_{c_j}^{1/3}) \qquad (5.13\text{-}19)$$

$$z_{c_{ij}} = 0.291 - 0.08\left(\frac{\omega_i + \omega_j}{2}\right) \qquad (5.13\text{-}20)$$

and

$$T_{c_{ij}} = (T_{c_i} T_{c_j})^{1/2}(1 - k_{ij}). \qquad (5.13\text{-}21)$$

The characteristic binary constant k_{ij} represents the deviation from the geometric mean for $T_{c_{ij}}$; it is, to a good approximation, a true molecular constant independent of temperature, composition and density. Since k_{ij} is characteristic of the i-j interaction, it must be obtained from some experimental data for the i-j mixture; a good source of such data is provided by second virial cross coefficients B_{ij} although other binary data may also be used.[35] Table 5-10 gives Chueh's estimates of k_{ij} for 118 binary sys-

tems; these estimates are subject to revision as new experimental data
become available.

In a typical application of Eq. (5.13-16), the molar volume of the mixture is not given but must be found by trial and error from Eq. (5.13-3)
using the known temperature, pressure, and composition. Calculations
for fugacity coefficients based on Eq. (5.13-16) are therefore best performed with an electronic computer.

Table 5-10

CHARACTERISTIC CONSTANT k_{12} FOR BINARY SYSTEMS[†]

$$\left[k_{12} = 1 - \frac{T_{c_{12}}}{(T_{c_{11}} T_{c_{22}})^{1/2}} \right]$$

System (1)	System (2)	$k_{12} \times 10^2$	System (1)	System (2)	$k_{12} \times 10^2$
Methane	Ethylene	1	Propylene	Propane	0
	Ethane	1	(or Propane)	n-Butane	0
	Propylene	2		i-Butane	0
	Propane	2		n-Pentane	1
	n-Butane	4		i-Pentane	0
	i-Butane	4		n-Hexane	(1)
	n-Pentane	6		Cyclohexane	(1)
	i-Pentane	6		n-Heptane	(2)
	n-Hexane	8		n-Octane	(3)
	Cyclohexane	8		Benzene	2
	n-Heptane	10		Toluene	(2)
	n-Octane	(12)[‡]			
	Benzene	(8)	n-Butane	i-Butane	0
	Toluene	(8)	(or i-Butane)	n-Pentane	0
	Naphthalene	14		i-Pentane	0
				n-Hexane	0
Ethylene	Ethane	0		Cyclohexane	0
(or Ethane)	Propylene	0		n-Heptane	0
	Propane	0		n-Octane	(1)
	n-Butane	1		Benzene	(1)
	i-Butane	1		Toluene	(1)
	n-Pentane	2			
	i-Pentane	2	n-Pentane	i-Pentane	0
	n-Hexane	3	(or i-Pentane)	n-Hexane	0
	Cyclohexane	3		Cyclohexane	0
	n-Heptane	4		n-Heptane	0
	n-Octane	(5)		n-Octane	0
	Benzene	3		Benzene	(1)
	Toluene	(3)		Toluene	(1)
	Naphthalene	8			

[†]P. L. Chueh and J. M. Prausnitz, *I&EC Fundamentals*, **6**, 492 (1967).
[‡]Numbers in parentheses are interpolated or estimated values.

Table 5-10—*Cont.*

System (1)	(2)	$k_{12} \times 10^2$	System (1)	(2)	$k_{12} \times 10^2$
n-Hexane	n-Heptane	0	Acetylene	Methane	(5)
(or Cyclohexane)	n-Octane	0		Ethylene	6
	Benzene	(1)		Ethane	8
	Toluene	1		Propylene	7
n-Heptane	n-Octane	0		Propane	9
	Benzene	(1)		n-Butane	(10)
	Toluene	(1)		i-Butane	(10)
				n-Pentane	(11)
n-Octane	Benzene	(1)		i-Pentane	(11)
	Toluene	(1)	Nitrogen	Methane	3
Benzene	Toluene	(0)		Ethylene	4
Carbon dioxide	Methane	(5 ± 2)		Ethane	5
	Ethylene	6		Propylene	(7)
	Ethane	8		Propane	(9)
	Propylene	10		n-Butane	12
	Propane	11 ± 1	Argon	Methane	2
	n-Butane	16 ± 2		Ethylene	3
	i-Butane	(16 ± 2)		Ethane	3
	n-Pentane	(18 ± 2)		Oxygen	1
	i-Pentane	(18 ± 2)		Nitrogen	0
	Naphthalene	24	Hydrogen	Nitrogen	0
Hydrogen sulfide	Methane	5 ± 1		Argon	0
	Ethylene	(5 ± 1)		Methane	3
	Ethane	6		n-Hexane	10
	Propylene	(7)	Tetra-	Methane	7
	Propane	8	fluoro methane	Nitrogen	2
	n-Butane	(9)		Helium	16 ± 2
	i-Butane	(9)	Neon	Methane	28
	n-Pentane	11 ± 1		Krypton	20 ± 2
	i-Pentane	(11 ± 1)	Krypton	Methane	1
	Carbon dioxide	8	Helium	Nitrogen	16
				Methane	(46)

Figure 5-28 shows fugacity coefficients as calculated by Eqs. (5.13-16) to (5.13-21) for carbon dioxide in a mixture containing 85 mol % n-butane at 340°F; for comparison, experimental fugacity coefficients are also shown. The test is a stringent one since the mole fraction of carbon dioxide is small and the temperature is close to the mixture's critical temperature. Also indicated is the poor result obtained when k_{ij} is assumed to be zero. Finally, Fig. 5-28 shows that the Lewis rule fails badly for carbon dioxide since butane is present in excess.

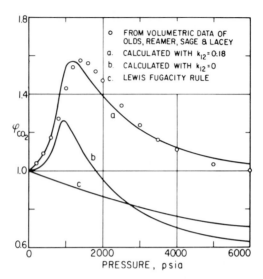

o FROM VOLUMETRIC DATA OF
   OLDS, REAMER, SAGE & LACEY
a. CALCULATED WITH $k_{12}=0.18$
b. CALCULATED WITH $k_{12}=0$
c. LEWIS FUGACITY RULE

Fig. 5-28   Fugacity coefficients of carbon dioxide in a mixture containing 85 mol % n-butane at 340°F. ($k_{12} = 0.18$ obtained from second-viral-coefficient data.)

Figure 5-29 gives calculated and experimental results for the saturated vapor of the n-pentane/hydrogen sulfide system. Also shown are results obtained using Redlich's original mixing rule, Eq. (5.13-7). While the compressibility factor is improved only moderately by Chueh's modification, this improvement has an appreciable effect on the fugacity coefficient of pentane above 600 psia when hydrogen sulfide is present in excess.

For quantum gases (helium, hydrogen, and neon) Chueh showed that good results can be obtained from the Redlich-Kwong equation provided effective critical constants are used in Eqs. (5.13-4) and (5.13-5). Since the configurational properties of low-molecular-weight molecules are described by quantum, rather than classical mechanics, effective critical constants depend on the molecular mass and on the temperature. By fitting the Redlich-Kwong equation (using $\Omega_a = 0.4278$ and $\Omega_b = 0.0867$) to experimental volumetric data, Chueh found that the effective critical constants are given by

$$T_c = \frac{T_c^0}{1 + (c_1/mT)} \qquad (5.13\text{-}22)$$

and

$$P_c = \frac{P_c^0}{1 + (c_2/mT)} \qquad (5.13\text{-}23)$$

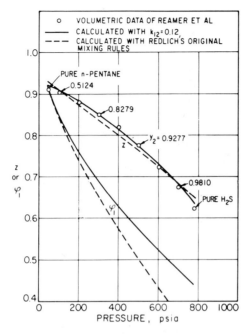

Fig. 5-29 Compressibility factors and fugacity coefficients for saturated vapor of n-pentane (1)/hydrogen sulfide (2) at 160°F. ($k_{12} = 0.12$ obtained from second-virial-coefficient data.)

where m is the molecular weight and c_1 and c_2 are constants. Equations (5.13-22) and (5.13-23) have also been used to include quantum gases in corresponding-states correlations.[54]

Further, by comparing experimental data with correlations for second and third virial coefficients, the effective critical volume is given by

$$v_c = \frac{v_c^0}{1 + (c_3/mT)}. \tag{5.13-24}$$

The constants c_1, c_2, and c_3 are the same for all quantum gases. They are (in degrees Kelvin): $c_1 = 21.8$, $c_2 = 44.2$, and $c_3 = -9.91$. In Eqs. (5.13-22) to (5.13-24), the superscript 0 designates the "classical" critical constant; as mT becomes large, $T_c \rightarrow T_c^0$, $P_c \rightarrow P_c^0$, and $v_c \rightarrow v_c^0$. Therefore, while Eqs. (5.13-22) to (5.13-24) are essentially empirical, they are consistent with the theoretical requirement that quantum corrections are important only for low molecular weights and at low temperatures. Table 5-11 gives effective "classical" critical constants for neon and for isotopes of hydrogen and helium.

Table 5-11

CLASSICAL CRITICAL CONSTANTS
FOR QUANTUM GASES

	$T_c^0\,(^\circ K)$	$P_c^0\,(atm)$	$v_c^0\,(cc/g\text{-}mol)$
Ne	45.5	26.9	40.3
He^4	10.47	6.67	37.5
He^3	10.55	5.93	42.6
H_2	43.6	20.2	51.5
HD	42.9	19.6	52.3
HT	42.3	19.1	52.9
D_2	43.6	20.1	51.8
DT	43.5	20.3	51.2
T_2	43.8	20.5	51.0

For mixtures containing one or more of the quantum gases, the mixing rules given by Eqs. (5.13-6), (5.13-8), and (5.13-17) should be used; however, Eqs. (5.13-18) to (5.13-21) must now be modified as follows:

$$T_{c_{ij}} = \frac{(T_{c_i}^0\, T_{c_j}^0)^{1/2}\,(1 - k_{ij})}{1 + (c_1/m_{ij}\, T)} \qquad (5.13\text{-}25)$$

and

$$P_{c_{ij}} = \frac{P_{c_{ij}}^0}{1 + (c_2/m_{ij}\, T)} \qquad (5.13\text{-}26)$$

where

$$P_{c_{ij}}^0 = \frac{z_{c_{ij}}^0\, R\,(T_{c_i}^0\, T_{c_j}^0)^{1/2}\,(1 - k_{ij})}{v_{c_{ij}}^0} \qquad (5.13\text{-}27)$$

$$v_{c_{ij}}^{0\,1/3} = \tfrac{1}{2}\,[(v_{c_i}^0)^{1/3} + (v_{c_j}^0)^{1/3}] \qquad (5.13\text{-}28)$$

$$z_{c_{ij}}^0 = 0.291 - 0.08 \left(\frac{\omega_i + \omega_j}{2}\right) \qquad (5.13\text{-}29)$$

$$\frac{1}{m_{ij}} = \frac{1}{2}\left(\frac{1}{m_i} + \frac{1}{m_j}\right). \qquad (5.13\text{-}30)$$

For all pure quantum gases, $\Omega_a = 0.4278$ and $\Omega_b = 0.0867$. Further, by assumption, for all pure quantum gases ω (effective) is zero; it therefore follows that

$$v_c^0 = 0.291\,\frac{R T_c^0}{P_c^0}. \qquad (5.13\text{-}31)$$

The "classical" critical volumes listed in Table 5-11 satisfy Eq. (5.13-31).

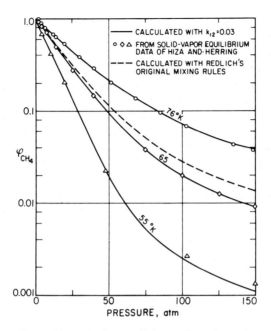

Fig. 5-30   Fugacity coefficients of methane in
hydrogen at saturation. (k_{12} = 0.03
obtained from second-virial-coefficient
data.)

Figure 5-30 presents calculated and experimental fugacity coefficients for saturated methane in hydrogen at 55–76°K. The calculations were made using Eq. (5.13-16) and the mixing rules just described; at these low temperatures it is important to include quantum corrections for hydrogen. Also shown are calculations at 65°K based on Redlich's original mixing rules, Eqs. (5.13-6), (5.13-7), and (5.13-8). The experimental results are based on measurements of Hiza and Herring[60] who obtained gas-phase solubility data for solid methane in compressed hydrogen.

5.14   Solubility of Solids and Liquids
   in Compressed Gases

Vapor-phase fugacity coefficients are required in any phase-equilibrium calculation wherein one of the phases is a gas at advanced pressure. These coefficients are particularly important for calculating the solubility of a solid or a high-boiling liquid in a gas at high pressures because in such cases, failure to include corrections for gas-phase nonideality can sometimes lead to serious errors. To illustrate, we now discuss calculations for the solubility of a condensed component in a dense gas.

First, we consider equilibrium between a compressed gas and a solid. This is a particularly simple case because the solubility of the gas in the solid is almost always negligible. The condensed phase may therefore be considered pure and thus all nonideal behavior in the system can be attributed entirely to the vapor phase. The case of gas-liquid equilibria, which we consider later, is a little more complicated because the solubility of the gas in the liquid, while it may be small, is usually not negligible.

In our discussion, let subscript 1 stand for the light (gaseous) component and let subscript 2 stand for the heavy (solid) component. We want to calculate the solubility of the solid component in the gas phase at temperature T and pressure P. We first write the general equation of equilibrium for component 2:

$$f_2^s = f_2^V, \tag{5.14-1}$$

where superscript s stands for the solid phase.

Since the solid phase is pure, the fugacity of component 2 is given by:

$$f_2^s = P_2^s \varphi_2^s \exp \int_{P_2^s}^{P} \frac{v_2^s dP}{RT} \tag{5.14-2}$$

as discussed in Sec. 3.3. In Eq. (5.14-2), P_2^s is the saturation (vapor) pressure of the pure solid, φ_2^s is the fugacity coefficient at saturation pressure P_2^s, and v_2^s is the solid molar volume, all at temperature T.

For the vapor-phase fugacity we introduce the fugacity coefficient φ_2 by recalling its definition:

$$\varphi_2 \equiv \frac{f_2^V}{y_2 P}. \tag{5.14-3}$$

Substituting and solving for y_2, we obtain for the solubility of the heavy component in the gas phase:

$$\boxed{y_2 = \frac{P_2^s}{P} \cdot E} \tag{5.14-4}$$

where

$$E \equiv \frac{\varphi_2^s \exp \int_{P_2^s}^{P} \frac{v_2^s dP}{RT}}{\varphi_2}. \tag{5.14-5}$$

The quantity E is nearly always greater than unity and it is therefore called the enhancement factor; that is, E is the correction factor which must be applied to the simple (ideal-gas) expression which is valid only at low pressure. The enhancement factor is a measure of the extent to which pressure enhances the solubility of the solid in the gas; as $P \rightarrow P_2^s$, $E \rightarrow 1$.

The enhancement factor contains three correction terms: φ_2^s, which takes into account nonideality of the pure saturated vapor; the Poynting correction, which gives the effect of pressure on the fugacity of the pure solid; and φ_2, the vapor-phase fugacity coefficient in the high-pressure gas mixture. Of these three correction terms, it is the last one which is by far the most important. In most practical cases, the saturation pressure P_2^s of the solid is quite small and thus φ_2^s is nearly equal to unity. The Poynting correction is not negligible, but it rarely accounts for an enhancement factor of more than 2 or 3 and frequently much less. However, the fugacity coefficient φ_2 can be so far removed from unity as to produce very large enhancement factors which, in some cases, can exceed 10^3 or more. Large enhancement factors are especially common at low temperatures; a dramatic example is given in Fig. 5-31, which shows experimentally observed enhancement factors for the oxygen-hydrogen system.[61] The solubility of (solid) oxygen in dense hydrogen is much larger than that calculated by a simple, ideal-gas computation because the fugacity coefficient of oxygen is much less than unity. This particular example is, no doubt, highly extreme. The magnitude 10^{12} is a difficult concept for any ordinary person who is not trained in astronomy or in governmental expenditures. (As a frame of reference, the U.S. federal budget in 1968 was close to 1.4×10^{11} dollars.)

The huge enhancement factors shown in Fig. 5-31 are perhaps somewhat misleading because at temperatures below the critical of hydrogen $(60°R)$ the observed solubilities are actually those in liquid, rather than gaseous, hydrogen. Nevertheless, even if attention is restricted to the left half of Fig. 5-31 where hydrogen is above its critical temperature, it is clear that in this type of phase equilibrium very large errors are incurred by failure to account for gas-phase nonideality.

Another example of enhanced solubility of a solid in a dense gas is given in Fig. 5-32, which shows the solubility of solid carbon dioxide in compressed air at $143°K$; the dashed line gives the calculated solubility with $\varphi_2 = 1$. In this calculation the solubility was calculated with the Poynting correction included but with the assumption of gas-phase ideality. The two lines at the top of the diagram were calculated from Eq. (5.4-2) based on the virial equation; one calculation included only second virial coefficients, while the other included second and third virial coefficients. In these calculations air was considered to be a mixture of oxygen

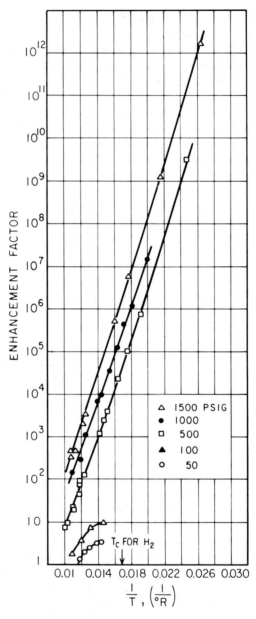

Fig. 5-31   Enhancement factor for the oxygen/
          hydrogen system.

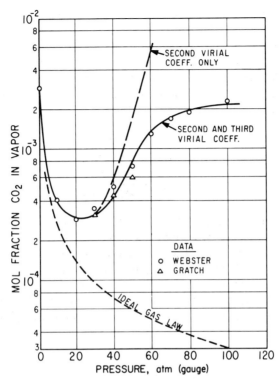

Fig. 5-32   Solubility of solid carbon dioxide in air at
143°K.

and nitrogen and therefore the results shown are those for a ternary system. The various virial coefficients were calculated from the Kihara
potential using only pure-component parameters and customary mixing
rules. The calculated results are compared with experimental values.[62]
When third virial coefficients are neglected, the calculations correctly
predict the minimum in the curve of log y_2 versus P, but at pressures
above that corresponding to the minimum mole fraction, third virial
coefficients are required. In this particular system, agreement between
calculated and experimental results at higher pressures is very good; unfortunately, such good agreement is not always obtained due to the difficulty of calculating third virial coefficients of mixtures. However, in the
moderate-pressure region where third (and higher) virial coefficients can
be neglected, it is almost always possible in nonpolar systems to make
good estimates of the vapor-phase solubility of a solid.[63] Such estimates
are frequently useful for engineering purposes; for example, in a freezing-
out process it is desirable to minimize the mole fraction of a heavy im-
purity in a gas stream or, in a gas-cooling process, it is necessary to know

the solubility of a heavy component in order to prevent plugging of the flow lines by precipitation of solid. Reuss and Beenakker[64] and also Hinckley and Reid[65] have shown that by differentiation and rearrangement of Eqs. (5.14-4) and (5.4-2) for a binary system it is possible to calculate rapidly the minimum mole fraction and its corresponding pressure. Neglecting φ_2^s, the Poynting correction, and all third and higher virial coefficients, the coordinates at the minimum mole fraction are:

$$y_{2\,(minimum)} = \frac{-5.44\,B_{12}\,P_2^s}{RT} \qquad (5.14\text{-}6)\dagger$$

$$P_{(at\,minimum)} = \frac{-RT}{B_{11} + 2B_{12}} \qquad (5.14\text{-}7)$$

where, as before, subscript 1 refers to the light (gaseous) component and subscript 2 to the heavy (solid) component. Equations (5.14-6) and (5.14-7) provide a rapid method for estimating the minimum solubility of solids in gases. Table 5-12 compares some experimental results with those calculated from Eqs. (5.14-6) and (5.14-7) and the agreement is most satisfying. In these calculations, second virial coefficients were obtained from the Lennard-Jones potential and customary mixing rules.

Table 5-12

SOLUBILITY OF SOLIDS IN DENSE GASES
VAPOR COMPOSITION AND PRESSURE FOR MINIMUM SOLUBILITY†

Solid	Gas	$T\,(^\circ K)$	Minimum mol fraction of heavy component		Pressure at minimum mol fraction (atm)	
			Calc.	Expt'l.	Calc.	Expt'l.
CO	H_2	60	2.7×10^{-3}	2.6×10^{-3}	23	21
		50	1.9×10^{-4}	1.8×10^{-4}	14	15
		40	1.0×10^{-5}	1.1×10^{-5}	7.4	8.5
N_2	H_2	60	6.2×10^{-3}	6.5×10^{-3}	24.5	22–24
		50	6.3×10^{-4}	6.3×10^{-4}	14.6	16–17
		40	2.0×10^{-5}	2.0×10^{-5}	7.9	8–9
		30	4.6×10^{-8}	$< 7 \times 10^{-7}$	3.6	< 6
CO_2	air	160	2.0×10^{-3}	1.9×10^{-3}	34	35
		140	1.7×10^{-4}	1.9×10^{-4}	23	23
		130	3.7×10^{-5}	4.0×10^{-5}	18.5	18–19
		125	1.6×10^{-5}	1.6×10^{-5}	17	16
		120	6.3×10^{-6}	7.6×10^{-6}	15	14

†R. B. Hinckley and R. C. Reid, *A. I. Ch. E. Journal*, **10**, 416 (1964).

†If $B_{12} > 0$, then no $y_{2\,(minimum)}$ exists.

At very high pressures the virial equation is not useful because almost no information is available on the fourth and higher virial coefficients and even the third virial coefficient can only be estimated with fair accuracy. Therefore, as discussed in the previous section, if the gas density is high (close to or exceeding the critical density) an essentially empirical equation of state must be used to relate the fugacity coefficient to the pressure, temperature, and gas-phase composition. The equation of Redlich and Kwong described earlier is useful for this purpose; to illustrate its use,

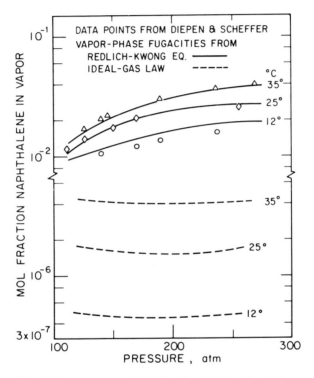

Fig. 5-33   Vapor-phase solubility of naphthalene in ethylene.

Fig. 5-33 compares calculated and experimental solubilities of naphthalene (melting point 80.2°C) in compressed ethylene gas in the temperature range 12–35°C. The Redlich-Kwong equation gives excellent results even up to very high pressures where the density of ethylene is above its critical value. For comparison, Fig. 5-33 also shows results calculated with the assumption that the fugacity coefficient of naphthalene is unity and it is evident that such an assumption leads to very large error. The solubility of naphthalene in compressed ethylene is remarkably high because at temperatures not far above its critical temperature (9.2°C) compressed ethylene gas has a solvent power comparable to that of a liquid hydro-carbon.

We now want to consider the solubility of a liquid in a compressed gas. This problem is somewhat more difficult to handle than that discussed above because a gas is soluble to an appreciable extent in a liquid whereas the solubility of a gas in a solid is usually negligible. The general problem of high-pressure, vapor-liquid equilibria is considered in more detail in Chap. 10; here we restrict ourselves to equilibrium between a high-boiling liquid and a sparingly soluble gas at conditions remote from critical. We are interested in the solubility of the heavy (liquid) component in the vapor phase, which contains the light (gaseous) component in excess; as before, we let 1 stand for the (light) gaseous solvent and 2 for the (heavy) solute in the gas phase. For the heavy component the equation of equilibrium is:

$$f_2^L = f_2^V = \varphi_2 y_2 P \qquad (5.14\text{-}8)$$

where superscripts V and L stand, respectively, for vapor and liquid and where φ_2 is the vapor-phase fugacity coefficient of component 2 in the gaseous mixture.

The fugacity coefficient φ_2 must again be calculated from an equation of state, but before we turn to that calculation, we must say something about the fugacity of component 2 in the liquid phase. If the gas is only sparingly soluble, the liquid-phase fugacity of component 2 can be calculated by assuming that the solubility of component 1 in the liquid is described by a pressure-corrected form of Henry's law (see Sec. 8.3): For component 1 we assume that in the liquid phase the fugacity f_1^L is related to the mole fraction x_1 by

$$f_1^L = H_{1,2} x_1 \exp \int_{P_2^s}^{P} \frac{\bar{v}_1^\infty}{RT} dP \qquad (5.14\text{-}9)$$

where $H_{1,2}$ is Henry's constant, $\bar{v}_1^\infty$ is the partial molar volume of component 1 in the liquid phase at infinite dilution, and P_2^s is the saturation pressure of pure liquid component 2, all at system temperature T. Then, from the Gibbs-Duhem equation, it can be shown that

$$f_2^L = (1 - x_1) P_2^s \varphi_2^s \exp \int_{P_2^s}^{P} \frac{v_2^L}{RT} dP \qquad (5.14\text{-}10)$$

where P_2^s is the saturation pressure, φ_2^s is the fugacity coefficient at saturation, and v_2^L is the molar volume, all of pure liquid 2 and all at temperature T.

Substituting Eq. (5.14-10) into Eq. (5.14-8), and solving for the desired quantity y_2, we obtain:

$$y_2 = \frac{(1 - x_1) P_2^s \varphi_2^s \exp \displaystyle\int_{P_2^s}^{P} \frac{v_2^L \, dP}{RT}}{\varphi_2 P},$$ (5.14-11)

where x_1, the solubility of gas 1 in liquid 2, is calculated from

$$x_1 = \frac{y_1 \varphi_1 P}{H_{1,2} \exp \displaystyle\int_{P_2^s}^{P} \frac{\bar{v}_1^{\infty} \, dP}{RT}}.$$ (5.14-12)

The solution of Eq. (5.14-11) necessarily requires a trial-and-error calculation because x_1 as well as φ_1 and φ_2 depend on the composition of the vapor. For the conditions under consideration, $y_1 \approx 1$ and thus, as discussed in Sec. 5.1, the Lewis fugacity rule is a good approximation for the fugacity of component 1 (although it is a very bad approximation for the fugacity of component 2). In many cases, therefore, the calculations can be simplified by replacing φ_1 in Eq. (5.14-12) by $\varphi_{\text{pure 1}}$ (evaluated at the temperature and pressure of the system) and by setting $y_2 = 0$ in Eq. (5.4-2) for φ_2.

The fugacity of pure liquid 2 at temperature T and pressure P is readily calculated from the volumetric properties of pure component 2 as shown in Sec. 3.3. Henry's constant H must be determined experimentally or else estimated from a suitable correlation (see Sec. 8.5); it need be known only approximately because under the conditions considered, $x_1 \ll 1$, and thus even a somewhat inaccurate estimate of x_1 does not seriously affect the value of y_2. A rough estimate of $\bar{v}_1^{\infty}$ is sufficient here since the exponential correction to H is usually small and often may be neglected entirely.

In Eq. (5.14-11) it is the fugacity coefficient φ_2 which accounts for the nonideal behavior of liquid solubility in compressed gases. To illustrate, Fig. 5-34 shows the solubility of decane (normal b.p. 174°C) in compressed nitrogen at 50°C. Curve (a) gives calculated results assuming ideal-gas behavior and curve (c) shows results obtained when the Lewis fugacity rule is used for calculating the fugacity coefficient of decane (component 2). Curve (b) is based on the virial equation with $B_{12} = -141$ cc/g-mol, and it gives excellent agreement with experimental data.[66] This is perhaps not surprising since the virial coefficient B_{12} used in the calculation was, in fact, back-calculated from the experimental solubility data. However, a reasonable estimate of B_{12} could probably have been made without any

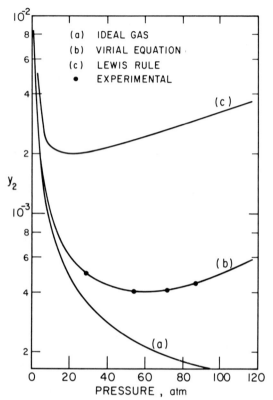

Fig. 5-34   Vapor-phase solubility of decane in nitrogen
at 50°C.

data for this binary system by using one of the generalized equations discussed in Sec. 5.7.

The nonideal behavior of the decane/nitrogen system at 50°C is not large because the fugacity coefficient of decane in this case is still of the order of magnitude of unity (or, in other words, the absolute value of B_{12} is not very large). This result follows from the fact that nitrogen is a very light gas with a potential energy which is much smaller than its kinetic energy at 50°C. Therefore, when a decane molecule escapes from the liquid into a space filled with gaseous nitrogen, it feels almost as if it were in a vacuum, or to put it more precisely, as if it were dissolved in an ideal gas. Thus, the calculation based on the ideal-gas assumption is not too seriously in error in this case; the solubility of decane in nitrogen is only a little larger than what it would be in an ideal-gas solvent because the attractive forces between solute and gaseous solvent are weak in this particular system at 50°C.

However, the ideal-gas assumption becomes increasingly bad for calculating the solubility of a liquid in a compressed gas when the potential energy of the gas is not small compared to its kinetic energy. For example, Fig. 5-35 shows fugacity coefficients of decane at saturation in three gases. It is evident that the vapor-phase fugacity coefficient of decane (and hence its vapor-phase solubility) is strongly dependent on the

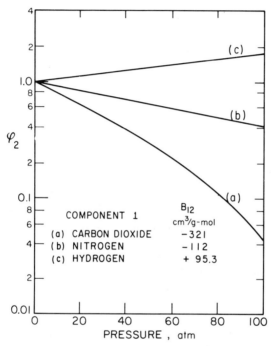

Fig. 5-35   Fugacity coefficient of decane in hydrogen, nitrogen, and carbon dioxide at 75°C.

nature of the gaseous component. Thus decane at 75°C and 100 atm is approximately ten times more soluble in carbon dioxide than in nitrogen and approximately forty times more soluble than in hydrogen. The relatively high solubility in carbon dioxide follows from the relatively large attractive forces between decane and carbon dioxide. A rough measure of the potential energy of a gaseous solvent is given by its critical temperature; as a result it is not surprising that carbon dioxide (T_c = 304.2°K) is a better solvent for decane than nitrogen (T_c = 126.2°K) which, in turn, is a better solvent than hydrogen (T_c = 33.3°K).

5.15   Summary

Efforts to describe volumetric properties of gases and gaseous mixtures extend over more than a hundred years and the scientific literature abounds with articles on this subject. In this chapter we have given some of the main ideas with special reference to those contributions which appear to be most useful for technical applications. At the end of such a long chapter it is helpful to recapitulate and to state briefly the current status of efforts toward the calculation of gas-phase fugacities.

The fugacity of a component in a gas mixture at any temperature and density can be calculated exactly if volumetric data can be obtained for the mixture at the system temperature and over a density range from zero up to the density of interest. But such data are rarely available and for practical purposes it is necessary to use approximations. A particularly simple approximation is provided by the Lewis fugacity rule; this rule is not generally reliable although it gives good results for a component present in excess (Sec. 5.1).

There are many empirical equations of state and new ones are continually appearing. These equations contain empirically determined constants and such constants have been reported for many common pure gases. To predict volumetric properties of a mixture, it is necessary to establish mixing rules which state how these constants depend on composition. For empirical equations of state, such mixing rules frequently cannot be established with confidence unless at least some data are available for the particular mixture of interest. Fugacity coefficients are usually sensitive to the mixing rules used.

A theoretically significant equation of state is the virial equation whose constants (virial coefficients) can be related to intermolecular potentials (Secs. 5.2, 5.3, and 5.4). The virial equation can readily be extended to mixtures; whereas for any pure component the virial coefficients depend only on the temperature and on the intermolecular potential for that component, for a mixture they depend also on the potentials between molecules of those different components which comprise the mixture. The fundamental advantage of the virial equation is that it directly relates fugacities in mixtures to intermolecular forces. The practical disadvantage of the virial equation follows from insufficient understanding of intermolecular forces. As a result, we can at present apply the virial equation only to certain restricted cases: We can use it only for mixtures at moderate pressures, and usually we can predict the required virial coefficients only for those mixtures whose components are nonpolar.

Several models for intermolecular potential functions have been proposed and from these, virial coefficients can be computed, provided we can estimate the potential parameters (Secs. 5.5 and 5.6). For practical purposes, however, it is simpler to calculate virial coefficients from correlations based on the theorem of corresponding states (Sec. 5.7).

For highly polar and associated systems, virial coefficients can be interpreted "chemically" by relating them to the enthalpy and entropy of formation of chemically stable complexes (Secs. 5.8, 5.9 and 5.10).

Despite the limited practical use of the virial equation, we have devoted much attention to potential functions in this chapter because it is almost certain that in the future, fundamental advances in the description of volumetric properties will be based on research in statistical mechanics.

At high densities, we must compute fugacity coefficients from empirical equations of state or from generalized correlations such as those based on the theorem of corresponding states (Secs. 5.11, 5.12 and 5.13). Research on the statistical mechanics of dense gases is progressing rapidly and we may expect practical results in the near future; however, at present empirical methods are more useful.

When applying an equation of state or a corresponding-states correlation to a gaseous mixture, flexibility in mixing rules is essential for consistently good results. A mixing rule which is good for one system may not be good for another; mixing rules should therefore contain one or two adjustable parameters which are determined, if possible, by a few data for the mixture under consideration. If no mixture data are available, they should be estimated by careful analysis of chemically similar systems for which mixture data are at hand. Good estimates of fugacity coefficients can be made for constituents of dense gas mixtures which contain nonpolar (or slightly polar) components. However, for mixtures containing polar gases, or for any mixture near critical conditions, calculated fugacity coefficients are not likely to be highly accurate.

In phase equilibria, the effect of gas-phase nonideality is most pronounced in those cases where a condensed component is sparingly soluble in a compressed gas. This solubility is strongly affected by the density and by the intermolecular forces between solute and gaseous solvent. In many cases, solubilities calculated with the assumption of ideal-gas behavior are in error by several orders of magnitude (Sec. 5.14).

We close this chapter by briefly repeating what we stressed at the beginning: Calculation of fugacities in gaseous mixtures is not a thermodynamic problem. The relations which express fugacity in terms of fundamental macroscopic properties are exact and well known; they easily lend themselves to numerical solution by electronic computers. The difficulty we face lies in our inability to characterize and to predict with sufficient accuracy the configurational (essentially volumetric) properties of pure and particularly of mixed fluids; this inability, in turn, is a consequence of insufficient knowledge concerning intermolecular forces. To increase our knowledge, we require on the one hand new results from theoretical molecular physics and on the other, more accurate experimental data on equilibrium properties of dense mixtures, especially for those containing one or more polar components.

PROBLEMS

1. Consider the following simple experiment: We have a container of fixed volume V; this container, kept at temperature T, has in it n_1 moles of gas 1. We now add isothermally to this container n_2 moles of gas 2 and we observe that the pressure rise is ΔP. Assume that the conditions are such that the volumetric properties of the gases and their mixture are adequately described by the virial equation neglecting the third and higher coefficients. The second virial coefficients of the pure gases are known. Find an expression which will permit calculation of B_{12}.

2. At $-100°C$, a gaseous mixture of one gram mole contains 1 mol % CO_2 and 99 mol % H_2. The mixture is compressed isothermally to 60 atm. Does any CO_2 precipitate? If so, approximately how much? [At $-100°C$, the saturation (vapor) pressure of pure solid CO_2 is 0.1374 atm and the molar volume of solid CO_2 is 27.6 cc/g-mol.]

3. Using the generalized correlations of Pitzer, calculate:
  (a) The energy of vaporization of acetylene at $0°C$.
  (b) The second virial coefficient of a mixture containing 80 mol % butane and 20 mol % nitrogen at $370°F$. (The experimental value is -172 cc/g-mol.)
  (c) The enthalpy of mixing (at $200°K$ and 100 atm) one gram mole of methane, one gram mole of nitrogen and one gram mole of hydrogen.

4. A gas stream at $25°C$ consists of 40 mol % nitrogen, 50 mol % hydrogen and 10 mol % propane. The propane is to be removed from this stream by high-pressure absorption using a mineral oil of very low volatility. At the bottom of the column (where the gas enters) what is the driving force for the rate of absorption? The total pressure is 40 atm. The driving force for a component i is given by:

$$\text{Driving force} = p_i - p_i^*$$

where p_i is the partial pressure of i in the gas bulk phase, and p_i^* is the partial pressure of i in the gas phase in equilibrium with the liquid phase.

In this column the liquid-to-gas flow ratio (L/G) is 5 moles of liquid per mole of gas. The column is to absorb 95% of the propane. Solubility of inert gases in the mineral oil is negligible.

Data: The compressibility factor of the entering gas mixture is 0.95.
         Solubility of propane in the mineral oil is given by $f_{c_3} = Hx_{c_3}$ with $H = 52.6$ atm at $25°C$.

5. Acetic acid vapor has a tendency to dimerize. Show that in an equilibrium mixture of monomer and dimer the fugacity of the dimer is proportional to the square of the fugacity of the monomer.

  6. The virial equation gives the compressibility factor as an expansion in the density

$$z = 1 + B\rho + C\rho^2 + \ldots . \tag{1}$$

Since it is more convenient to use pressure as an independent variable, another form of the virial equation is given by

$$z = 1 + B'P + C'P^2 + \ldots . \tag{2}$$

(a) Relate the virial coefficients B' and C' to the constants a and b which appear in the Redlich-Kwong equation, which is

$$P = \frac{RT}{v - b} - \frac{a}{T^{1/2} v(v + b)} .$$

(b) Consider a binary gas mixture of ethylene and nitrogen having mole fraction $y_{C_2H_4} = 0.2$. Starting with Eq. (2), calculate the fugacity of ethylene at $50°C$ and 50 atm; neglect virial coefficients beyond the second. For the second virial coefficient use the relation found in (a) for the Redlich-Kwong equation with the mixing rules

$$a_{mixture} = y_1^2 a_1 + y_2^2 a_2 + 2 y_1 y_2 (a_1 a_2)^{1/2}$$
$$b_{mixture} = y_1 b_1 + y_2 b_2 .$$

For a pure component, a and b in the Redlich-Kwong equation are related to the critical temperature and pressure by

$$a = \frac{0.4278 R^2 T_c^{2.5}}{P_c}$$

$$b = \frac{0.0867 R T_c}{P_c} .$$

(From experimental volumetric data, the fugacity of ethylene, under the conditions specified, is 8.76 atm.)

7. One mole of a binary gas mixture containing ethylene and argon at $40°C$ is confined to a constant-volume container. Compute the composition of this mixture which gives the maximum pressure.

Assume that virial coefficients beyond the second may be neglected.

8. Benham and Katz† report the following equilibrium data for the propane/methane/hydrogen system at $-200°$ F:

Total Pressure (psia)	Liquid Mole Percent			Vapor Mole Percent		
	H_2	CH_4	C_3H_8	H_2	CH_4	C_3H_8
500	0.789	8.59	90.62	95.73	3.79	0.48
1000	1.79	10.14	88.07	99.0	0.37	0.63

From these data the K values for propane at $-200°$ F are:

$$500 \text{ psia}, \quad K = \frac{0.48}{90.62} = 5.29 \times 10^{-3}$$

$$1000 \text{ psia}, \quad K = \frac{0.63}{88.07} = 7.15 \times 10^{-3}$$

†A. I. Ch. E. Journal, **3**, 33 (1957).

Since these K values are much larger than those calculated from the Benedict-Webb-Rubin equation some doubt has been expressed about the validity of this equation of state.† On the other hand, are we sure that the data are reliable? Devise a test for the internal consistency of the K values and compare the results of Katz and Benham with K values computed on the basis of the modern theory of intermolecular forces.

9. 100 moles of H_2 initially at 85° F and 300 psia are mixed adiabatically and isobarically with 2000 moles of ethylene initially at 7.5° F and 300 psia. What is the temperature of the final mixture?

REFERENCES

1. Cullen, E. J., and K. A. Kobe, *A. I. Ch. E. Journal*, **1**, 452 (1955).

2. Douslin, D. R., in *Progress in International Research on Thermodynamic and Transport Properties* (J. F. Masi and D. H. Tsai, eds.), ASME. New York: Academic Press Inc., 1962.

3. Michels, A., W. De Graaff and C. A. Ten Seldam, *Physica*, **26**, 393 (1960).

4. Guggenheim, E. A., and M. L. McGlashan, *Proc. Roy. Soc.* (London), **A255**, 456 (1960); R. J. Munn, *J. Chem. Phys.*, **40**, 1439 (1964); R. J. Munn and F. J. Smith, *J. Chem. Phys.*, **43**, 3998 (1965); *Disc. Faraday Soc.*, **40** (1965).

5. See, for example, T. L. Hill, *Introduction to Statistical Thermodynamics.* Reading, Mass.: Addison-Wesley Publishing Co., 1960.

6. Hirschfelder, J. O., C. F. Curtiss and R. B. Bird, *Molecular Theory of Gases and Liquids.* New York: John Wiley & Sons, Inc., 1954.

7. Michels, A., J. M. Levelt and W. De Graaff, *Physica*, **24**, 659 (1958).

8. Sherwood, A. E., and J. M. Prausnitz, *J. Chem. Phys.*, **41**, 429 (1964).

9. Sherwood, A. E., and J. M. Prausnitz, *J. Chem. Phys.*, **41**, 413 (1964).

10. Kihara, T., *Advan. Chem. Phys.*, **1**, 276 (1958) and **5**, 147 (1963).

11. Kihara, T., *Rev. Mod. Phys.*, **25**, 831 (1953).

12. Connolly, J. F., and G. A. Kandalic, *Phys. Fluids*, **3**, 463 (1960).

13. Prausnitz, J. M., and R. N. Keeler, *A. I. Ch. E. Journal*, **7**, 399 (1961).

14. Prausnitz, J. M., and A. L. Myers, *A. I. Ch. E. Journal*, **9**, 5 (1963).

15. Myers, A. L., and J. M. Prausnitz, *Physica*, **28**, 303 (1962).

16. Rossi, J. C., and F. Danon, *J. Phys. Chem.*, **70**, 942 (1966).

17. O'Connell, J. P., and J. M. Prausnitz, *J. Phys. Chem.*, **72**, 632 (1968).

18. Rowlinson, J. S., *Trans. Faraday Soc.*, **45**, 974 (1949).

†R. L. Motard and E. I. Organick, *A. I. Ch. E. Journal*, **6**, 39 (1960).

19. Rowlinson, J. S., *Disc. Faraday Soc.*, **40**, 19 (1965).

20. Graben, H. W., and R. D. Present, *Phys. Rev. Letters*, **9**, 247 (1962).

21. Axilrod, B. M., and E. Teller, *J. Chem. Phys.*, **11**, 299 (1943).

22. Sherwood, A. E., A. G. De Rocco and E. A. Mason, *J. Chem. Phys.*, **44**, 2984 (1966).

23. Dymond, J. H., and B. J. Alder, *Chem. Phys. Letters*, **2**, No. 1, 54 (1968).

24. Orentlicher, M., and J. M. Prausnitz, *Can. J. Chem.*, **45**, 373 (1967).

25. McGlashan, M. L., and D. J. B. Potter, *Proc. Roy. Soc.*, (London) **A267**, 478 (1962).

26. Crain, R. W., Jr., and R. E. Sonntag, *Advan. Cryog. Eng.*, **11**, 379 (1965).

27. McGlashan, M. L., and C. J. Wormald, *Trans. Faraday Soc.*, **60**, 646 (1964).

28. Meissner, H. P., and R. Seferian, *Chem. Eng. Progr.*, **47**, 579 (1951).

29. Riedel, L., *Chem. Ing. Tech.*, **28**, 557 (1956), and earlier papers.

30. Rowlinson, J. S., *Trans. Faraday Soc.*, **51**, 1317 (1955).

31. Pitzer, K. S., and R. F. Curl, Jr., *J. Am. Chem. Soc.*, **79**, 2369 (1957).

32. Reid, R. C., and T. W. Leland, Jr., *A. I. Ch. E. Journal*, **11**, 229 (1965).

33. Prausnitz, J. M., and R. D. Gunn, *A. I. Ch. E. Journal*, **4**, 430 (1958).

34. Prausnitz, J. M., and P. R. Benson, *A. I. Ch. E. Journal*, **5**, 301 (1959).

35. Chueh, P. L., and J. M. Prausnitz, *I&EC Fundamentals*, **6**, 492 (1967).

36. Black, C., *Ind. Eng. Chem.*, **50**, 391 (1958).

37. O'Connell, J. P., and J. M. Prausnitz, *I&EC Process Design Develop.*, **6**, 245 (1967).

38. Chueh, P. L., and J. M. Prausnitz, *A. I. Ch. E. Journal*, **13**, 896 (1967).

39. MacDougall, F. H., *J. Am. Chem. Soc.*, **58**, 2585 (1936).

40. MacDougall, F. H., *J. Am. Chem. Soc.*, **63**, 3420 (1941).

41. Christian, S. D., *J. Phys. Chem.*, **61**, 1441 (1957).

42. Lambert, J. D., *Disc. Faraday Soc.*, **15**, 226 (1953).

43. Prausnitz, J. M., and W. B. Carter, *A. I. Ch. E. Journal*, **6**, 611 (1960).

44. Pauling, L., *The Nature of the Chemical Bond.* Ithaca, N. Y.: Cornell University Press, 1945.

45. Cheh, H. Y., J. P. O'Connell, and J. M. Prausnitz, *Can. J. Chem.*, **44**, 429 (1966).

46. Kay, W. B., *Ind. Eng. Chem.*, **28**, 1014 (1936).

47. Reid, R. C., and T. W. Leland, *A. I. Ch. E. Journal*, **11**, 229 (1965); T. W. Leland and P. S. Chappelear, *Ind. Eng. Chem.*, **60**, No. 7, 15 (1968).

48. Joffe, J., *Ind. Eng. Chem.*, **40**, 738 (1948).

49. Gamson, B. W., and K. M. Watson, *Natl. Petrol. News*, **36**, R623 (1944).

50. Dodge, B. F., *Chemical Engineering Thermodynamics*, pp. 104–107. New .York: McGraw-Hill Book Company, 1944.

51. Lewis, G. N., and M. Randall, *Thermodynamics* (second edition by K. S. Pitzer and L. Brewer). New York: McGraw-Hill Book Company, 1961, Appendix 1.

52. Pitzer, K. S., and G. O. Hultgren, *J. Am. Chem. Soc.*, **80**, 4793 (1958).

53. Olds, R. H., H. H. Reamer, B. H. Sage and W. N. Lacey, *Ind. Eng. Chem.*, **41**, 475 (1949).

54. Gunn, R. D., P. L. Chueh, and J. M. Prausnitz, *A. I. Ch. E. Journal*, **12**, 937 (1966).

55. Redlich, O., F. J. Ackerman, R. D. Gunn, M. Jacobson, and S. Lau, *I&EC Fundamentals*, **4**, 369 (1965).

56. Redlich, O., and J. N. S. Kwong, *Chem. Rev.*, **44**, 233 (1949).

57. Robinson, R. L., and R. H. Jacoby, *Hydrocarbon Process. Petrol. Refiner*, **44**, No. 4, 141 (1965).

58. Wilson, G. M., *Advan. Cryog. Eng.*, **9**, 168 (1964).

59. Joffe, J., and D. Zudkevitch, *I & EC Fundamentals*, **5**, 455 (1966).

60. Hiza, M. J., and R. N. Herring, *Int. Advan. Cryog. Engineering* (K. D. Timmerhaus, ed.), **10**, 182 (1965).

61. McKinley, C., J. Brewer, and E. S. J. Wang, *Advan. Cryog. Eng.*, **7**, 114 (1961).

62. Webster, T. J., *Proc. Roy. Soc.* (London), **A214**, 61 (1952).

63. Ewald, A. H., W. B. Jepson, and J. S. Rowlinson, *Disc. Faraday Soc.*, **15**, 238 (1953).

64. Reuss, J. and J. J. M. Beenakker, *Physica*, **22**, 869 (1956).

65. Hinckley, R. B., and R. C. Reid, *A. I. Ch. E. Journal*, **10**, 416 (1964).

66. Prausnitz, J. M., and P. R. Benson, *A. I. Ch. E. Journal*, **5**, 161 (1959).

67. Rowlinson, J. S., *J. Chem. Phys.*, **19**, 827 (1951).

Fugacities in Liquid Mixtures: Excess Functions

6

Calculation of fugacities from volumetric properties was discussed in Chap. 3; many of the relations derived there [in particular Eqs. (3.1-14) and (3.4-16)] are general and may be applied to condensed phases as well as to the gas phase. However, it is not practical to do so because the necessary integrations require that volumetric data be available at constant temperature and constant composition over the entire density range from the ideal-gas state (zero density) to the density of the condensed phase, including the two-phase region. It is a tedious task to obtain such data for liquid mixtures, and very few data of this type have been reported. A more useful alternate technique, therefore, is needed for calculation of fugacities in liquid solutions. Such a technique is obtained by defining an ideal liquid solution and by describing deviations from ideal behavior in terms of excess functions; these functions yield the familiar activity coefficients which give a quantitative measure of departure from ideal behavior.

The fugacity of component i in a liquid solution is most conveniently related to the mole fraction x_i by an equation of the form

$$f_i^L = \gamma_i x_i f_i^0, \tag{1}$$

where γ_i is the activity coefficient and f_i^0 is the fugacity of i at some arbitrary condition known as the standard state. At any composition, the activity coefficient depends on the choice of standard state and the numerical value of γ_i has no significance unless the numerical value of f_i^0 is also specified.†

Since the choice of standard state is arbitrary, it is convenient to choose f_i^0 such that γ_i assumes values close to unity and when, for a range of conditions, γ_i is exactly equal to unity, we say that the solution is ideal. However, because of the intimate relation between the activity coefficient and the standard-state fugacity, the definition of solution ideality ($\gamma_i = 1$) is not complete unless the choice of standard state is clearly indicated. Either of two choices is frequently used. One of these leads to an ideal solution in the sense of Raoult's law and the other leads to an ideal solution in the sense of Henry's law.

6.1 The Ideal Solution

The history of modern science has shown repeatedly that a quantitative description of nature can often be achieved most successfully by first idealizing natural phenomena, i.e., by setting up a simplified model, either physical or mathematical, which crudely describes the essential behavior while neglecting details. (In fact, one of the outstanding characteristics of great contributors to modern science has been their ability to distinguish between what is essential and what is incidental.) The behavior of nature is then related to the idealized model by various correction terms which can be interpreted physically and which sometimes can be related quantitatively to those details in nature which were neglected in the process of idealization.

An ideal liquid solution is one where, at constant temperature and pressure, the fugacity of every component is proportional to some suitable measure of its concentration which is usually taken to be the mole fraction. That is, at some constant temperature and pressure, for any component i in an ideal solution,

$$f_i^L = \mathcal{R}_i x_i, \qquad (6.1\text{-}1)$$

---

† The interdependence of γ_i and f_i^0 is an essential element in the thermodynamics of solutions, although it has not been sufficiently stressed in most thermodynamics texts. It is similar to the moral of an old folk song:

> "Love and marriage, love and marriage
> Go together like a horse and carriage
> This I tell you, brother:
> You can't have one without the other."

where $\Re_i$ is a proportionality constant dependent on temperature and pressure but independent of x_i.

We notice at once from Eq. (1) of the introductory section that if we let $f_i^0 = \Re_i$, then $\gamma_i = 1$. If Eq. (6.1-1) holds for the entire range of composition (from $x_i = 0$ to $x_i = 1$), the solution is ideal in the sense of Raoult's law. For such a solution it follows from the boundary condition at $x_i = 1$ that the proportionality constant $\Re_i$ is equal to the fugacity of pure liquid i at the temperature of the solution.† For this case, if the fugacities are set equal to partial pressures, we then obtain the familiar relation known as Raoult's law.

In many cases the simple proportionality between f_i^L and x_i holds only over a small range of composition. If x_i is near zero, it is still possible to have an ideal solution according to Eq. (6.1-1), without, however, equating $\Re_i$ to the fugacity of pure liquid i. We call such a solution an ideal dilute solution leading to the familiar relation known as Henry's law.

The strict definition of an ideal solution requires that Eq. (6.1-1) must hold not only at a special temperature and pressure of interest but also at temperatures and pressures in their immediate vicinity. It is this feature which leads to an interesting conclusion concerning heat effects and volume changes of mixing for a solution ideal in the sense of Raoult's law. For such a solution, we have, at any T and P,

$$f_i(T, P) = f_{\text{pure } i}(T, P)x_i, \tag{6.1-2}$$

where, for convenience, we have deleted the superscript L. We now utilize two sets of exact thermodynamic relations:

$$\left(\frac{\partial \ln f_i}{\partial T}\right)_{P, x} = -\frac{(\bar{h}_i - h_i^+)}{RT^2}; \quad \left(\frac{\partial \ln f_{\text{pure } i}}{\partial T}\right)_P = -\frac{(h_i - h_i^+)}{RT^2} \tag{6.1-3}$$

and

$$\left(\frac{\partial \ln f_i}{\partial P}\right)_{T, x} = \frac{\bar{v}_i}{RT}; \quad \left(\frac{\partial \ln f_{\text{pure } i}}{\partial P}\right)_T = \frac{v_i}{RT}, \tag{6.1-4}$$

where $\bar{h}_i$ is the partial molar enthalpy of component i in the liquid phase, h_i is the enthalpy of pure liquid i, h_i^+ is the enthalpy of pure i in the ideal-gas state, $\bar{v}_i$ is the partial molar volume of i, and v_i is the molar volume of pure i, both in the liquid phase, all at system temperature T and pressure P.

---

†The standard-state fugacity of pure liquid i at system temperature is usually taken either at P_i^s, the saturation pressure of pure i, or else at P, the total pressure of the mixture. The latter choice is more common, especially at low or moderate pressures. At high pressures, special care must be taken in specifying the pressure of the standard state. See Chapter 10.

Upon substitution of Eq. (6.1-2) we find that:

$$\bar{h}_i = h_i \qquad (6.1\text{-}5)$$

and

$$\bar{v}_i = v_i. \qquad (6.1\text{-}6)$$

Since the partial molar enthalpy and partial molar volume of component i in an ideal solution are, respectively, the same as the molar enthalpy and molar volume of pure i at the same temperature and pressure, it follows that the formation of an ideal solution occurs without evolution or absorption of heat and without change of volume.

Mixtures of real fluids do not form ideal solutions although mixtures of similar liquids often exhibit behavior close to ideality. However, all solutions of chemically stable nonelectrolytes behave as ideal dilute solutions in the limit of very large dilution.

The correction terms which relate the properties of real solutions to those of ideal solutions are called excess functions.

6.2 Fundamental Relations of Excess Functions

Excess functions are thermodynamic properties of solutions which are in excess of those of an ideal (or ideal dilute) solution at the same conditions of temperature, pressure, and composition.† For an ideal solution all excess functions are zero. For example, G^E, the excess Gibbs energy, is defined by:

$$G^E \equiv G_{\left(\substack{\text{actual solution} \\ \text{at } T,\, P,\, \text{and } x}\right)} - G_{\left(\substack{\text{ideal solution at} \\ \text{same } T,\, P,\, \text{and } x}\right)}. \qquad (6.2\text{-}1)$$

Similar definitions hold for excess volume V^E, excess entropy S^E, excess enthalpy H^E, excess internal energy U^E, and excess Helmholtz energy A^E. Relations between these excess functions are exactly the same as those between the total functions:

$$H^E = U^E + PV^E \qquad (6.2\text{-}2)$$

$$G^E = H^E - TS^E \qquad (6.2\text{-}3)$$

$$A^E = U^E - TS^E. \qquad (6.2\text{-}4)$$

Also, partial derivatives of extensive excess functions are analogous to those of the total functions. For example,

$$\left(\frac{\partial G^E}{\partial T}\right)_{P,\,x} = -S^E \qquad (6.2\text{-}5)$$

---

†Most excess functions, but not all, are extensive. However, it follows from the definition that we cannot have an excess pressure, temperature or composition. See R. Missen, *I & EC Fundamentals* (February 1969).

$$\left(\frac{\partial G^E/T}{\partial T}\right)_{P,\,x} = -\frac{H^E}{T^2} \tag{6.2-6}$$

$$\left(\frac{\partial G^E}{\partial P}\right)_{T,\,x} = V^E. \tag{6.2-7}$$

Excess functions may be positive or negative; when the excess Gibbs energy of a solution is greater than zero the solution is said to exhibit positive deviations from ideality whereas if it is less than zero the deviations from ideality are said to be negative.

Partial excess functions are defined in a manner completely analogous to that used for partial molar thermodynamic properties. If M is an extensive thermodynamic property, then $\bar{m}_i$, the partial molar M of component i, is defined by:

$$\bar{m}_i \equiv \left(\frac{\partial M}{\partial n_i}\right)_{T,\,P,\,n_j}, \tag{6.2-8}$$

where n_i is the number of moles of i and where the subscript n_j designates that the number of moles of all components other than i are kept constant. Similarly,

$$\bar{m}_i^E \equiv \left(\frac{\partial M^E}{\partial n_i}\right)_{T,\,P,\,n_j}. \tag{6.2-9}$$

Also, from Euler's theorem we have that:

$$M = \sum_i n_i \bar{m}_i. \tag{6.2-10}$$

It then follows that

$$M^E = \sum_i n_i \bar{m}_i^E. \tag{6.2-11}$$

For our purposes, an extensive excess property is a homogeneous function of the first degree in the mole numbers.†

For phase-equilibrium thermodynamics the most useful partial excess property is the partial excess Gibbs energy which is directly related to the

†Many, but not all, extensive excess properties can be defined in this way. See O. Redlich, "Fundamental Thermodynamics Since Carathéodory," *Revs. Mod. Phys.*, **40**, 556 (1968).

activity coefficient. The partial excess enthalpy and partial excess volume are related, respectively, to the temperature and pressure derivatives of the activity coefficient. These relations are summarized in the next section.

6.3 Activity and Activity Coefficients

The activity of component i at some temperature, pressure and composition is defined as the ratio of the fugacity of i at these conditions to the fugacity of i in the standard state, which is a state at the same temperature as that of the mixture and at some specified condition of pressure and composition:

$$a_i(T, P, x) \equiv \frac{f_i(T, P, x)}{f_i(T, P^0, x^0)}, \tag{6.3-1}$$

where P^0 and x^0 are, respectively, an arbitrary but specified pressure and composition.

The activity coefficient γ_i is the ratio of the activity of i to some convenient measure of the concentration of i which is usually taken to be the mole fraction:

$$\gamma_i \equiv \frac{a_i}{x_i}. \tag{6.3-2}$$

The relation between partial excess Gibbs energy and activity coefficient is obtained by first recalling the definition of fugacity. At constant temperature and pressure, for a component i in solution,

$$\bar{g}_{i(\text{real})} - \bar{g}_{i(\text{ideal})} = RT \left[\ln f_{i(\text{real})} - \ln f_{i(\text{ideal})}\right]. \tag{6.3-3}$$

Next, we introduce the partial excess function $\bar{g}_i^E$ by differentiation of Eq. (6.2-1) at constant T, P and n_j:

$$\bar{g}_i^E = \bar{g}_{i(\text{real})} - \bar{g}_{i(\text{ideal})}. \tag{6.3-4}$$

Substitution then gives

$$\bar{g}_i^E = RT \ln \frac{f_{i(\text{real})}}{f_{i(\text{ideal})}}, \tag{6.3-5}$$

and substituting Eq. (6.1-1) we obtain

$$\bar{g}_i^E = RT \ln \frac{f_i}{\mathcal{R} x_i}. \tag{6.3-6}$$

It follows from Eq. (6.1-1) that an ideal solution is one where the activity is equal to the mole fraction; if we set the standard-state fugacity f_i^0 equal to $\mathcal{R}_i$ we then have

$$\mathfrak{a}_i = \gamma_i x_i = \frac{f_i}{\mathcal{R}_i}. \qquad (6.3\text{-}7)$$

But for an ideal solution [Eq. (6.1-1)], f_i is equal to $\mathcal{R}_i x_i$ and therefore, for an ideal solution, $\gamma_i = 1$ and $\mathfrak{a}_i = x_i$. Substitution of Eq. (6.3-7) into Eq. (6.3-6) gives the important and useful result:

$$\boxed{\bar{g}_i^E = RT \ln \gamma_i.} \qquad (6.3\text{-}8)\dagger$$

Substitution into Eq. (6.2-11) gives the equally important relation

$$\boxed{g^E = RT \sum_i x_i \ln \gamma_i,} \qquad (6.3\text{-}9)$$

where g^E is the molar excess Gibbs energy. Equations (6.3-8) and (6.3-9) are used repeatedly in the remainder of this chapter as well as in later chapters.

We now want to consider the temperature and pressure derivatives of the activity coefficient. Let us first discuss the case where the ideal solution is ideal over the entire range of composition, in the sense of Raoult's law. In this case,

$$\mathcal{R}_i = f_i \text{ (pure liquid } i \text{ at } T \text{ and } P \text{ of solution)} \qquad (6.3\text{-}10)$$

and

$$\ln \gamma_i = \ln f_i - \ln x_i - \ln f_{\text{pure } i}. \qquad (6.3\text{-}11)$$

Differentiation with respect to temperature at constant P and x gives:

$$\left(\frac{\partial \ln \gamma_i}{\partial T} \right)_{P,x} = \frac{h_{\text{pure } i} - \bar{h}_i}{RT^2} = -\frac{\bar{h}_i^E}{RT^2}, \qquad (6.3\text{-}12)$$

---

†A shorter but equivalent derivation of Eq. (6.3-8) follows from writing

$$\bar{g}_i^E \equiv g_i - g_{i(\text{ideal})} \quad \text{and} \quad \bar{g}_i^E = RT \ln \frac{f_i}{f_{i(\text{ideal})}}.$$

Since $f_i = \gamma_i x_i f_i^0$ and $f_{i(\text{ideal})} = x_i f_i^0$, we obtain $\bar{g}_i^E = RT \ln \gamma_i$.

where $\bar{h}_i^E$ is the partial molar enthalpy of i minus the molar enthalpy of pure liquid i at the same temperature and pressure. Differentiation with respect to pressure at constant T and x gives

$$\left(\frac{\partial \ln \gamma_i}{\partial P}\right)_{T,x} = \frac{\bar{v}_i - v_{\text{pure } i}}{RT} = \frac{\bar{v}_i^E}{RT} \qquad (6.3\text{-}13)\dagger$$

where $\bar{v}_i^E$ is the partial molar volume of i minus the molar volume of pure liquid i at the same temperature and pressure.

Now let us consider an ideal dilute solution. It is useful to define excess functions relative to an ideal dilute solution whenever the liquid mixture cannot exist over the entire composition range, as happens for example in a liquid mixture containing a gaseous solute. If the critical temperature of solute 2 is lower than the temperature of the mixture, then a liquid phase cannot exist as $x_2 \rightarrow 1$, and relations based on an ideal mixture in the sense of Raoult's law can be used only by introducing a hypothetical standard state for solute 2. However, relations based on an ideal dilute solution avoid this difficulty. The proportionality constant $\mathcal{R}_2$ is not determined from the pure-component boundary condition $x_2 = 1$, but rather from the boundary condition of the infinitely dilute solution, i.e., $x_2 \rightarrow 0$. For an ideal dilute solution, we have for the solute 2:

$$\mathcal{R}_2 = \lim_{x_2 \to 0} \frac{f_2}{x_2} = H_{2,1}, \qquad (6.3\text{-}14)\ddagger$$

where $H_{2,1}$ is Henry's constant for solute 2 in solvent 1.

However, for the solvent (component 1, which is present in excess) we obtain the same result as before:

$$\mathcal{R}_1 = \lim_{x_2 \to 0} \frac{f_1}{x_1} = f_{\text{pure liquid 1}}. \qquad (6.3\text{-}15)$$

---

†In some cases $\mathcal{R}_i$ is set equal to the fugacity of pure liquid i at temperature T and at its own saturation pressure P_i^s (rather than the total pressure P, which may vary with the composition in an isothermal mixture). In that case Eq. (6.3-12) is not affected significantly because the enthalpy of a pure liquid is usually a weak function of pressure. However, Eq. (6.3-13) becomes:

$$\left(\frac{\partial \ln \gamma_i}{\partial P}\right)_{T,x} = \frac{\bar{v}_i}{RT}. \qquad (6.3\text{-}13a)$$

‡For any binary system, Henry's constant $H_{2,1}$ depends on both temperature and pressure. Unless the pressure is large, however, the effect of pressure on $H_{2,1}$ is negligible. See Sec. 8.3.

For the solute, the activity coefficient is given by

$$\gamma_2 = \frac{f_2}{x_2 H_{2,1}}. \tag{6.3-16}$$

Substituting into Eq. (6.3-8) and differentiating with respect to temperature we obtain, as before,

$$\left(\frac{\partial \ln \gamma_2}{\partial T}\right)_{P,x} = -\frac{\bar{h}_2^E}{RT^2}. \tag{6.3-17}$$

However, $\bar{h}_2^E$ now has a different meaning; it is given by

$$\bar{h}_2^E = \bar{h}_2 - \bar{h}_2^\infty, \tag{6.3-18}$$

where $\bar{h}_2^\infty$ is the partial molar enthalpy of solute 2 in an infinitely dilute solution.

The effect of pressure on the activity coefficient of the solute is given by

$$\left(\frac{\partial \ln \gamma_2}{\partial P}\right)_{T,x} = \frac{\bar{v}_2}{RT}, \tag{6.3-19}$$

where $\bar{v}_2$ is the partial molar volume of solute 2.†

The derivatives with respect to temperature and pressure of the activity coefficient for component 1 (the solvent) are the same as those given by Eqs. (6.3-12) and (6.3-13).

For all components in a mixture, the partial excess Gibbs energies (and the activity coefficients) are related to one another by a fundamental relation known as the *Gibbs-Duhem equation* which is discussed in Sec. 6.6.

---

†Notice that the derivation of Eq. (6.3-19) requires that Henry's constant in Eq. (6.3-14) be evaluated at the temperature of the solution and at some pressure which usually is the saturation pressure of pure 1. The relationship between Henry's constant at system pressure P and Henry's constant at pressure P_1^s is given by

$$H_{2,1}(P) = H_{2,1}(P_1^s) \exp \int_{P_1^s}^{P} \frac{\bar{v}_2^\infty}{RT} dP. \tag{6.3-20}$$

If, in Eq. (6.3-14), $H_{2,1}(P)$ is used rather than $H_{2,1}(P_1^s)$, then Eq. (6.3-19) is not valid but must be replaced by

$$\left(\frac{\partial \ln \gamma_2}{\partial P}\right)_{T,x} = \frac{\bar{v}_2 - \bar{v}_2^\infty}{RT} \tag{6.3-19a}$$

where $\bar{v}_2^\infty$ is the partial molar volume of solute 2 at infinite dilution.

6.4 Normalization of Activity Coefficients

As mentioned in the previous section, it is convenient to define activity in such a way that for an ideal solution activity is equal to the mole fraction or, equivalently, that the activity coefficient is equal to unity. Since we have distinguished between two types of ideality (one leading to Raoult's law and the other leading to Henry's law), it follows that activity coefficients may be normalized (that is, become unity) in two different ways.

If activity coefficients are defined with reference to an ideal solution in the sense of Raoult's law [Eq. (6.3-10)], then for each component i the normalization is

$$\gamma_i \longrightarrow 1 \quad \text{as} \quad x_i \longrightarrow 1. \tag{6.4-1}$$

Since this normalization holds for both solute and solvent, Eq. (6.4-1) is called the symmetric convention for normalization.

However, if activity coefficients are defined with reference to an ideal dilute solution [Eq. (6.3-14)], then

$$\left.\begin{array}{ll} \gamma_1 \longrightarrow 1 & \text{as} \quad x_1 \longrightarrow 1 \quad \text{(solvent)} \\ \gamma_2 \longrightarrow 1 & \text{as} \quad x_2 \longrightarrow 0 \quad \text{(solute)} \end{array}\right\}. \tag{6.4-2}$$

Since solute and solvent are not normalized in the same way, Eq. (6.4-2) gives the unsymmetric convention for normalization. In order to differentiate between symmetrically and unsymmetrically normalized activity coefficients, it is useful to denote with an asterisk (*) the activity coefficient of a component which approaches unity as its mole fraction goes to zero. With this notation, Eq. (6.4-2) becomes:

$$\left.\begin{array}{ll} \gamma_1 \longrightarrow 1 & \text{as} \quad x_1 \longrightarrow 1 \; \text{(solvent)} \\ \gamma_2^* \longrightarrow 1 & \text{as} \quad x_2 \longrightarrow 0 \; \text{(solute)} \end{array}\right\}. \tag{6.4-2a}$$

The two methods of normalization are illustrated in Fig. 6-1. In the dilute region ($x_2 \ll 1$), $\gamma_2^* = 1$ and the solution is ideal;† however, $\gamma_2 \neq 1$ and therefore, while the dilute solution is ideal in the sense of Henry's law, it is not ideal in the sense of Raoult's law.

Symmetric normalization of activity coefficients is easily extended to solutions containing more than two components; the activity coefficient of any component approaches unity as its mole fraction goes to unity. However, extension to multicomponent solutions of unsymmetric normalization of activity coefficients requires more care. This extension is discussed

---

†It can readily be shown from the Gibbs-Duhem equation that for the region where $\gamma_2^* = 1, \gamma_1 = 1$ also.

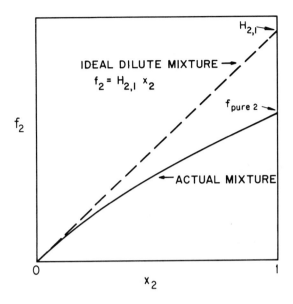

Fig. 6-1   Normalization of activity coefficients.

$$\gamma_2 \equiv \frac{f_2}{x_2 f_{\text{pure 2}}} \quad \text{and} \quad \gamma_2^* \equiv \frac{f_2}{x_2 H_{2,1}}$$

$$\begin{array}{cc} \gamma_2 \rightarrow 1 & \gamma_2^* \rightarrow 1 \\ \text{as } x_2 \rightarrow 1 & \text{as } x_2 \rightarrow 0 \end{array}$$

$$H_{2,1} \equiv \lim_{x_2 \rightarrow 0} \frac{f_2}{x_2}$$

further in Sec. 8.6, where we consider the solubility of gases in mixed solvents, and in Chap. 10, which is concerned with high-pressure equilibria.

In binary mixtures, activity coefficients which are normalized symmetrically are easily related to activity coefficients which are normalized unsymmetrically. The definitions of γ_2 and γ_2^* are:

$$\gamma_2 = \frac{f_2}{x_2 f_{\text{pure 2}}} \tag{6.4-3}$$

$$\gamma_2^* = \frac{f_2}{x_2 H_{2,1}}. \tag{6.4-4}$$

Therefore

$$\frac{\gamma_2}{\gamma_2^*} = \frac{H_{2,1}}{f_{\text{pure 2}}}. \tag{6.4-5}$$

Now, since

$$\lim_{x_2 \to 0} \gamma_2^* = 1, \qquad (6.4\text{-}6)$$

we obtain

$$\lim_{x_2 \to 0} \gamma_2 = \frac{H_{2,1}}{f_{\text{pure 2}}}. \qquad (6.4\text{-}7)$$

Substitution in Eq. (6.4-5) gives

$$\frac{\gamma_2}{\gamma_2^*} = \lim_{x_2 \to 0} \gamma_2. \qquad (6.4\text{-}8)$$

By a similar argument, we can also show that

$$\frac{\gamma_2^*}{\gamma_2} = \lim_{x_2 \to 1} \gamma_2^*. \qquad (6.4\text{-}9)$$

Both Eqs. (6.4-8) and (6.4-9) relate to each other the two activity co-efficients of the solute, one normalized by the symmetric convention and the other by the unsymmetric convention. However, Eq. (6.4-8) is much more useful than Eq. (6.4-9) because the limit given on the right side of Eq. (6.4-8) corresponds to a real physical situation whereas the limit on the right side of Eq. (6.4-9) corresponds to a situation which is hypothetical (physically unreal) whenever component 2 cannot exist as a pure liquid at the temperature of the solution.

6.5 Activity Coefficients from Excess
Functions in Binary Mixtures

At a fixed temperature, the molar excess Gibbs energy g^E of a mixture depends on the composition of the mixture and, to a smaller extent, on pressure. At low or moderate pressures, well removed from critical conditions, the effect of pressure is sufficiently small to be negligible, and it is therefore not considered in this section.†

We now consider a binary mixture where the excess properties are taken with reference to an ideal solution wherein the standard state for each component is the pure liquid at the temperature and pressure of the mixture. In that case, any expression for the molar excess Gibbs energy

---

†Excess functions for liquid mixtures under high-pressure conditions are discussed in Sec. 8.3 and in Chap. 10.

must obey the two boundary conditions:

$$g^E = 0 \quad \text{when} \quad x_1 = 0$$
$$g^E = 0 \quad \text{when} \quad x_2 = 0.$$

The simplest nontrivial expression which obeys these boundary conditions is

$$g^E = Ax_1 x_2, \tag{6.5-1}$$

where A is an empirical constant with units of energy, characteristic of components 1 and 2, which depends on the temperature but not on composition.

Equation (6.5-1) immediately gives expressions for activity coefficients γ_1 and γ_2 by substitution in the relation between activity coefficient and excess Gibbs energy [Eq. (6.3-8)]:

$$RT \ln \gamma_i = \bar{g}_i^E = \left(\frac{\partial n_T g^E}{\partial n_i} \right)_{T, P, n_j}, \tag{6.5-2}$$

where n_i is the number of moles of i and n_T is the total number of moles. Remembering that $x_1 = n_1/n_T$ and $x_2 = n_2/n_T$ we obtain:

$$\ln \gamma_1 = \frac{A}{RT} x_2^2 \tag{6.5-3}$$

$$\ln \gamma_2 = \frac{A}{RT} x_1^2 \tag{6.5-4}$$

Equations (6.5-3) and (6.5-4), often called the two-suffix Margules equations, provide a good representation for many simple liquid mixtures, i.e., for mixtures of molecules which are similar in size, shape, and chemical nature. The two equations are symmetrical: When $\ln \gamma_1$ and $\ln \gamma_2$ are plotted against x_2 (or x_1) the two curves are mirror images. At infinite dilution the activity coefficients of both components are equal:

$$\gamma_1^\infty \equiv \lim_{x_1 \to 0} \gamma_1 = \exp \frac{A}{RT} \tag{6.5-5}$$

$$\gamma_2^\infty \equiv \lim_{x_2 \to 0} \gamma_2 = \exp \frac{A}{RT}. \tag{6.5-6}$$

The coefficient A may be positive or negative and while it is in general a function of temperature, it frequently happens that for simple systems over a small temperature range, A is nearly constant. For example, vapor-liquid equilibrium data of Staveley, et. al.,[1] for argon/oxygen are well represented by the two-suffix Margules equations as shown in Fig. 6-2. At $83.8°K$, A is equal to 35.4 cal/g-mol and at $89.6°K$, A is equal to 33.7 cal/g-mol. For this simple system, A is a weak function of temperature.

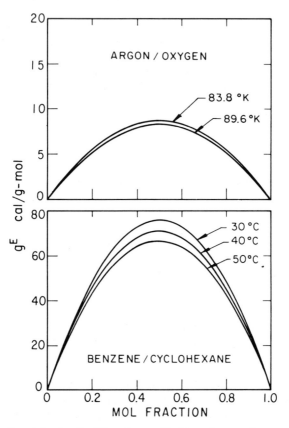

Fig. 6-2   Applicability of two-suffix Margules equation to simple binary mixtures.

Another binary system whose excess Gibbs energy is well represented by the two-suffix Margules equation is the benzene/cyclohexane system studied by Scatchard, et al.[2] Their results are also shown in Fig. 6-2. The variation of A with temperature is again not large although certainly not negligible; at 30, 40, and 50°C, the values of A are, respectively, 303, 283, and 266 cal/g-mol. In nonpolar solutions A frequently falls with rising temperature.

Equation (6.5-1) is a very simple relation and in the general case a more complex equation is needed to represent adequately the excess Gibbs energy of a binary solution. Since the boundary conditions given just before Eq. (6.5-1) must be obeyed regardless of the complexity of the solution, one convenient extension of Eq. (6.5-1) is to write a series expansion:

$$g^E = x_1 x_2 [A + B(x_1 - x_2) + C(x_1 - x_2)^2 + D(x_1 - x_2)^3 + \ldots]$$

$$(6.5\text{-}7)$$

where $B, C, D, \ldots$ are additional, temperature-dependent parameters which must be determined from experimental data. Equation (6.5-7) is commonly known as the Redlich-Kister expansion and, upon using Eq. (6.5-2), we obtain these expressions for the activity coefficients

$$RT \ln \gamma_1 = a^{(1)} x_2^2 + b^{(1)} x_2^3 + c^{(1)} x_2^4 + d^{(1)} x_2^5 + \ldots \qquad (6.5\text{-}8)$$

$$RT \ln \gamma_2 = a^{(2)} x_1^2 + b^{(2)} x_1^3 + c^{(2)} x_1^4 + d^{(2)} x_1^5 + \ldots \qquad (6.5\text{-}9)$$

where:

$$
\begin{array}{ll}
a^{(1)} = A + 3B + 5C + 7D & a^{(2)} = A - 3B + 5C - 7D \\
b^{(1)} = -4(B + 4C + 9D) & b^{(2)} = 4(B - 4C + 9D) \\
c^{(1)} = 12(C + 5D) & c^{(2)} = 12(C - 5D) \\
d^{(1)} = -32D & d^{(2)} = 32D
\end{array}
$$

The number of parameters $(A, B, C, \ldots)$ which should be used to represent the experimental data depends on the molecular complexity of the solution, on the quality of the data, and on the number of data points available. Typical vapor-liquid equilibrium data reported in the literature justify no more than two or at most three constants; very accurate and extensive data are needed to warrant the use of four or more empirical parameters.

The Redlich-Kister expansion provides a flexible algebraic expression for representing the excess Gibbs energy of a liquid mixture. The first term in the expansion is symmetric in x and gives a parabola when g^E is plotted against x. The odd-powered correction terms [first (B), third (D), ...] are asymmetric in x and therefore tend to skew the parabola either to the left or right. The even-powered correction terms [second (C), fourth (E), ...] are symmetric in x and tend to flatten or sharpen the parabola. To illustrate, we show in Fig. 6-3 the first three terms of the Redlich-Kister expansion for unit value of the coefficients $A, B,$ and C.

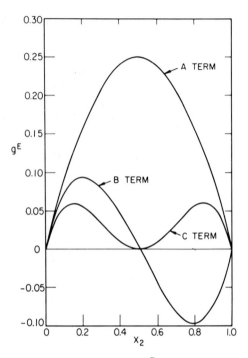

Fig. 6-3   Contributions to g^E in Redlich-Kister
equation (for $A = B = C = 1$).

The Redlich-Kister equations provide not only a convenient method for representing liquid-phase activity coefficients, but also for classifying different types of liquid solutions. From experimental data, γ_1 and γ_2 are calculated and these are then plotted as log γ_1/γ_2 versus x_1, as shown in Figs. 6-4, 6-5, and 6-6 taken from the paper of Redlich, Kister and Turnquist.[3] Strictly speaking, Eqs. (6.5-7), (6.5-8), and (6.5-9) apply to isothermal data, but in fact they are often applied to isobaric data and, provided the temperature does not change much with composition, this practice does not introduce large errors. From Eqs. (6.3-8) and (6.5-7) we obtain, after some rearrangement,

$$RT \ln \frac{\gamma_1}{\gamma_2} = A(x_2 - x_1) + B(6x_1 x_2 - 1) + C(x_1 - x_2)(8x_1 x_2 - 1)$$

$$+ D(x_1 - x_2)^2(10x_1 x_2 - 1) + \ldots . \qquad (6.5\text{-}10)$$

For simple solutions, $B = C = D = \cdots = 0$, and a plot of log γ_1/γ_2 versus x_1 gives a straight line, as shown in Fig. 6-4 for the system n-hexane/toluene. In this case a good representation of the data is obtained with $A/RT = 0.352$.

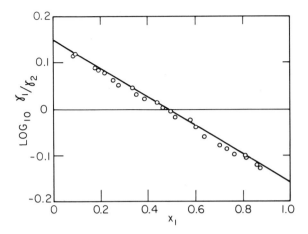

Fig. 6-4   Activity-coefficient ratio for a simple mixture. Experimental data for *n*-hexane (1)/toluene (2) at 1 atm. The line is drawn so as to satisfy the area (consistency) test described in Sec. 6.10.

Data for a somewhat more complicated solution are shown in Fig. 6-5. In this case the plot has some curvature and two parameters are required to represent it adequately; they are $A/RT = 0.433$ and $B/RT = 0.104$. Although the systems in Figs. 6-4 and 6-5 are similar, both being mixtures

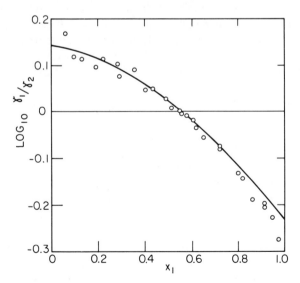

Fig. 6-5   Activity-coefficient ratio for a mixture of intermediate complexity.   Experimental data for benzene (1)/isooctane (2) at total pressures ranging from 736 to 760 mm Hg.

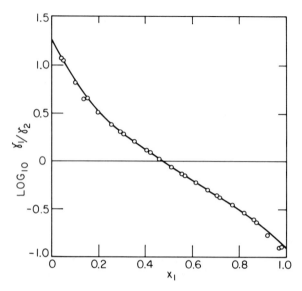

Fig. 6-6   Activity-coefficient ratio for a highly complex
mixture. Experimental data for ethanol (1)/
methylcyclohexane (2) in the region 30–35°C.

of a paraffinic and an aromatic hydrocarbon, it may appear surprising
that two parameters are needed for the second system while only one is
required for the first. This result is most probably related to the fact that
the difference in the molecular sizes of the two components is much larger
in the isooctane/benzene system than it is in the n-hexane/toluene system.
At 25°C, the ratio of molar volumes (paraffin-to-aromatic) in the system
containing benzene is 1.86, while it is only 1.23 in the system containing
toluene. The effect of molecular size on the representation of activity
coefficients is shown more clearly by Wohl's expansion discussed in
Sec. 6.13.

Finally, data for a highly complex solution are shown in Fig. 6-6. The
plot is not only curved but has a point of inflection, and four parameters
are required to give an adequate representation. This mixture, containing
an alcohol and a saturated hydrocarbon, is a complex one because the degree of hydrogen bonding of the alcohol is strongly dependent on the
composition, especially in the region dilute with respect to alcohol.

The number of parameters in Eq. (6.5-10) required to represent activity
coefficients of a binary mixture gives an indication of the apparent complexity of the mixture, thereby providing a means for classification. If the
number of required parameters is large (4 or more parameters) then the
mixture is classified as a complex solution and if it is small (1 parameter)
then the mixture is classified as a simple solution. Most solutions of non-

electrolytes commonly encountered in chemical engineering are of intermediate complexity, requiring 2 or 3 parameters in the Redlich-Kister
expansion.

Classification of solutions as determined by the number of parameters
is necessarily arbitrary and just because a good bit of experimental data is
obtained with one or two parameters, we cannot necessarily conclude that
the mixture is truly simple in a molecular sense; in many cases a one- or
two-parameter equation may fortuitously provide an adequate representation. For example, binary mixtures of two alcohols (such as methanol/
ethanol) are certainly highly complex because of the various types of
hydrogen bonds which can exist in such mixtures; nevertheless, it is
frequently possible to represent the excess Gibbs energies of such solutions with a one-or two-parameter equation.

Numerous equations have been proposed for expressing analytically
the composition-dependence of the excess Gibbs energies of binary
mixtures. Most of these equations are empirical and we mention several
of the better known ones in Sec. 6.13, but some have at least a little
theoretical basis and we discuss a few of these in Chap. 7. However,
before continuing our presentation of equations for representing excess
Gibbs energies of nonideal liquid mixtures, we first want to discuss one
of the most fundamental relations in phase-equilibrium thermodynamics,
known as the Gibbs-Duhem equation.

6.6 The Gibbs-Duhem Equation

In a mixture, the partial molar properties of the components are related
to one another by one of the most useful equations in thermodynamics,
the Gibbs-Duhem equation. This equation says that at constant temperature and pressure,

$$\sum_i x_i \, d\bar{m}_i = 0, \tag{6.6-1}$$

where $\bar{m}_i$ is any partial molar property. Equation (6.6-1) holds for ideal
as well as real solutions and it can be rewritten in terms of excess partial
properties:

$$\sum_i x_i \, d\bar{m}_i^E = 0. \tag{6.6-2}$$

We want to discuss here some applications of the Gibbs-Duhem
equation. A derivation and more detailed discussion of this equation is
given in App. IV (p. 468).

There are two important applications of Eq. (6.6-2). First, in the absence of complete experimental data on the properties of a mixture, Eq. (6.6-2) frequently may be used to calculate additional properties; for example, in a binary solution, if experimental measurements over a range of concentration yield activity coefficients of only one component, activity coefficients of the other component can be computed for the same concentration range. Second, if experimental data are available for a directly measured partial molar property for each component over a range of composition, it is then possible to check the data for thermodynamic consistency, by which we mean that if the data satisfy the Gibbs-Duhem equation, it is likely that they are reliable, but if they do not, it is certain that they are incorrect.

While the Gibbs-Duhem equation is applicable to all partial excess properties, it is most useful for the partial excess Gibbs energy, which is directly related to the activity coefficient by Eq. (6.3-8). In terms of activity coefficients, Eq. (6.6-2) is:

$$\sum_i x_i \, d \ln \gamma_i = 0 \qquad \text{(constant } T \text{ and } P\text{).} \qquad (6.6\text{-}3)$$

We now consider in more detail some applications of Eq. (6.6-3).

6.7   Activity Coefficients for One
Component from Data for Those
of the Other Components

The Gibbs-Duhem equation is a differential relation between the activity coefficients of all the components in the solution. Hence, in a solution containing m components, data for activity coefficients of $m - 1$ components may be used to compute the activity coefficient of the mth component. To illustrate, consider the simplest case, viz., a binary solution for which isothermal data are available for one component and where the pressure is sufficiently low to permit neglect of the effect of pressure on the liquid-phase activity coefficient. For this case, the Gibbs-Duhem equation may be written:

$$x_1 \frac{d \ln \gamma_1}{dx_1} = x_2 \frac{d \ln \gamma_2}{dx_2} . \qquad (6.7\text{-}1)$$

Assume now that data have been obtained for γ_1 at various values of x_1. By graphical integration of Eq. (6.7-1) one may then obtain values of

γ_2. A simpler procedure, which is less exact in principle but much easier to use, is to curve-fit the data for γ_1 to an algebraic expression using x_1 (or x_2) as the independent variable. Once such an expression has been obtained, the integration of Eq. (6.7-1) may be performed analytically. For this purpose it is convenient to rewrite Eq. (6.7-1):

$$\frac{d \ln \gamma_1/\gamma_2}{dx_2} = \frac{1}{x_2} \frac{d \ln \gamma_1}{dx_2} . \tag{6.7-2}$$

Assume now, as first suggested in 1895 by Margules, that the data for γ_1 can be represented by an empirical equation of the form

$$\ln \gamma_1 = \sum_k \alpha_k x_2^{\beta_k} , \qquad (\beta_k > 1) \tag{6.7-3}\dagger$$

where α_k and β_k are empirical constants to be determined from the data. Substituting (6.7-3) into (6.7-2) yields:

$$\frac{d \ln \gamma_1/\gamma_2}{dx_2} = \sum_k \alpha_k \beta_k x_2^{\beta_k - 2} . \tag{6.7-4}$$

Integrating Eq. (6.7-4) gives

$$\ln \gamma_2 = \ln \gamma_1 - \sum_k \frac{\alpha_k \beta_k}{\beta_k - 1} x_2^{\beta_k - 1} - I, \tag{6.7-5}$$

where I is a constant of integration. To eliminate $\ln \gamma_1$ in Eq. (6.7-5), we substitute Eq. (6.7-3):

$$\ln \gamma_2 = \sum_k \alpha_k x_2^{\beta_k} - \sum_k \frac{\alpha_k \beta_k}{\beta_k - 1} x_2^{\beta_k - 1} - I. \tag{6.7-6}$$

To evaluate I it is necessary to impose a suitable boundary condition. If component 2 can exist as a pure liquid at the temperature of the solution, then it is common to use pure liquid component 2 at that temperature as the standard state for γ_2; in that case,

$$\gamma_2 = 1 \quad \text{when} \quad x_2 = 1.$$

The constant of integration then is

$$I = \sum_k \alpha_k - \sum_k \frac{\alpha_k \beta_k}{\beta_k - 1} , \tag{6.7-7}$$

†We need $\beta_k > 1$ in order to avoid singularities in $\ln \gamma_2$ when $x_2 = 0$.

and the expression for $\ln \gamma_2$ becomes:

$$\ln \gamma_2 = \sum_k \alpha_k x_2^{\beta_k} - \sum_k \frac{\alpha_k}{\beta_k - 1} (\beta_k x_2^{\beta_k - 1} - 1). \qquad (6.7\text{-}8)$$

Equation (6.7-8) is a general relation and it is strictly for convenience that one customarily uses only positive integers for β_k. To illustrate the use of Eq. (6.7-8), suppose that the data for γ_1 can be adequately represented by Eq. (6.7-3) terminated after the fourth term ($k = 2, 3, 4$) with $\beta_k = k$. Equation (6.7-3) becomes:

$$\ln \gamma_1 = \alpha_2 x_2^2 + \alpha_3 x_2^3 + \alpha_4 x_2^4, \qquad (6.7\text{-}9)$$

which is known as the four-suffix Margules equation.† The coefficients α_2, α_3, and α_4 must be found from the experimental data which give γ_1 as a function of the mole fraction. When this four-suffix Margules equation for $\ln \gamma_1$ is substituted into Eq. (6.7-8), the result for γ_2 is:

$$\ln \gamma_2 = (\alpha_2 + \tfrac{3}{2} \alpha_3 + 2\alpha_4)x_1^2 - (\alpha_3 + \tfrac{8}{3} \alpha_4)x_1^3 + \alpha_4 x_1^4. \qquad (6.7\text{-}10)$$

The important feature of Eq. (6.7-10) is that γ_2 is given in terms of constants which are determined exclusively from data for γ_1.

In a binary system, calculating the activity coefficient of one component from data for the other, is a common practice whenever the two components in the solution differ markedly in volatility. In that event the experimental data frequently give the activity coefficient of only the more volatile component and the activity of the less volatile component is found from the Gibbs-Duhem equation. For example, if one wished to have information on the thermodynamic properties of some high-boiling liquid (such as a polymer) dissolved in, say, benzene, near room temperature, then the easiest procedure would be to measure the activity (partial pressure) of the benzene in the solution and to compute the activity of the other component from the Gibbs-Duhem equation as outlined above; in this case it would not be practical to measure the extremely small partial pressure of the high-boiling component over the solution.

While the binary case is the simplest, the method just discussed can be extended to systems of any number of components. In these more complicated cases the computational work is greater but the theoretical principles are exactly the same.

In carrying out numerical work it is important that activity coefficients be calculated as rigorously as possible from the equilibrium data. We recall the definition of activity coefficient:

$$\gamma_i \equiv \frac{f_i}{x_i f_i^0}. \qquad (6.7\text{-}11)$$

---

†The n-suffix Margules equation gives $\ln \gamma_1$ as a polynomial in x_2 of degree n.

The quantities f_i and f_i^0 must be computed with care; in vapor-liquid equilibria, gas-phase corrections for both of these quantities are frequently important. For solutions of liquids, it is useful to use the pure component as the standard state and, to simplify calculations, it is common practice to evaluate γ_i from the data by the simplified expression

$$\gamma_i = \frac{y_i P}{x_i P_i^s},\tag{6.7-12}$$

where P_i^s is the saturation (vapor) pressure of pure i. Equation (6.7-12), however, is only an approximate form of Eq. (6.7-11). In some cases the approximation is justified but before Eq. (6.7-12) is used it is important to inquire if the simplifying assumptions apply to the case under consideration. For mixtures of strongly polar or hydrogen-bonding components, or for mixtures at low temperatures, gas-phase corrections may be significant even at pressures near 1 atmosphere.

6.8   Partial Pressures from Isothermal Total-Pressure Data

Complete description of vapor-liquid equilibria for a system gives equilibrium compositions of both phases as well as temperature and total pressure. In a typical experimental investigation temperature or total pressure is held constant; in a system of m components then, complete measurements require that for each equilibrium point data must be obtained for either the temperature or pressure and for $2(m - 1)$ mole fractions. Even in a binary system this represents a significant experimental effort; it is advantageous to reduce this effort by utilizing the Gibbs-Duhem equation for calculation of at least some of the desired information.

   First, we consider a procedure for calculating partial pressures from isothermal total pressure data for binary systems. According to this procedure, total pressures are measured as a function of the composition of one of the phases (usually the liquid phase) and no measurements are made of the composition of the other phase. Instead, the composition of the other phase is calculated from the total-pressure data with the help of the Gibbs-Duhem equation. The necessary experimental work is thereby much reduced.

   Numerous techniques have been proposed for making this kind of calculation and it is not necessary to review all of them here. However, to indicate the essentials of these techniques one representative and useful procedure is given below.

A numerical method due to Barker[4]

The total pressure for a binary system is written:

$$P = \gamma_1 x_1 P_1^{s'} + \gamma_2 x_2 P_2^{s'},\tag{6.8-1}$$

where $P_i^{s'}$ is the "corrected" vapor pressure of component i:

$$P_1^{s'} \equiv P_1^s \exp\left[\frac{(v_1^L - B_{11})(P - P_1^s) - P\delta_{12}y_2^2}{RT}\right]\tag{6.8-2}$$

$$P_2^{s'} \equiv P_2^s \exp\left[\frac{(v_2^L - B_{22})(P - P_2^s) - P\delta_{12}y_1^2}{RT}\right]\tag{6.8-3}$$

where δ_{12} is related to the second virial coefficients by

$$\delta_{12} \equiv 2B_{12} - B_{11} - B_{22}.\tag{6.8-4}$$

At constant temperature, activity coefficients γ_1 and γ_2 are functions only of composition.

Equation (6.8-1) is rigorous provided one assumes that the vapor phase of the mixture, as well as the vapors in equilibrium with the pure components, are adequately described by the volume-explicit virial equation terminated after the second virial coefficient; that the pure-component liquid volumes are incompressible over the pressure range in question; and that the liquid partial molar volume of each component is invariant with composition. The standard states for the activity coefficients in Eq. (6.8-1) are the pure components at the same temperature and pressure as those of the mixture.†

Barker's method is used to reduce experimental data which give the variation of total pressure with liquid composition at constant temperature. One further relation is needed in addition to Eqs. (6.8-1) to (6.8-4), and that is an equation relating the activity coefficients to mole fractions. This relation may contain any desired number of undetermined numerical coefficients which are then found from the total pressure data as shown below. For example, suppose we assume that

$$\ln \gamma_1 = \alpha x_2^2 + \beta x_2^3,\tag{6.8-7}$$

†From these assumptions it follows that

$$\ln \gamma_1 = \ln \frac{y_1 P}{x_1 P_1^s} + \frac{(B_{11} - v_1^L)(P - P_1^s)}{RT} + \frac{Py_2^2\delta_{12}}{RT}\tag{6.8-5}$$

and

$$\ln \gamma_2 = \ln \frac{y_2 P}{x_2 P_2^s} + \frac{(B_{22} - v_2^L)(P - P_2^s)}{RT} + \frac{Py_1^2\delta_{12}}{RT}.\tag{6.8-6}$$

where α and β are unknown constants. Then, from the Gibbs-Duhem equation [see Eq. (6.7-10)] it follows that

$$\ln \gamma_2 = (\alpha + \tfrac{3}{2} \beta) x_1^2 - \beta x_1^3. \tag{6.8-8}$$

Equations (6.8-1) to (6.8-8) contain only two unknowns, viz., α and β. (It is assumed that values for the quantities v^L, B, P^s, and δ_{12} are available. It is true that y is also unknown but once α and β are known, y can be determined.)

In principle, Eqs. (6.8-1) to (6.8-8) could yield α and β using only two points on the experimental P-x curve. In practice, however, more than just two points are required; we prefer to utilize all reliable experimental points and then optimize the values of α and β in order to give the best agreement between the observed total pressure curve and that calculated with the parameters α and β.

The calculations are tedious because y_1 and y_2 can only be calculated after α and β have been determined; the method of successive approximations must be used. In the first approximation y_1 and y_2 are set equal to zero in Eqs. (6.8-2) and (6.8-3). Then α and β are found, and immediately thereafter y_1 and y_2 are computed (from the first approximation of the parameters α and β) using Eqs. (6.8-5) to (6.8-8). The entire calculation is then repeated except that the new values of y_1 and y_2 are now used in Eqs. (6.8-2) and (6.8-3). One proceeds in this way until the assumed and calculated values of y_1 and y_2 are in agreement; usually 3 or 4 successive approximations are sufficient.

The form of Eqs. (6.8-7) and (6.8-8) is arbitrary and one may use any desired set of equations with any desired number of constants, provided that the two equations satisfy the Gibbs-Duhem equation.

Although Barker's numerical method is too complicated for manual calculation, it can easily be programmed for a digital electronic computer which is capable of rapidly transforming isothermal P-x data to isothermal y-x data.

An illustration of Barker's method is provided by the work of Hermsen,[5] who measured isothermal total vapor pressures for the benzene/cyclopentane system at 25, 35, and 45°C. Hermsen assumed that the excess Gibbs energy of this system is described by a two-parameter expansion of the Redlich-Kister type:

$$\frac{g^E}{RT} = x_1 x_2 [A' + B'(x_1 - x_2)], \tag{6.8-9}†$$

where subscript 1 refers to benzene and subscript 2 to cyclopentane. Activity coefficients are obtained from Eq. (6.8-9) by differentiation ac-

---

†The coefficients A' and B' are dimensionless.

cording to Eq. (6.3-8). They are:

$$\ln \gamma_1 = (A' + 3B')x_2^2 - 4B'x_2^3 \qquad (6.8\text{-}10)$$

and

$$\ln \gamma_2 = (A' - 3B')x_1^2 + 4B'x_1^3. \qquad (6.8\text{-}11)$$

As indicated by Eqs. (6.8-2) and (6.8-3), Barker's method requires the molar liquid volumes of the pure liquids and the three second virial coefficients B_{11}, B_{22}, and B_{12}. For the benzene/cyclopentane system these are given in Table 6-1.

Table 6-1

SECOND VIRIAL COEFFICIENTS AND LIQUID MOLAR VOLUMES
FOR BENZENE (1) AND CYCLOPENTANE (2)†

$(cm^3/g\text{-}mol)$

Temp (°C)	v_1	v_2	B_{11}	B_{22}	B_{12}	δ_{12}
25.00	89.39	94.71	−1314	−1054	−1176	16
35.00	90.49	95.98	−1224	− 983	−1096	15
45.00	91.65	97.29	−1143	− 919	−1024	14

†R. W. Hermsen and J. M. Prausnitz, *Chem. Eng. Sci.*, **18**, 485 (1963).

Hermsen's calculated and experimental results are given in Table 6-2. At a given temperature, the total pressure and the liquid composition were determined. Vapor-phase compositions were not measured. From the experimental measurements, optimum values of A' and B' were found such that the calculated total pressures [Eq. (6.8-1)] reproduce as closely as possible the experimental ones. The optimum values are:

	25°C	35°C	45°C
A'	0.45598	0.42463	0.40085
B'	−0.01815	−0.01627	−0.02186

In the benzene/cyclopentane system, deviations from ideality are not large and only weakly asymmetric and therefore, a two-parameter expression for the excess Gibbs energy is sufficient. For other systems, where the excess Gibbs energy is large or strongly asymmetric, it is advantageous to use a more complicated expression for the excess Gibbs energy, or else to include higher terms in the Redlich-Kister expansion. For example, Orye[6] has measured total pressures for five binary systems containing a hydrocarbon and a polar solvent and he has reduced these

Table 6-2

TOTAL PRESSURE DATA AND CALCULATED RESULTS FOR THE SYSTEM
BENZENE (1)/CYCLOPENTANE (2)†

x_1	y_1 (calc)	P (exp) (mm Hg)	P (calc) (mm Hg)	γ_1	γ_2	g^E (cal/g-mol)
			25.00°C			
0.1417	0.0655	294.11	294.09	1.408	1.010	33.8
0.2945	0.1324	268.36	268.53	1.253	1.043	57.0
0.4362	0.1984	243.29	243.47	1.151	1.095	66.8
0.5166	0.2410	228.33	228.14	1.108	1.135	67.4
0.5625	0.2682	218.98	218.86	1.087	1.160	66.2
0.8465	0.5510	148.07	148.19	1.010	1.380	34.1
			35.00°C			
0.1417	0.0684	430.51	430.48	1.375	1.009	32.5
0.2945	0.1391	394.04	393.80	1.234	1.040	54.9
0.4362	0.2091	357.56	357.72	1.140	1.088	64.4
0.5166	0.2543	335.47	335.66	1.100	1.125	65.0
0.5625	0.2829	322.43	322.37	1.080	1.148	63.7
0.8465	0.5732	222.12	222.14	1.009	1.350	32.9
			45.00°C			
0.1417	0.0697	612.14	612.27	1.353	1.009	32.1
0.2945	0.1421	559.86	560.37	1.219	1.039	53.9
0.4362	0.2142	508.77	509.18	1.130	1.085	62.7
0.5166	0.2607	478.08	477.89	1.092	1.119	63.2
0.5625	0.2903	459.83	459.04	1.074	1.141	62.0
0.8465	0.5862	317.92	318.36	1.008	1.325	31.7

†R. W. Hermsen and J. M. Prausnitz, *Chem. Eng. Sci.*, **18**, 485 (1963).

Table 6-3

EXCESS GIBBS ENERGIES OF FIVE BINARY SYSTEMS OBTAINED
FROM TOTAL PRESSURE MEASUREMENTS AT 45°C†

$$\frac{g^E}{RT} = x_1 x_2 [A' + B'(x_1 - x_2) + C'(x_1 - x_2)^2]$$

System	A'	B'	C'
Toluene (1)/Acetonitrile (2)	1.17975	−0.05992	0.12786
Toluene (1)/2,3-Butanedione (2)	0.79810	0.01763	−0.01023
Toluene (1)/Acetone (2)	0.66365	−0.00477	0.00227
Toluene (1)/Nitroethane (2)	0.76366	0.07025	0.06190
Methylcyclohexane (1)/Acetone (2)	1.69070	−0.00010	0.18324

†R. V. Orye and J. M. Prausnitz, *Trans. Faraday Soc.*, **61**, 1338 (1965).

data with Barker's method, using a three-parameter Redlich-Kister expansion for the excess Gibbs energy. Orye's results at 45°C are given in Table 6-3, and a typical plot of total and partial pressures is shown for one of the systems in Fig. 6-7. In order to justify three parameters the number

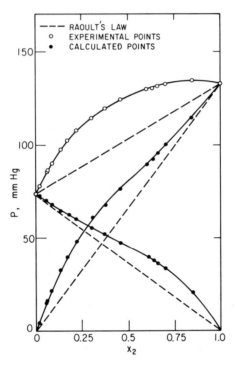

Fig. 6-7   Total and partial pressures of toluene (1)/2, 3-butanedione (2) at 45°C.

of experimental determinations of total pressure must necessarily be larger than that required to justify only two parameters. Whereas measurements for six compositions at any one temperature were sufficient for Hermsen to fix two parameters for the moderately nonideal benzene/cyclopentane system, Orye's measurements comprised about 15 compositions in order to specify three parameters for the more strongly nonideal hydrocarbon/polar-solvent systems.

Total pressure measurements at constant temperature are particularly convenient for obtaining excess Gibbs energies of those binary liquid mixtures whose components have similar volatilities. Such measurements can be made rapidly, and with some care and experience, they can be very accurate.

6.9   Partial Pressures from Isobaric Boiling-Point Data

As indicated in the previous section, the Gibbs-Duhem equation can be used to convert isothermal P-x data for a binary system into y-x data; similarly, it can be used to convert isobaric T-x data into y-x data. However, the latter calculation is often less useful because the Gibbs-Duhem equation for an isobaric, nonisothermal system (see App. IV) contains a correction term proportional to the heat of mixing and this correction term is not always negligible. The isothermal, nonisobaric Gibbs-Duhem equation also contains a correction term (which is proportional to the volume change on mixing) but in mixtures of two liquids at low pressures this term may safely be neglected.

Rigorous reduction of isobaric T-x data, therefore, requires data on the heat of mixing at the boiling point of the solution. Such data are almost never at hand and if the object of a particular study is to obtain accurate isobaric y-x data, then it is usually easier to measure the y-x data directly in an equilibrium still than to obtain heat-of-mixing data in addition to the T-x data. However, for approximate results, sufficient for some practical applications, boiling-point determinations may be useful because of experimental simplicity; it is a simple matter to place a liquid mixture in a monostated flask and to measure the boiling temperature. We now discuss briefly how isobaric T-x data may be reduced to yield an isobaric y-x diagram.

We assume that in the Gibbs-Duhem equation the correction term for nonisothermal conditions may be neglected. Further, we assume for simplicity that the gas phase is ideal and that the two-suffix Margules equation is adequate for the relation between activity coefficient and mole fraction:

$$RT \ln \gamma_1 = Ax_2^2. \tag{6.9-1}$$

We assume that A is a constant independent of temperature, pressure, and composition. The Gibbs-Duhem equation then gives:

$$RT \ln \gamma_2 = Ax_1^2. \tag{6.9-2}$$

The problem now is to find the parameter A from the T-x data. Once A is known, it is a simple matter to calculate the y-x diagram. To be consistent with the approximate nature of this calculation we here use the simplified definition of the activity coefficient as given by Eq. (6.7-12).

To find A we write:

$$P = \text{constant} = x_1 P_1^s \exp\left[\frac{A}{RT} x_2^2\right] + x_2 P_2^s \exp\left[\frac{A}{RT} x_1^2\right]. \tag{6.9-3}$$

From the T-x data and from the vapor-pressure curves of the pure components, everything in Eq. (6.9-3) is known except A. Unfortunately, Eq. (6.9-3) is not explicit in A but for any point on the T-x diagram, a value of A may be found by trial and error. Thus, in principle, the boiling point for one particular mixture of known composition is sufficient to determine A. However, to obtain a more representative value it is preferable to measure boiling points for several compositions of the mixture, to calculate a value for A for each boiling point, and then either to use an optimum average value in the subsequent calculations or, if the data warrant doing so, to use a two- (or three-) parameter equation for the activity coefficients.

To illustrate, we consider boiling-point data for the di-isopropyl ether/ 2-propanol system obtained at atmospheric pressure. Table 6-4 gives experimental boiling points. From Eq. (6.9-3) we find an average value of

Table 6-4

BOILING POINTS OF DI-ISOPROPYL ETHER/
2-PROPANOL MIXTURES AT 1 ATM PRESSURE†

Mol % Ether in Liquid	T (°C)	Mol % Ether in Liquid	T (°C)
0	82.3	58.4	66.77
8.40	76.02	73.2	66.20
18.00	72.48	75.4	66.18
28.2	69.93	84.6	66.31
38.5	68.18	89.1	66.33
43.6	67.79	91.8	66.77
47.7	67.56	98.9	67.73
52.0	67.19	100.0	68.5

†H. C. Miller and H. Bliss, *Ind. Eng. Chem.*, **32**, 123 (1940).

$A = 760 \pm 30$ cal/g-mol. When this average value is used, a y-x diagram is obtained, as shown by the line in Fig. 6-8; the points represent the experimental y-x data of Miller and Bliss.[7] In this case agreement between the observed and calculated y-x diagram is good but one should not assume that this will always be the case.

The method outlined above is quite rough and there are many refinements one can make; for example, one may use additional coefficients in the expressions for the activity coefficients or one may correct for vapor-phase nonideality. But these refinements are frequently not worth while as long as the temperature correction in the Gibbs-Duhem equation is neglected. It appears to be unavoidable that unless the heat of mixing can

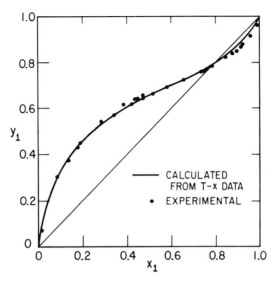

Fig. 6-8   Vapor-liquid equilibria for the di-isopropyl
ether (1)/2-propanol (2) system at 1 atm.

be estimated with at least fair accuracy, the boiling-point method for obtaining the y-x diagram is necessarily an approximation.

Techniques for estimating the isothermal or isobaric y-x diagram with
the help of the Gibbs-Duhem equation have received a large amount of
attention in the chemical literature and many variations on this theme
have been proposed. One particularly popular topic in this subject is concerned with the use of azeotropic data; if a binary system has an azeotrope, and if one knows its composition, temperature and pressure, one
can compute the two constants of any two-constant equation for the
activity coefficients and thereby calculate the entire x-y diagram.[8] This
method assumes validity of the isothermal, isobaric Gibbs-Duhem equation. Good results are often obtained provided the azeotropic mole fraction is in the interval 0.3–0.7.

Van Ness[9] has given a good discussion of the relation between experimental P-x data and the Gibbs-Duhem equation. An interesting variation
has been proposed by Christian[10] who describes apparatus and calculations for measuring P-y rather than P-x data to obtain the y-x diagram.
A method for computing equilibrium phase compositions from dew point
(T-y) data has been given by Bellemans;[11] and calculations using (P-x)
data for ternary systems have been described by McDermott and Ellis.[12]
All of these techniques have but one aim: to reduce the experimental effort
which is needed to characterize liquid-mixture properties. Such techniques are useful but they have one serious limitation: Results obtained

by data reduction with the Gibbs-Duhem equation cannot be checked for thermodynamic consistency since the method of calculation already forces the results to be thermodynamically consistent, regardless of how faulty the data may be. Thus, the Gibbs-Duhem equation may be used either to extend limited data, or else to test critically more complete data. In the next three sections of this chapter we briefly discuss some techniques for performing such tests.

6.10　Testing Equilibrium Data for Thermodynamic Consistency

The Gibbs-Duhem equation interrelates activity coefficients of all components in a mixture. Therefore, if data are available for all of the activity coefficients, these data should obey the Gibbs-Duhem equation; if they do not, then the data cannot be correct. If they do obey the Gibbs-Duhem equation, then the data are probably, although not necessarily, correct; it is conceivable that a given set of incorrect data may fortuitously satisfy the Gibbs-Duhem equation, but this is not likely. Unfortunately, there are many phase-equilibrium data in the chemical literature which do not satisfy the Gibbs-Duhem equation and which therefore must be incorrect.

To illustrate, we consider the simplest case: a binary solution of two liquids at low pressures for which isothermal activity-coefficient data have been obtained. For this case, the Gibbs-Duhem equation can be written:

$$x_1 \frac{d \ln \gamma_1}{dx_1} = x_2 \frac{d \ln \gamma_2}{dx_2} . \qquad (6.10\text{-}1)$$

A theoretically simple technique is to test the data directly with Eq. (6.10-1); that is, plots are prepared of $\ln \gamma_1$ vs. x_1 and $\ln \gamma_2$ vs. x_2 and slopes are measured. These slopes are then substituted into Eq. (6.10-1) at various compositions to see if the Gibbs-Duhem equation is satisfied. While this test appears to be both simple and exact, it is of little practical value since it is difficult to measure slopes with sufficient accuracy. Hence, the "slope method" provides at best a rough measure of thermodynamic consistency which can only be applied in a semiquantitative manner. For example, if, at a given composition, $d \ln \gamma_1/dx_1$ is positive, then $d \ln \gamma_2/dx_2$ must also be positive, and likewise if $d \ln \gamma_1/dx_1$ is zero, then $d \ln \gamma_2/dx_2$ must also be zero. The slope method can therefore be used easily to detect serious errors in the equilibrium data.

Integral (area) test

For quantitative purposes it is much easier to use an integral rather than a differential (slope) test. The integral test is not as stringent because it tests the data as a whole rather than point by point, but it has the important advantage that it can easily be carried out quantitatively. The integral test was proposed by Redlich and Kister[13] and also by Herington[14] and is derived below.

The molar excess Gibbs energy is written:

$$\frac{g^E}{RT} = x_1 \ln \gamma_1 + x_2 \ln \gamma_2. \tag{6.10-2}$$

Differentiating with respect to x_1 at constant temperature and pressure gives:

$$\frac{d(g^E/RT)}{dx_1} = x_1 \frac{\partial \ln \gamma_1}{\partial x_1} + \ln \gamma_1 + x_2 \frac{\partial \ln \gamma_2}{\partial x_1} + \ln \gamma_2 \frac{dx_2}{dx_1}. \tag{6.10-3}$$

Noting that $dx_1 = -dx_2$ and substituting the Gibbs-Duhem equation [Eq. (6.10-1)], we obtain

$$\frac{d(g^E/RT)}{dx_1} = \ln\left(\frac{\gamma_1}{\gamma_2}\right). \tag{6.10-4}$$

Integrating with respect to x_1,

$$\int_0^1 \frac{d(g^E/RT)}{dx_1} dx_1 = \int_0^1 \ln \frac{\gamma_1}{\gamma_2} dx_1 = \frac{g^E}{RT} \text{ (at } x_1 = 1) - \frac{g^E}{RT} \text{ (at } x_1 = 0). \tag{6.10-5}$$

If the pure liquids at the temperature of the mixture are used as the standard states,

$$\left.\begin{array}{ll} \ln \gamma_1 \longrightarrow 0 & \text{as} \quad x_1 \longrightarrow 1 \\ \ln \gamma_2 \longrightarrow 0 & \text{as} \quad x_1 \longrightarrow 0 \end{array}\right\} \tag{6.10-6}$$

and

$$\left.\begin{array}{l} \dfrac{g^E}{RT} \text{ (at } x_1 = 1) = 0 \\[2mm] \dfrac{g^E}{RT} \text{ (at } x_1 = 0) = 0. \end{array}\right\} \tag{6.10-7}$$

Equation (6.10-5) therefore becomes:

$$\boxed{\int_0^1 \ln \frac{\gamma_1}{\gamma_2} dx_1 = 0.} \tag{6.10-8}$$

Equation (6.10-8) provides what is called an *area test* of phase-equilibrium data. A plot of $\ln(\gamma_1/\gamma_2)$ versus x_1 is prepared; a typical plot of this type is shown in Fig. 6-5. Since the integral on the left-hand side of Eq. (6.10-8) is given by the area under the curve shown in the figure, the requirement of thermodynamic consistency is met if that area is zero, i.e., if the area above the x-axis is equal to that below the x-axis. These areas can be measured easily and accurately with a planimeter and thus the area test is a particularly simple one to carry out.

In performing the area test on real data, the integral in Eq. (6.10-8) will not come out to be exactly zero. How, then, can one decide whether a set of data "passes" or "fails" the area test? There is no absolute answer to this question since all real data have some error and thus the answer depends on just how much error one is willing to accept. As a practical guide, however, for systems of moderate nonideality, it is usually agreed that a given set of data are thermodynamically consistent if

$$0.02 > \left| \frac{\text{Area above } x\text{-axis} - \text{Area below } x\text{-axis}}{\text{Area above } x\text{-axis} + \text{Area below } x\text{-axis}} \right|. \quad (6.10\text{-}9)\dagger$$

Equation (6.10-9) is necessarily arbitrary and the quantity on the left-hand side of this equation may be raised or lowered depending on how liberal or how conservative one wishes to be in interpreting the meaning of thermodynamic consistency. For systems which exhibit only slight deviations from ideality (say $|g_{max}^E| < 50$ cal/g-mol), the arbitrary value of 0.02 is probably too low.

In applying the area test, or any of the other tests for thermodynamic consistency, it is important that the activity coefficients be calculated as rigorously as possible from the equilibrium data. It is especially important to consider the effect of vapor-phase corrections.

The area test for binary systems can be extended to ternary systems as shown by Herington[15] and to systems containing any number of components, as discussed by Snider.[16] These extensions are for isothermal data and, as before, they assume that the pressure is sufficiently low to neglect the effect of pressure on liquid-phase activity coefficients.

6.11   Integral Tests for Nonisobaric
or Nonisothermal Data

As indicated by the phase rule, vapor-liquid equilibrium data for a binary system cannot be both isobaric and isothermal. Therefore, the thermo-

---

†In Eq. (6.10-9) all areas are taken as positive.

dynamic consistency test of Redlich-Kister and Herington is, strictly speaking, an approximation although the degree of approximation in a given case may be extremely good, especially if the data are isothermal.

For a binary system it is a simple matter to write down rigorous expressions for the area test under nonisobaric or nonisothermal conditions. In these cases the forms of the Gibbs-Duhem equation to be used are Eqs. (IV.1-20) and (IV.1-26) of Appendix IV rather than Eq. (6.10-1); it follows then that the criteria for thermodynamic consistency are:

Isobaric but
nonisothermal
data:
$$\int_0^1 \ln \frac{\gamma_1}{\gamma_2} \, dx_1 = \int_{x_1=0}^{x_1=1} \frac{h^E}{RT^2} \, dT \qquad (6.11\text{-}1)$$

Isothermal but
nonisobaric
data:
$$\int_0^1 \ln \frac{\gamma_1}{\gamma_2} \, dx_1 = -\int_{x_1=0}^{x_1=1} \frac{v^E}{RT} \, dP \qquad (6.11\text{-}2)$$

These equations are derived in exactly the same manner as that used to obtain Eq. (6.10-8) except that the complete form rather than the approximate form of the Gibbs-Duhem equation is used. In Eqs. (6.11-1) and (6.11-2) the standard states are the pure liquids at the same temperature and pressure as those of the solution.

The integrals in Eqs. (6.11-1) and (6.11-2) can be evaluated by plotting the appropriate quantities versus x_1, P, or T as indicated, and then measuring the areas underneath the curves. If the relations between these areas are as given by Eqs. (6.11-1) and (6.11-2), the data are thermodynamically consistent. In most cases, however, the data required for the integrals on the right-hand sides of these equations are not available and the simplest approximation is to set these integrals equal to zero. In the case of Eq. (6.11-2) this is usually a very good approximation. However, for isobaric, nonisothermal data the right-hand integral of Eq. (6.11-1) often cannot be neglected. If the boiling points of the two components at the prevailing pressure are close together and no azeotrope is formed, then dT/dx must be small; if, in addition, the two components are similar chemically, h^E is probably small also and then it might be reasonable to neglect the right-hand integral. But in general the nonisothermal correction term in the Gibbs-Duhem equation cannot be ignored if an accurate test of the data is to be obtained.

A semiempirical test for the thermodynamic consistency of binary isobaric, nonisothermal equilibrium data has been proposed by Herington.[17] This proposal uses an area test as given by Eq. (6.11-1), and the heart of it lies in an estimate of the right-hand side of this equation.

Herington proposes that, as suggested by Eq. (6.11-1), we plot $\ln (\gamma_1/\gamma_2)$ vs. x_1 for the range $x_1 = 0$ to $x_1 = 1$. From this plot we calculate the percent deviation D, defined by

$$D \equiv \left| \frac{\text{Area above } x\text{-axis} \; - \; \text{Area below } x\text{-axis}}{\text{Area above } x\text{-axis} \; + \; \text{Area below } x\text{-axis}} \right|. \qquad (6.11\text{-}3)\dagger$$

If the data were isothermal as well as isobaric, then D should be zero for consistent data. However, for nonisothermal, consistent data, D is not zero and thus the practical problem is essentially one of making a good estimate of what D should be in the common situation where no data on the heat of mixing are available. Herington suggests that the quantity D should be compared with another quantity J which is found as follows:

Let τ be the total boiling-point range in degrees centigrade. If no azeotrope is formed,

$$\tau = |T_1 - T_2|, \qquad (6.11\text{-}4)$$

where T_1 is the boiling point of pure component 1, and T_2 is the boiling point of pure component 2. If an azeotrope is formed, then τ is found as shown in Fig. 6-9.

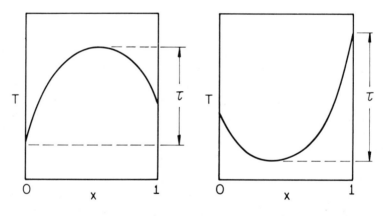

MAXIMUM-BOILING MIXTURE   MINIMUM-BOILING MIXTURE

Fig. 6-9 Maximum boiling temperature difference τ in azeotrope-forming binary mixtures at constant pressure.

Let $T_{\min}$ be the lowest boiling point observed in the composition range $x_1 = 0$ to $x_1 = 1$, in degrees Kelvin. The quantity J is now defined by:

$$J \equiv 150 \frac{\tau}{T_{\min}}. \qquad (6.11\text{-}5)$$

†In Eq. (6.11-3) all areas are taken as positive.

The constant 150 is empirical, based on Herington's analysis of typical heat-of-mixing data. Herington suggests that:

If $(D - J) < 10$, the data are probably consistent
If $(D - J) > 10$, the data are probably inconsistent.

Herington's method is necessarily an approximation but in the absence of heat-of-mixing data it is a useful semiempirical technique for testing isobaric vapor-liquid equilibrium data. However, we must emphasize that since Herington's empirical constant was determined from mixtures of typical organic liquids not far from room temperature, it is probably not applicable to mixtures which are very much different from those included in his study. It is an interesting feature of Herington's work that when he chose isobaric data for 24 systems taken from the third edition of Perry's *Handbook for Chemical Engineers*,† he found that the data for 9 of these systems were inconsistent according to his criterion.

6.12   The Differential (Composition-Resolution) Test of Van Ness and Mrazek

The accuracy of the area test suffers from its integral nature; it would be more desirable to check the thermodynamic consistency of each separate datum. An integral (area) test has the disadvantage that it does not represent a point-by-point test of the data and thus it is possible, because of cancellation of errors, that an integral test is satisfied even though the slope test at various concentrations is not. A differential test provides a much stricter examination of thermodynamic consistency than does an integral test. If the data satisfy the differential test at every point they must also satisfy the integral test, but the converse of this statement is unfortunately not true.

Van Ness and Mrazek[18, 19] have discussed a useful technique for overcoming the limitations of the area test for binary systems called the *composition-resolution* test. The essence of their technique lies in the fact that whereas a given curve for the excess Gibbs energy vs. mole fraction can be computed from an infinite number of sets of experimentally determined values of γ_1 and γ_2, only one consistent set of γ_1 and γ_2 can be calculated from a given curve for the excess Gibbs energy of mixing.

For a binary system, γ_1 and γ_2 are related by the Gibbs-Duhem equation and hence, from a curve of g^E/RT vs. x_1 only one *consistent* set of γ_1 and γ_2 can be calculated from any individual point on the curve. If these calculated values of γ_1 and γ_2 agree with those observed directly

†New York: McGraw-Hill Book Company, 1950.

from the equilibrium data, then the data are thermodynamically consistent.

In the derivation below, the standard state for each activity coefficient is the pure liquid at the temperature and pressure of the solution.

We start with the general relation

$$\frac{g^E}{RT} = x_1 \ln \gamma_1 + x_2 \ln \gamma_2. \qquad (6.12\text{-}1)$$

Next, we differentiate Eq. (6.12-1) with respect to x_1 at either constant temperature or pressure and substitute the nonisothermal or nonisobaric Gibbs-Duhem equation derived in Appendix IV. We then obtain:

$$\frac{d(g^E/RT)}{dx_1} = \ln \gamma_1 - \ln \gamma_2 + \mathfrak{D}, \qquad (6.12\text{-}2)$$

where

$$\mathfrak{D} \equiv \left(\frac{v^E}{RT}\right)\frac{dP}{dx_1} \quad \text{for isothermal data,}$$

or

$$\mathfrak{D} \equiv -\left(\frac{h^E}{RT^2}\right)\frac{dT}{dx_1} \quad \text{for isobaric data.}$$

Figure 6-10(a) shows a typical plot of g^E/RT vs. x_1 for data obtained at either constant temperature or constant pressure. The slope of a tangent drawn to the curve at some arbitrary point is $d(g^E/RT)/dx_1$, and the tangent is shown to intersect the ordinate at $x_1 = 1$ and $x_1 = 0$ at points labeled a and b respectively. From inspection of Fig. 6-10(a), it follows that the points a and b are given by

$$a = \frac{g^E}{RT} + x_2 \frac{d(g^E/RT)}{dx_1} \qquad (6.12\text{-}3)$$

and

$$b = \frac{g^E}{RT} - x_1 \frac{d(g^E/RT)}{dx_1}. \qquad (6.12\text{-}4)$$

Combining Eqs. (6.12-3) and (6.12-4) with Eqs. (6.12-1) and (6.12-2), we see that the intercepts are related to the activity coefficients at the point of tangency by

$$a = \ln \gamma_1 + x_2 \mathfrak{D} \qquad (6.12\text{-}5)$$

and

$$b = \ln \gamma_2 - x_1 \mathfrak{D}. \qquad (6.12\text{-}6)$$

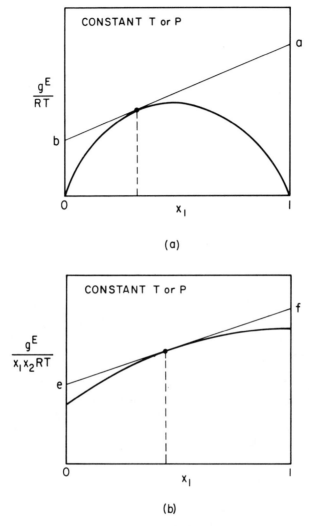

Fig. 6-10   (a) Classical method of tangent intercepts for isothermal or isobaric data.   (b) Method of tangent intercepts of Van Ness and Mrazek for isothermal or isobaric data.

Figure 6-10(a) illustrates the classical method of tangent intercepts. The procedure for testing for thermodynamic consistency is as follows: From the equilibrium data, prepare a plot of g^E/RT vs. x_1. Then, for a given point, calculate a and b by construction of a tangent as shown in Fig. 6-10(a). Compare these graphically calculated values of a and b with the values calculated directly from the equilibrium data for the point in question as indicated by Eqs. (6.12-5) and (6.12-6). If the comparison is satisfactory, then the data at the point in question are thermodynamically consistent.

A practical disadvantage of the method of tangent intercepts is that it is difficult to draw tangents with accuracy. To overcome this disadvantage at least partly, Van Ness and Mrazek have suggested that instead of plotting g^E/RT vs. x_1, it is better to plot g^E/RTx_1x_2 vs. x_1, as shown in Fig. 6-10(b) and to construct a tangent having intercepts e and f. The desired quantities a and b can now be found from the exact relations

$$a = x_2^2 \left[\frac{2g^E}{x_1x_2RT} - e \right] \qquad (6.12\text{-}7)$$

and

$$b = x_1^2 \left[\frac{2g^E}{x_1x_2RT} - f \right]. \qquad (6.12\text{-}8)$$

There is an operational advantage to using the plot shown in Fig. 6-10(b) rather than the one shown in (a); when Eqs. (6.12-7) and (6.12-8) are used to compute a and b, the results are much less sensitive to errors in tangent construction than are those obtained by the classical method of tangent intercepts.

For rigorous testing, the procedure of Van Ness and Mrazek requires enthalpy-of-mixing data for the isobaric case and volumetric data for the isothermal case, as indicated by Eqs. (6.12-5) and (6.12-6). This may be a serious limitation since the required data are rarely available, but it is a limitation which is clearly unavoidable and which pertains to all tests for thermodynamic consistency. For low-pressure isothermal data remote from critical conditions, it is usually a safe approximation to set $\mathfrak{D}$ equal to zero, but for isobaric data this simplification is often not justified.[†]

6.13 Wohl's Expansion for the Excess Gibbs Energy

In Sec. 6.5 we discussed briefly some expressions for the excess Gibbs energy of binary solutions. We now continue this discussion with a general method for expressing excess Gibbs energies as proposed by Wohl.[20] One of the main advantages of this method is that some ap-

---

[†] Whenever the boiling points of the two (chemically similar) components in the mixture are close to each other and when no azeotrope is formed, isobaric data can often be analyzed with $\mathfrak{D} = 0$. See Francesconi and Trevissoi, *Chem. Eng. Sci.*, **21**, 123 (1966).

proximate physical significance can be assigned to the parameters which appear in the equations; as a result, and as shown later on in Sec. 6.17, Wohl's expansion can be extended systematically to multicomponent solutions.

Wohl expresses the excess Gibbs energy of a binary solution as a power series in z_1 and z_2, the effective volume fractions of the two components:

$$\frac{g^E}{RT(x_1 q_1 + x_2 q_2)} = 2a_{12} z_1 z_2 + 3a_{112} z_1^2 z_2 + 3a_{122} z_1 z_2^2$$

$$+ 4a_{1112} z_1^3 z_2 + 4a_{1222} z_1 z_2^3 + 6a_{1122} z_1^2 z_2^2$$

$$+ \ldots \tag{6.13-1}$$

where

$$z_1 \equiv \frac{x_1 q_1}{x_1 q_1 + x_2 q_2} \quad \text{and} \quad z_2 \equiv \frac{x_2 q_2}{x_1 q_1 + x_2 q_2}.$$

Wohl's equation contains two types of parameters, q's and a's. The q's are effective volumes, or cross-sections of the molecules; q_i is a measure of the size of molecule i, or of its "sphere of influence" in the solution. A large molecule has a larger q than a small one and, in solutions of non-polar molecules of similar shape, it is often a good simplifying assumption that the ratio of the q's is the same as the ratio of the pure-component liquid molar volumes. The a's are interaction parameters whose physical significance, while not precise, is in a rough way similar to that of virial coefficients. The parameter a_{12} is a constant characteristic of the interaction between molecule 1 and molecule 2; the parameter a_{112} is a constant characteristic of the interaction between three molecules, two of component 1 and one of component 2, and so on. The probability that any pair of two molecules consists of one molecule of component 1 and one molecule of component 2 is assumed to be $2z_1 z_2$; similarly the probability that a triplet of three nearest-neighbor molecules consists of molecules 1, 1, and 2 is assumed to be $3z_1^2 z_2$, and so forth. Thus there is a very rough analogy between Wohl's equation and the virial equation of state but it is no more than an analogy because, while the virial equation has an exact theoretical basis, Wohl's equation cannot be derived from any rigorous theory without drastic simplifying assumptions.

When, as in Eq. (6.13-1), the excess Gibbs energy is taken with reference to an ideal solution in the sense of Raoult's law, only interactions involving at least two dissimilar molecules are contained in Eq. (6.13-1); that is, terms of the type $z_1^2, z_1^3, \ldots$ and $z_2^2, z_2^3, \ldots$ do not explicitly appear in the expansion. This is a necessary consequence of the boundary condition that g^E must vanish as x_1 or x_2 becomes zero.

However, if g^E is taken relative to an ideal dilute solution which is dilute in, say, component 2, then Wohl's expansion takes the form

$$\frac{g^{E^*}}{RT(x_1 q_1 + x_2 q_2)} = -a_{22} z_2^2 - a_{222} z_2^3 - a_{2222} z_2^4 - \dots \quad (6.13\text{-}2)\dagger$$

In this case a_{22} is the self-interaction coefficient characteristic of the interaction between two molecules of component 2, a_{222} is the self-interaction coefficient characteristic of the interaction between three molecules of component 2, and so on. Since the ideal solution to which g^{E^*} in Eq. (6.13-2) refers is one very dilute in component 2, it is not interaction between molecules 2 and molecules 1 but rather interaction between molecules of component 2 which cause deviation from ideal behavior and hence a nonvanishing g^{E^*}.

Equation (6.13-1) is a well-known expression for mixtures whose components can exist as pure liquids at the solution temperature. Equation (6.13-2) is hardly known at all but it is useful for solutions of gases or solids in liquids. (See Chaps. 8 and 10.)

To illustrate the generality of Eq. (6.13-1) we consider first the case of a binary solution of two components which are not strongly dissimilar chemically but which have different molecular size. An example might be a solution of benzene (molar volume 89 cc/g-mol at 25°C) and iso-octane (molar volume 166 cc/g-mol at 25°C). We make the simplifying assumption that interaction coefficients $a_{112}, a_{122}, \dots$ and higher may be neglected; i.e., Wohl's expression is truncated after the first term. In that case Eq. (6.13-1) becomes:

$$\frac{g^E}{RT} = \frac{2a_{12} x_1 x_2 q_1 q_2}{(x_1 q_1 + x_2 q_2)}. \quad (6.13\text{-}3)$$

From Eq. (6.3-8) expressions for the activity coefficients can be found. They are:

$$\ln \gamma_1 = \frac{A'}{\left[1 + \dfrac{A'}{B'} \dfrac{x_1}{x_2}\right]^2} \quad (6.13\text{-}4)\ddagger$$

---

†The minus signs before the coefficients are arbitrarily introduced for convenience. For most mixtures a_{22} in Eq. (6.13-2) is a positive number. The asterisk on g^{E^*} indicates that g^{E^*} is taken relative to an *ideal dilute* solution.

‡A' and B' are dimensionless.

$$\ln \gamma_2 = \frac{B'}{\left[1 + \dfrac{B'\,x_2}{A'\,x_1}\right]^2} \qquad (6.13\text{-}5)$$

where $A' = 2q_1 a_{12}$; $B' = 2q_2 a_{12}$.

Equations (6.13-4) and (6.13-5) are the familiar van Laar equations which are commonly used to represent activity-coefficient data. These equations include two empirical constants, A' and B'; the ratio of A' to B' is the same as the ratio of the effective volumes q_1 and q_2 and it is also equal to the ratio of $\ln \gamma_1^\infty$ to $\ln \gamma_2^\infty$. Whereas Eqs. (6.13-4) and (6.13-5) contain only two parameters, Eq. (6.13-3) appears to be a three-parameter equation. However, from the empirically determined values of A' and B' it is not possible to find a value of the interaction coefficient a_{12} unless some independent assumption is made concerning the value of q_1 or q_2. For practical purposes it is never necessary to know the values of q_1 and q_2 separately because it is only their ratio which is important.

Figure 6-11 gives activity coefficients for the benzene/isooctane system at 45°C. The data of Weissman and Wood[21] are well represented by the van Laar equations with $A' = 0.419$ and $B' = 0.745$.

The derivation of the van Laar equations suggests that they should be used for solutions of relatively simple, preferably nonpolar liquids but empirically it has been found that these equations are frequently able to

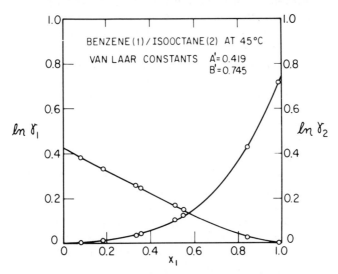

Fig. 6-11 Application of van Laar's equations to a mixture whose components differ appreciably in molecular size.

represent activity coefficients of more complex mixtures. In such mixtures the physical significance of the van Laar constants is even more obscure, and they must be regarded as essentially little more than empirical parameters in a thermodynamically consistent equation. The van Laar equations are widely used in the chemical literature; they have become popular for applied work because of their flexibility and because of their mathematical simplicity relative to the many other equations which have been proposed.†

Whenever Wohl's expansion is used as the basis for an expression giving the the activity coefficient as a function of composition, the resulting equation holds for a fixed temperature and pressure; that is, the constants in the equation are not dependent on composition but instead are functions of temperature and pressure. Since the effect of pressure on liquid-phase properties is usually small (except at high pressures and at conditions near critical) the pressure-dependence of the constants can often be neglected; however, the temperature-dependence is often not negligible. Since many industrial operations (e.g., distillation) are conducted at constant pressure rather than at constant temperature, there is a strong temptation to assume that the constants in equations such as those of van Laar are temperature-independent. Physically this assumption is most unreasonable; as indicated by Eq. (6.3-12), the activity coefficient of a component in solution is independent of temperature only in an athermal solution, i.e., one where the components mix isothermally and isobarically without evolution or absorption of heat. For practical applications, however, this assumption is often quite tolerable provided the temperature range in question is not large. For example, Fig. 6-12 shows activity coefficients for the propanol/water system calculated from the data of Gadwa[22] at a constant pressure of 1 atmosphere. These activity coefficients are well represented by the van Laar equations with $A' = 2.60$ and $B' = 1.13$. In this particular case the assumption of temperature-invariant constants appears to be a good approximation because at a constant pressure of 1 atmosphere the boiling temperature varies only from 87.8 to 100°C.

When, as frequently happens, experimental data are insufficient to specify the temperature-dependence of the activity coefficient, either one of two simplifying approximations is usually made. The first one, mentioned in the previous paragraph, is to assume that at constant composition the activity coefficient is invariant with temperature; the second one is to assume that at constant composition $\ln \gamma$ is proportional to the reciprocal of the absolute temperature. The first assumption is equivalent to assuming that the solution is athermal ($h^E = 0$), and the second one is equivalent to assuming that the solution is regular ($s^E = 0$). Actual

---

†As pointed out by Van Ness,[25] the van Laar equation implies that when the experimental data are plotted in the form $x_1 x_2 / g^E$ vs. x_1, a straight line should result.

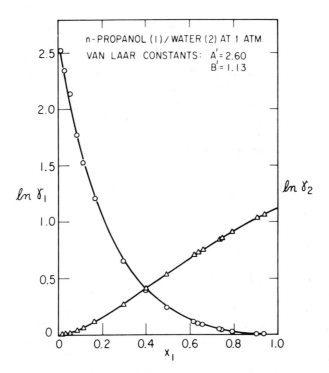

n-PROPANOL (1)/WATER (2) AT 1 ATM.

VAN LAAR CONSTANTS: $A' = 2.60$
$B' = 1.13$

Fig. 6-12   Application of van Laar's equations to an isobaric
system. In this system the temperature varies only
from 87.8 to 100°C.

solutions are neither athermal nor regular but more often than not, the
assumption of regularity provides a better approximation to the effect of
temperature on the activity coefficient than does the assumption of
athermal behavior.

Next we consider a binary solution of two components whose molecular sizes are not much different. In that case we assume that $q_1 = q_2$ in
Wohl's expansion. Neglecting terms higher than the fourth power in the
mole fraction, and again using Eq. (6.3-8) to obtain expressions for the
activity coefficients, we now obtain:

$$\ln \gamma_1 = A'x_2^2 + B'x_2^3 + C'x_2^4 \qquad (6.13\text{-}6)$$

$$\ln \gamma_2 = (A' + \tfrac{3}{2}B' + 2C')x_1^2 - (B' + \tfrac{8}{3}C')x_1^3 + C'x_1^4, \qquad (6.13\text{-}7)$$

where

$$A' = q(2a_{12} + 6a_{112} - 3a_{122} + 12a_{1112} - 6a_{1122})$$

$$B' = q(6a_{122} - 6a_{112} - 24a_{1112} - 8a_{1222} + 24a_{1122})$$

$$C' = q(12a_{1112} + 12a_{1222} - 18a_{1122}).$$

To simplify matters, and because experimental data are usually limited, it is common to truncate the expansions after the cubic terms, i.e., to set $C' = 0$; in that case the equation is called the three-suffix Margules equation and it has two parameters. Only in those cases where the data are sufficiently precise and plentiful is the expansion truncated after the quartic terms and we then have a four-suffix Margules equation with three parameters as given in Eqs. (6.13-6) and (6.13-7). On the other hand, if the mixture is a simple one, containing similar components, it is sometimes sufficient to retain only the quadratic term (two-suffix Margules equation).

It can be shown by simple algebra that the expansion of Redlich and Kister, Eq. (6.5-7), is equivalent to the Margules expansion. Further, a little algebra shows that the two-suffix Margules equation becomes identical to the van Laar equation for the special case where the van Laar constants A' and B' are equal.

Although the assumption $q_1 = q_2$ suggests that the Margules equations should be used only for mixtures whose components have similar molar volumes, it is nevertheless used frequently for all sorts of liquid mixtures, regardless of the relative sizes of the different molecules. The primary value of the Margules and van Laar equations lies in their ability to serve as simple empirical equations for representing experimentally determined activity coefficients with only a few constants. When, as is often the case, experimental data are scattered or scarce, these equations can be used to smooth the data and, more important, they serve as an efficient tool for interpolation and extrapolation with respect to composition.†

The three-suffix Margules equations have been used to reduce experimental vapor-liquid equlibrium data for many systems; to illustrate, we show in Fig. 6-13 results for three binaries at 50°C: acetone/methanol, acetone/chloroform and chloroform/methanol.[23] Each of these binaries has an azeotrope at 50°C.

Figure 6-13 shows that the thermodynamic properties of these three systems differ markedly from one another; in the acetone/methanol system there are strong positive deviations from ideality, while in the acetone/chloroform system there are equally strong negative deviations; in the chloroform/methanol system there are very large positive deviations at the chloroform-rich end, and at the methanol-rich end the activity coefficient of chloroform exhibits rather unusual behavior since it goes through a maximum.‡ Despite these large differences, the three-

---

†As pointed out by Van Ness,[25] the three-suffix Margules equation implies that when the experimental data are plotted in the form $g^E/x_1 x_2$ vs. x_1, a straight line should result.

‡One of the advantages of the three-suffix Margules equations is that they are capable of representing maxima or minima; the two-suffix Margules equations and the van Laar equations cannot do so. However, maxima and minima are only rarely observed in isothermal plots of activity coefficient versus mole fraction.

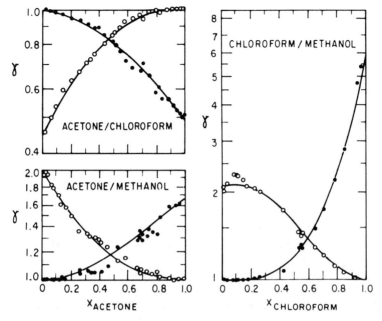

Fig. 6-13   Activity coefficients for three binary systems at 50°C. Lines calculated from three-suffix Margules equations.

suffix Margules equations give a good representation of the data for all three systems. The Margules constants,† as obtained from the experimental data at 50°C, are:

	A'	B'
Acetone(1)/Chloroform(2)	−0.553	−0.276
Acetone (1)/Methanol (2)	0.334	0.368
Chloroform (1)/Methanol (2)	2.89	−2.17

For one additional example of the flexibility of Wohl's expansion we consider the case where the series is truncated after the third-order terms but where, instead of assuming that $q_1 = q_2$, we assume that

$$\frac{q_1}{q_2} = \frac{v_1}{v_2},$$

where v_1 and v_2 are, respectively, the molar volumes of the pure liquids at the temperature of the solution. Truncating Wohl's expansion after the cubic terms, we then obtain expressions for the activity coefficients

†These constants pertain to Eqs. (6.13-6) and (6.13-7) with $C' = 0$.

first proposed by Scatchard and Hamer.[24] They are:

$$\ln \gamma_1 = A'z_2^2 + B'z_2^3 \qquad (6.13\text{-}8)$$

$$\ln \gamma_2 = (A' + \tfrac{3}{2}B')\left(\frac{v_2}{v_1}\right)z_1^2 - B'\left(\frac{v_2}{v_1}\right)z_1^3, \qquad (6.13\text{-}9)$$

where

$$A' = v_1(2a_{12} + 6a_{112} - 3a_{122}),$$

and

$$B' = v_1(6a_{122} - 6a_{112}).$$

The Scatchard-Hamer equations use only two adjustable parameters but they have not received much attention in the extensive literature on vapor-liquid equilibria, apparently because they are mathematically more complex than the popular van Laar equations and the three-suffix Margules equations. This is unfortunate because in the general case the assumptions of Scatchard and Hamer appear more reasonable than those of van Laar or those in Margules' equation truncated after the cubic

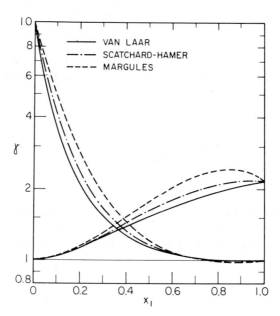

Fig. 6-14   Activity coefficients according to three, two-parameter equations with $\gamma_1^\infty = 10$ and $\gamma_2^\infty = 2.15$. For the Scatchard-Hamer equation, $v_2/v_1 = \tfrac{2}{3}$.

terms. In the age of computers, mathematical complexity is no longer a major obstacle.

The behavior of the Scatchard-Hamer equations may be considered as intermediate between that of the van Laar equations and that of the three-suffix Margules equations. All three equations contain two adjustable parameters, and if for each of these equations we arbitrarily determine these parameters from values of γ_1^∞ and γ_2^∞, we can then compare the three equations as shown in Fig. 6-14.

Many other equations have been proposed for the relation between activity coefficients and mole fractions and new ones appear every year.[†] Some, but not all, of these can be derived from Wohl's general method. In particular, two equations, due to Wilson and Renon, which appear to be promising for many practical calculations, cannot be obtained by Wohl's formulation.

6.14 Equations of Wilson and Renon

On the basis of molecular considerations which are discussed in Sec. 7.8, Wilson[26, 27] derived the following expression for the excess Gibbs energy of a binary solution:

$$\frac{g^E}{RT} = -x_1 \ln (x_1 + \Lambda_{12}x_2) - x_2 \ln (x_2 + \Lambda_{21}x_1). \qquad (6.14\text{-}1)$$

The activity coefficients derived from this equation are given by

$$\ln \gamma_1 = -\ln (x_1 + \Lambda_{12}x_2) + x_2 \left[\frac{\Lambda_{12}}{x_1 + \Lambda_{12}x_2} - \frac{\Lambda_{21}}{\Lambda_{21}x_1 + x_2} \right] \qquad (6.14\text{-}2)$$

$$\ln \gamma_2 = -\ln (x_2 + \Lambda_{21}x_1) - x_1 \left[\frac{\Lambda_{12}}{x_1 + \Lambda_{12}x_2} - \frac{\Lambda_{21}}{\Lambda_{21}x_1 + x_2} \right]. \qquad (6.14\text{-}3)$$

In Eq. (6.14-1) the excess Gibbs energy is defined with reference to an ideal solution in the sense of Raoult's law; Eq. (6.14-1) obeys the boundary condition that g^E vanishes as either x_1 or x_2 becomes zero.

Wilson's equation has two adjustable parameters, Λ_{12} and Λ_{21}. In Wilson's derivation, these are related to the pure-component molar vol-

---

†See, for example, E. Hála, J. Pick, V. Fried & O. Vilim, *Vapor-Liquid Equilibrium*, tr. G. Standart, 2nd ed. (Oxford: Pergamon Press Ltd., 1967), Part 1; and C. Black, *A. I. Ch. E. Journal*, **5**, 249 (1959). Also, a particularly simple but unusual equation has been proposed by H. Mauser, *Z. Elektrochem.*, **62**, 895 (1958).

umes and to characteristic energy differences by

$$\Lambda_{12} \equiv \frac{v_2}{v_1} \exp - \frac{(\lambda_{12} - \lambda_{11})}{RT} \qquad (6.14\text{-}4)$$

$$\Lambda_{21} \equiv \frac{v_1}{v_2} \exp - \frac{(\lambda_{12} - \lambda_{22})}{RT} \qquad (6.14\text{-}5)$$

where v_i is the molar liquid volume of pure component i and where the λ's are energies of interaction between the molecules designated in the subscripts. To a fair approximation, the differences in the characteristic energies are independent of temperature, at least over modest temperature intervals. As a result, Wilson's equation gives not only an expression for the activity coefficients as a function of composition but also an estimate of the variation of the activity coefficients with temperature. This is an important practical advantage in isobaric calculations where the temperature varies as the composition changes. For accurate work, $(\lambda_{12} - \lambda_{11})$ and $(\lambda_{12} - \lambda_{22})$ should be considered temperature-dependent but in many cases this dependence can be neglected without serious error.

Wilson's equation appears to provide a good representation of excess Gibbs energies for a variety of miscible mixtures. It is particularly useful for highly asymmetric systems such as solutions of polar or associating components (e.g., alcohols) in nonpolar solvents. The three-suffix Margules equation and the van Laar equation are usually not adequate for such solutions. For a good data-fit an equation of the Margules type or a modification of van Laar's equation[28] may be used, but such equations require at least three parameters and, more important, these equations are not readily generalized to multicomponent solutions without further assumptions or ternary parameters.

A study of Wilson's equation by Orye[27] shows that for approximately one hundred miscible binary mixtures of various chemical types, activity coefficients were well represented by the Wilson equation; in essentially all cases this representation was as good as, and in many cases better than, the representation given by the three-suffix (two-constant) Margules equation and by the van Laar equation.

To illustrate, Table 6-5 gives calculated and experimental vapor compositions for the nitromethane/carbon tetrachloride system. The calculations were made twice, once using the van Laar equation and once using Wilson's equation; in both cases required parameters were found from a least-squares computation using experimental $P\text{-}x$ data at 45°C given by Brown and Smith.[29] In both calculations the average error in the predicted vapor compositions is not very high but for the calculation based on van Laar's equation it is almost three times as large as that based on Wilson's equation.

Table 6-5

CALCULATED VAPOR COMPOSITIONS FROM
FIT OF P-x DATA AT 45°C
[Nitromethane (1)/Carbon Tetrachloride (2)]

Experimental†			Calculated y_1	
x_1	P (atm)	y_1	Wilson	van Laar
0.0459	0.3782	0.130	0.147	0.117
0.0918	0.3910	0.178	0.191	0.183
0.1954	0.3986	0.222	0.225	0.247
0.2829	0.3981	0.237	0.236	0.262
0.3656	0.3966	0.246	0.243	0.264
0.4659	0.3932	0.253	0.251	0.261
0.5366	0.3906	0.260	0.258	0.259
0.6065	0.3859	0.266	0.266	0.259
0.6835	0.3778	0.277	0.279	0.266
0.8043	0.3482	0.314	0.318	0.304
0.9039	0.2824	0.408	0.410	0.411
0.9488	0.2249	0.528	0.524	0.540

Error   ±0.004   ±0.011

$\Lambda_{12} = 0.1156$ $A' = 2.230$

$\Lambda_{21} = 0.2879$ $B' = 1.959$

†I. Brown and F. Smith, *Australian J. Chem.*, **10,** 423 (1957).

A similar calculation is shown in Fig. 6-15 for the ethanol/isooctane system. Wilson and van Laar parameters were calculated from the isothermal vapor-pressure data of Kretschmer.[30]   In this case the Wilson equation is much superior to the van Laar equation, which erroneously predicts an immiscible region for this system at 50°C.

For isothermal solutions which do not exhibit large or highly asymmetric deviations from ideality, Wilson's equation does not offer any particular advantages over the more familiar three-suffix Margules or van Laar equations although it appears to be as good as these. For example, isothermal vapor-liquid equilibrium data for the cryogenic systems argon/nitrogen and nitrogen/oxygen are represented equally well by all three equations.

Wilson's equation has two disadvantages which are not serious for many applications. First, Eqs. (6.14-2) and (6.14-3) are not strictly applicable for systems where the logarithms of the activity coefficients, when plotted against x, exhibit maxima or minima. (The van Laar equations are also not strictly applicable for this case.) Such systems, however, are not common. The second and more serious disadvantage of Wilson's equation lies in its inability to predict limited miscibility. When Wilson's

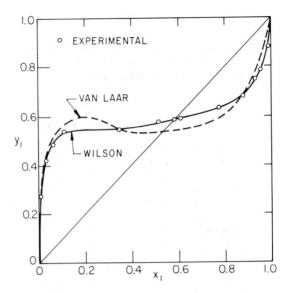

Fig. 6-15   Vapor-liquid equilibria in the ethanol (1)/
isooctane (2) system at 50°C. Lines calculated from P-x data.

equation is substituted into the equations of thermodynamic stability for a binary system (see next section), no parameters Λ_{12} and Λ_{21} can be found which indicate the existence of two stable liquid phases.† Wilson's equation, therefore, should be used only for liquid systems which are completely miscible or else for those limited regions of immiscible systems where only one liquid phase is present.

The basic idea in Wilson's derivation of Eq. (6.14-1) follows from the concept of local composition which is discussed further in Sec. 7.8. This concept was also used by Renon[31] in his derivation of the NRTL (Non-Random, Two-Liquid) equation [see Sec. 7.10]; however, Renon's equation, unlike Wilson's, is applicable to partially miscible as well as completely miscible systems. Renon's equation for the excess Gibbs energy is:

$$\frac{g^E}{RT} = x_1 x_2 \left[\frac{\tau_{21} G_{21}}{x_1 + x_2 G_{21}} + \frac{\tau_{12} G_{12}}{x_2 + x_1 G_{12}} \right] \qquad (6.14\text{-}6)$$

where

$$\tau_{12} = \frac{(g_{12} - g_{22})}{RT}; \qquad \tau_{21} = \frac{(g_{12} - g_{11})}{RT} \qquad (6.14\text{-}7)$$

$$G_{12} = \exp\left(-\alpha_{12}\tau_{12}\right); \quad G_{21} = \exp\left(-\alpha_{12}\tau_{21}\right). \qquad (6.14\text{-}8)$$

†For partially miscible systems, Wilson[26] suggested that the right-hand side of Eq. (6.14-1) be multiplied by a constant.

The significance of g_{ij} is similar to that of λ_{ij} in Wilson's equation; g_{ij} is a (Gibbs) energy parameter characteristic of the i-j interaction. The parameter α_{12} is related to the nonrandomness in the mixture; when α_{12} is zero, the mixture is completely random and Eq. (6.14-6) reduces to the two-suffix Margules equation.   The NRTL equation contains three parameters but reduction of experimental data for a large number of binary systems indicates that α_{12} varies from about 0.20 to 0.47; when experimental data are scarce, the value of α_{12} can be set equal to that found to be best for the type of mixture under consideration.[31]

From Eq. (6.14-6), the activity coefficients are

$$\ln \gamma_1 = x_2^2 \left[\tau_{21} \left(\frac{G_{21}}{x_1 + x_2 G_{21}} \right)^2 + \frac{\tau_{12} G_{12}}{(x_2 + x_1 G_{12})^2} \right] \qquad (6.14\text{-}9)$$

$$\ln \gamma_2 = x_1^2 \left[\tau_{12} \left(\frac{G_{12}}{x_2 + x_1 G_{12}} \right)^2 + \frac{\tau_{21} G_{21}}{(x_1 + x_2 G_{21})^2} \right]. \qquad (6.14\text{-}10)$$

For moderately nonideal systems, the NRTL equation offers no advantages over the simpler van Laar and three-suffix Margules equations. However, for strongly nonideal mixtures, and especially for partially immiscible systems,† the NRTL equation provides a good representation of experimental data.   For example, consider the nitroethane/isooctane system studied by Renon; below 30°C, this system has a miscibility gap. Reduction of liquid-liquid equilibrium data below 30°C and vapor-liquid equilibrium data at 25 and 45°C, gave the results shown in Fig. 6-16.   The parameters $(g_{12} - g_{22})$ and $(g_{21} - g_{11})$ appear to be linear functions of temperature, showing no discontinuities in the region of the critical solution temperature.

Renon's and Wilson's equations are readily generalized to multi-component mixtures as discussed in Sec. 6.18.

6.15 Excess Functions and Partial Miscibility

In the previous sections we have been concerned with mixtures of liquids which are completely miscible.   We now want to consider briefly the thermodynamics of binary liquid systems wherein the components are only partially miscible.‡

At a fixed temperature and pressure, a stable state is that which has a minimum Gibbs energy.   Thermodynamic stability analysis tells us that a

†See App. V (p. 473).
‡See also App. V.

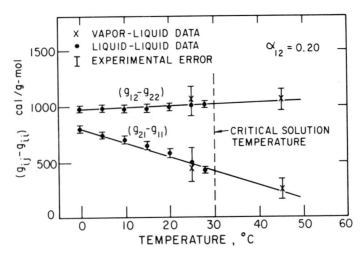

Fig. 6-16   Parameters in Renon's equation for the nitroethane (1)/
isooctane (2) system calculated from vapor-liquid and
liquid-liquid equilibrium data.

liquid mixture splits into two separate liquid phases if upon doing so, it can lower its Gibbs energy. To fix ideas, let us consider a mixture of two liquids, 1 and 2, whose Gibbs energy of mixing at constant temperature and pressure is given by the heavy line in Fig. 6-17. If the composition of the mixture is that corresponding to point a, then the molar Gibbs

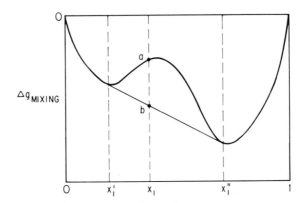

Fig. 6-17   Molar Gibbs energy of mixing of a binary,
partially miscible liquid mixture at constant
temperature and pressure.

energy of a mixture is:

$$g_{mixture \atop (at\ a)} = x_1 g_{pure\ 1} + x_2 g_{pure\ 2} + \Delta g_a. \qquad (6.15\text{-}1)$$

However, if the mixture splits into two separate liquid phases, one having mole fraction x_1' and the other having mole fraction x_1'', then the Gibbs energy change upon mixing is given by point b and the molar Gibbs energy of the two-phase mixture is:

$$g_{mixture \atop (at\ b)} = x_1 g_{pure\ 1} + x_2 g_{pure\ 2} + \Delta g_b. \qquad (6.15\text{-}2)$$

The mole fractions x_1 and x_2 in Eq. (6.15-2) represent the overall composition and they are the same as those in Eq. (6.15-1).

It is evident from Fig. 6-17 that point b represents a lower Gibbs energy of the mixture than does point a, and as a result the liquid mixture having overall composition x_1 splits into two liquid phases having mole fractions x_1' and x_1''. Point b represents the lowest possible Gibbs energy which the mixture may attain subject to the restraints of fixed temperature, pressure, and overall composition x_1.

A decrease in the Gibbs energy of a binary liquid mixture due to the formation of another liquid phase can occur only if a plot of the Gibbs energy change of mixing against mole fraction is, in part, concave downward. Therefore, the condition for instability of a binary liquid mixture is:

$$\left(\frac{\partial^2 g_{mixture}}{\partial x^2}\right)_{T,\ P} < 0, \qquad (6.15\text{-}3)\dagger$$

or

$$\left(\frac{\partial^2 \Delta g_{mixing}}{\partial x^2}\right)_{T,\ P} < 0. \qquad (6.15\text{-}4)\dagger$$

Let us now introduce an excess function into Eq. (6.15-3). As before, we define the excess Gibbs energy of a mixture relative to the Gibbs energy of an ideal mixture in the sense of Raoult's law:

$$g^E \equiv g_{mixture} - RT[x_1 \ln x_1 + x_2 \ln x_2]$$
$$- x_1 g_{pure\ 1} - x_2 g_{pure\ 2}. \qquad (6.15\text{-}5)$$

†In Eqs. (6.15-3) and (6.15-4) x stands for either x_1 or x_2.

Substituting into Eq. (6.15-3) we obtain for instability:

$$\left(\frac{\partial^2 g^E}{\partial x_1^2}\right)_{T,\,P} + RT\left(\frac{1}{x_1} + \frac{1}{x_2}\right) < 0. \qquad (6.15\text{-}6)$$

For an ideal solution $g^E = 0$ for all x and in that event the inequality is never obeyed for any values of x_1 and x_2 in the interval zero to one. Therefore, we conclude that an ideal solution is always stable and cannot exhibit phase splitting.

Suppose now that the excess Gibbs energy is not zero but is given by the simple expression

$$g^E = A x_1 x_2, \qquad (6.15\text{-}7)$$

where A is a constant dependent on temperature. Then

$$\left(\frac{\partial^2 g^E}{\partial x_1^2}\right)_{T,\,P} = -2A, \qquad (6.15\text{-}8)$$

and substitution in Eq. (6.15-6) gives

$$-2A < -RT\left(\frac{1}{x_1} + \frac{1}{x_2}\right). \qquad (6.15\text{-}9)$$

Multiplying both sides by minus one inverts the inequality sign, and the condition for instability becomes

$$2A > RT\left(\frac{1}{x_1} + \frac{1}{x_2}\right) = \frac{RT}{x_1 x_2}. \qquad (6.15\text{-}10)$$

The smallest value of A which satisfies inequality (6.15-10) is

$$A = 2RT \qquad (6.15\text{-}11)$$

and, therefore, instability occurs whenever

$$\frac{A}{RT} > 2. \qquad (6.15\text{-}12)$$

The borderline between stability and instability of a liquid mixture is called *incipient instability*. This condition corresponds to a critical state and it occurs when the two points of inflection shown in Fig. 6-17 merge into a single point. Incipient instability, therefore, is characterized by the two equations

$$\left(\frac{\partial^2 g_{\text{mixture}}}{\partial x^2}\right)_{T,\,P} = 0 \tag{6.15-13}$$

and

$$\left(\frac{\partial^3 g_{\text{mixture}}}{\partial x^3}\right)_{T,\,P} = 0. \tag{6.15-14}$$

An equivalent, but more useful characterization of incipient instability is provided by introducing into Eqs. (6.15-13) and (6.15-14) the activity function given by

$$g_{\text{mixture}} = RT\,[x_1 \ln a_1 + x_2 \ln a_2]$$
$$+ x_1 g_{\text{pure 1}} + x_2 g_{\text{pure 2}}. \tag{6.15-15}$$

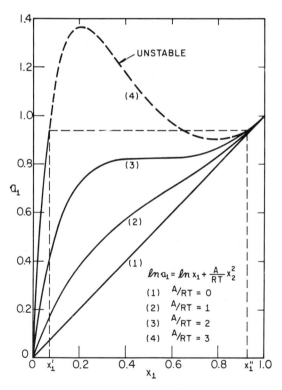

Fig. 6-18 Activity of component 1 in a binary liquid solution for different values of A/RT. Curve (3) shows incipient instability.

We then obtain for incipient instability

$$\left(\frac{\partial \ln a_1}{\partial x_1}\right)_{T, P} = 0 \tag{6.15-16}$$

and

$$\left(\frac{\partial^2 \ln a_1}{\partial x_1^2}\right)_{T, P} = 0. \tag{6.15-17}$$

We now want to illustrate graphically instability, incipient instability, and stability in a binary liquid mixture. Figure 6-18 gives a plot of activity versus mole fraction as calculated from the simple excess Gibbs energy expression given by Eq. (6.15-7); the activity is given by

$$\ln a_1 = \ln \gamma_1 + \ln x_1 = \frac{A}{RT} x_2^2 + \ln x_1. \tag{6.15-18}$$

When $A/RT > 2$, the curve has a maximum and a minimum; for this case there are two stable liquid phases whose compositions are given by x_1' and x_1'' as shown schematically in Fig. 6-18. When $A/RT = 2$, the maximum and minimum points coincide and we have incipient instability. For $A/RT < 2$, only one liquid phase is stable. A plot similar to that in Fig. 6-18 is shown in Fig. 6-19, which presents experimental results re-

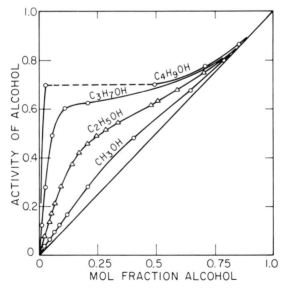

Fig. 6-19   Activities of four alcohols in aqueous solution at 25°C.

ported by Butler[32] for four binary aqueous systems containing methyl, ethyl, n-propyl or n-butyl alcohol at 25°C. Methyl and ethyl alcohols are completely miscible with water and when their activities are plotted against their mole fractions, there is no point of inflection. Propyl alcohol is also completely miscible with water at this temperature but just barely so; the plot of activity versus mole fraction almost shows a point of inflection. Butyl alcohol, however, is miscible with water over only small ranges of concentration, and if a continuous line were drawn through the experimental points shown in Fig. 6-19 it would necessarily have to go through a maximum and a minimum, in complete analogy to an equation-of-state isotherm on P-V coordinates in the two-phase region.

6.16 Upper and Lower Consolute Temperatures

As indicated in the previous section, the condition for instability of a binary liquid mixture depends on the nonideality of the solution and on the temperature. In the simplest case, when the excess Gibbs energy is given by a one-parameter equation such as Eq. (6.15-7), the temperature T^c at which incipient instability occurs is given by

$$T^c = \frac{A}{2R}. \qquad (6.16\text{-}1)$$

The temperature T^c is called the *consolute temperature*;† when the excess Gibbs energy is given by a temperature-independent one-parameter Margules equation, T^c is always a maximum but in general, it may be a maximum (upper) or a minimum (lower) temperature on a T-x diagram as shown in Fig. 6-20. Some binary systems have both upper and lower consolute temperatures. Upper critical solution temperatures are much more common than lower critical solution temperatures although the latter are sometimes observed in mixtures of components which form hydrogen bonds with one another (e.g., aqueous mixtures of amines). In many simple liquid mixtures, the parameter A is a weak function of temperature and therefore vapor-liquid equilibrium measurements obtained at some temperature not too far removed from T^c may be used to estimate the consolute temperature. However, it sometimes happens that A depends on the temperature in a complicated way and, unless data are taken very near T^c, Eq. (6.16-1) provides only an approximation even in those cases where the simple one-parameter equation is adequate for the excess Gibbs energy. Figure 6-21 illustrates three possible cases of phase stability corresponding to three types of temperature dependence

†Or, alternatively, the critical solution temperature.

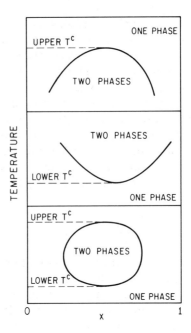

Fig. 6-20   Phase stability in three binary liquid mixtures.

for the parameter A. In the first case T^c is a maximum and in the second case it is a minimum; in the third case there is first a minimum and then a maximum. These three cases are not the only ones possible; for example, a plot of A/RT may cross the dotted line ($A/RT = 2$) twice by going through a minimum rather than a maximum as shown. In that event we again have an upper and lower consolute temperature but the interval between these two limiting temperatures now corresponds to a region of complete miscibility, with incomplete miscibility at temperatures above and below this interval.

When the excess Gibbs energy is given by Eq. (6.15-7), we always find that the composition corresponding to the consolute temperature is $x_1 = x_2 = \frac{1}{2}$. However, when the excess Gibbs energy is given by a function which is not symmetric with respect to x_1 and x_2, the coordinates of the consolute point are not at the composition midpoint.

For example, if the excess Gibbs energy is given by van Laar's equation in the form

$$g^E = \frac{x_1 x_2 A}{\dfrac{A}{B} x_1 + x_2}, \qquad (6.16\text{-}2)$$

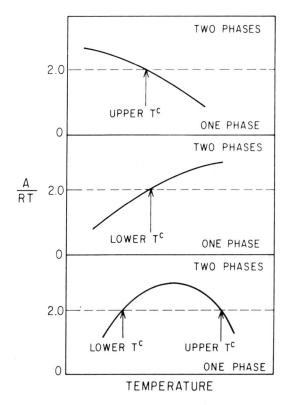

Fig. 6-21 Phase stability in three binary liquid mixtures whose excess Gibbs energy is given by a two-suffix Margules equation.

we find, upon substitution into Eqs. (6.15-16) and (6.15-17) that the coordinates of the consolute point are:

$$
\left.
\begin{aligned}
T_c &= \frac{2x_1 x_2 \dfrac{A^2}{B}}{R\left(\dfrac{A}{B} x_1 + x_2\right)^3} \\[2em]
x_1 &= \frac{\left[\left(\dfrac{A}{B}\right)^2 + 1 - \dfrac{A}{B}\right]^{1/2} - \dfrac{A}{B}}{1 - \dfrac{A}{B}}.
\end{aligned}
\right\} \qquad (6.16\text{-}3)
$$

In the previous section we indicated that when the excess Gibbs energy is assumed to follow a two-suffix Margules expression as given by Eq.

(6.15-7), incipient instability occurs when $A = 2RT$. A two-suffix Margules equation, however, is only a rough approximation for many real mixtures and a more accurate description is given by including higher terms. If we write the excess Gibbs energy in the three-parameter Redlich-Kister series:

$$g^E = x_1 x_2 [A + B(x_1 - x_2) + C(x_1 - x_2)^2],\qquad (6.16\text{-}4)$$

and substitute this series into the equations of incipient instability, we obtain the results given in Figs. 6-22 and 6-23 taken from the work of

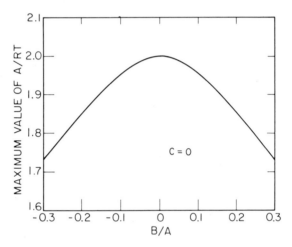

Fig. 6-22   Maximum values of A/RT for complete miscibility.

Shain.[33] Figure 6-22 gives the effect of the coefficient B (when $C = 0$) on the maximum value of A/RT for complete miscibility. The coefficient B reflects the asymmetry of the excess Gibbs energy function (see Sec. 6.5), and we see that the maximum permissible value of A/RT decreases below 2 as the asymmetry rises. Figure 6-23 gives the effect of both coefficients B and C on the maximum value of A/RT for complete miscibility. As discussed in Sec. 6.5, the third term in the Redlich-Kister expansion is symmetric in the mole fraction but affects the flatness or steepness of the excess Gibbs energy curve. Positive values of C make the excess Gibbs energy curve flatter, and from Fig. 6-23 we see that for small values of B, positive values of C increase the maximum permissible value of A/RT beyond 2 up to 2.4.

Shain's calculations show that whereas large positive values of the constant A favor limited miscibility, values of the constant B of either sign increase the tendency toward limited miscibility for a given value

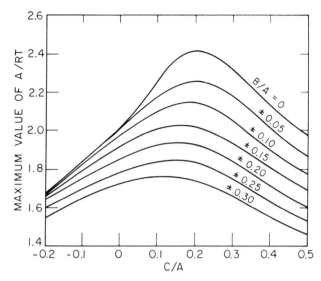

Fig. 6-23   Maximum values of A/RT for complete miscibility.

of A; however, small positive values of the constant C tend to decrease the tendency for phase separation. These calculations offer useful guidelines but must not be taken too seriously since they are based on a twofold differentiation of the excess Gibbs energy function. Extremely accurate data are required in order to assign precise quantitative significance to a function based on the second derivative of such data.

In concluding this section we should note that the thermodynamics of phase stability in binary liquid systems does not require that instability can occur only if the excess Gibbs energy is positive. In principle, it is possible that a binary liquid mixture may be only partially miscible even though it has a negative excess Gibbs energy. However, such behavior is unlikely since the composition-dependence of a negative excess Gibbs energy would have to be very unusual in order to satisfy the condition of instability, Eq. (6.15-6).

Some further discussion on liquid-liquid equilibria is given in App. V (p. 473).

6.17   Excess Functions for
Multicomponent Mixtures

So far in this chapter, we have primarily considered binary mixtures. We now turn to a discussion of mixtures containing more than two components, with particular attention to their excess Gibbs energies.

One of the main uses of excess functions for describing the thermo-dynamic properties of liquid mixtures lies in establishing thermo-dynamically consistent relationships for multicomponent mixtures containing any desired number of components; from these relationships we can then calculate activity coefficients needed to find liquid-phase fugacities. Expressions for the excess functions require a number of constants and many of these can be evaluated from binary data alone; in some cases it is possible to obtain all the required constants from binary data. Application of excess functions to mixtures of more than two components, therefore, is a labor-saving device which minimizes the experimental data required to describe a mixture of many components.

To illustrate the utility of excess functions, we consider first a ternary mixture as treated by Wohl's method (see Sec. 6.13). Extension to systems containing more than three components will then be evident.

As in the case of a binary solution, the Gibbs energy is again a summation of two-body, three-body, etc., interactions. When the excess Gibbs energy is relative to an ideal solution in the sense of Raoult's law, we have:

$$\frac{g^E}{RT(x_1 q_1 + x_2 q_2 + x_3 q_3)} = 2a_{12} z_1 z_2 + 2a_{13} z_1 z_3 + 2a_{23} z_2 z_3$$

$$+ 3a_{112} z_1^2 z_2 + 3a_{122} z_1 z_2^2 + 3a_{113} z_1^2 z_3$$

$$+ 3a_{133} z_1 z_3^2 + 3a_{223} z_2^2 z_3 + 3a_{233} z_2 z_3^2$$

$$+ 6a_{123} z_1 z_2 z_3 + \dots \qquad (6.17\text{-}1)$$

First, consider a simple case: Suppose that components 1, 2, and 3 are chemically similar and of approximately the same size. We assume that $q_1 = q_2 = q_3$ and that all three-body terms (and higher) may be neglected. Equation (6.17-1) simplifies to:

$$\frac{g^E}{RT} = 2q a_{12} x_1 x_2 + 2q a_{13} x_1 x_3 + 2q a_{23} x_2 x_3. \qquad (6.17\text{-}2)$$

The activity coefficients follow from differentiation as given by Eq. (6.3-8). They are:

$$\ln \gamma_1 = A'_{12} x_2^2 + A'_{13} x_3^2 + (A'_{12} + A'_{13} - A'_{23}) x_2 x_3 \qquad (6.17\text{-}3)$$

$$\ln \gamma_2 = A'_{12} x_1^2 + A'_{23} x_3^2 + (A'_{12} + A'_{23} - A'_{13}) x_1 x_3 \qquad (6.17\text{-}4)$$

$$\ln \gamma_3 = A'_{13} x_1^2 + A'_{23} x_2^2 + (A'_{13} + A'_{23} - A'_{12}) x_1 x_2 \qquad (6.17\text{-}5)$$

where $A'_{12} = 2q a_{12}$, $A'_{13} = 2q a_{13}$, and $A'_{23} = 2q a_{23}$.

Equations (6.17-3), (6.17-4), and (6.17-5) possess a great advantage: All the constants may be obtained from binary data without further assumptions. The constant $2qa_{12}$ is given by data for the 1-2 binary; the constants $2qa_{13}$ and $2qa_{23}$ are given, respectively, by data for the 1-3 and 2-3 binaries. Thus, by assumption, no ternary data at all are required to calculate activity coefficients for the ternary mixture.

A somewhat more realistic, but still simplified, model of a ternary solution is provided by again assuming that all three-body terms (and higher) in Eq. (6.17-1) may be neglected, but this time we do not assume equality of all the q terms. This treatment leads to the van Laar equations for a ternary mixture. The molar excess Gibbs energy is given by:

$$\frac{g^E}{RT} = \frac{2q_2 a_{12} x_1 x_2 + 2q_3 a_{13} x_1 x_3 + 2 \dfrac{q_2 q_3}{q_1} a_{23} x_2 x_3}{x_1 + \dfrac{q_2}{q_1} x_2 + \dfrac{q_3}{q_1} x_3}.$$ (6.17-6)

To simplify notation, let

$$A'_{12} = 2q_1 a_{12} \qquad A'_{21} = 2q_2 a_{12}$$
$$A'_{13} = 2q_1 a_{13} \qquad A'_{31} = 2q_3 a_{13}$$
$$A'_{23} = 2q_2 a_{23} \qquad A'_{32} = 2q_3 a_{23}$$

Upon differentiation according to Eq. (6.3-8) the activity coefficient of component 1 is:

$$\ln \gamma_1 = \frac{x_2^2 A'_{12}\left(\dfrac{A'_{21}}{A'_{12}}\right)^2 + x_3^2 A'_{13}\left(\dfrac{A'_{31}}{A'_{13}}\right)^2 + x_2 x_3 \dfrac{A'_{21}}{A'_{12}} \dfrac{A'_{31}}{A'_{13}}\left(A'_{12} + A'_{13} - A'_{32}\dfrac{A'_{13}}{A'_{31}}\right)}{\left(x_1 + x_2 \dfrac{A'_{21}}{A'_{12}} + x_3 \dfrac{A'_{31}}{A'_{13}}\right)^2}.$$ (6.17-7)

Expressions for γ_2 and γ_3 are of exactly the same form as that for γ_1. To obtain γ_2, Eq. (6.17-7) should be used with this change of all subscripts on the right-hand side: Replace 1 with 2; replace 2 with 3; and replace 3 with 1. To obtain γ_3, replace 1 with 3; replace 2 with 1; and replace 3 with 2. Notice that if $q_1 = q_2 = q_3$, then Eq. (6.17-7) reduces to Eq. (6.17-3).

All the parameters in Eq. (6.17-7) may be obtained from binary data as indicated by Eqs. (6.13-4) and (6.13-5).

The three-suffix Margules equations can also be extended to a ternary mixture by Wohl's method but now all the constants cannot be found from binary data alone unless an additional assumption is made. We assume that the q's are all equal to one another but we retain three-body terms in the expansion for excess Gibbs energy; higher-body terms are neglected. Equation (6.17-1) becomes:

$$\frac{g^E}{RT} = 2qa_{12}x_1x_2 + 2qa_{13}x_1x_3 + 2qa_{23}x_2x_3 + 3qa_{112}x_1^2x_2$$
$$+ 3qa_{122}x_1x_2^2 + 3qa_{113}x_1^2x_3 + 3qa_{133}x_1x_3^2 + 3qa_{223}x_2^2x_3$$
$$+ 3qa_{233}x_2x_3^2 + 6qa_{123}x_1x_2x_3. \tag{6.17-8}$$

All the constants appearing in Eq. (6.17-8) can be obtained from binary data except the last one, qa_{123}. This last constant is characteristic of the interaction between three different molecules: one of component 1, one of component 2, and one of component 3. It is a true ternary constant and, in principle, can be obtained only from ternary data.

To simplify notation, let:

$$A'_{12} = q(2a_{12} + 3a_{122}) \qquad A'_{31} = q(2a_{13} + 3a_{113})$$
$$A'_{21} = q(2a_{12} + 3a_{112}) \qquad A'_{23} = q(2a_{23} + 3a_{233})$$
$$A'_{13} = q(2a_{13} + 3a_{133}) \qquad A'_{32} = q(2a_{23} + 3a_{223})$$

and

$$Q' = \frac{3q}{2}[a_{122} + a_{112} + a_{133} + a_{113} + a_{233} + a_{223} - 4a_{123}].$$

All constants of the type A'_{ij} can be determined from binary data alone.† However, the constant Q' requires information on the ternary mixture because it is a function of a_{123}.

The activity coefficient for component 1 is given by:

$$\ln \gamma_1 = A'_{12}x_2^2(1 - 2x_1) + 2A'_{21}x_1x_2(1 - x_1) + A'_{13}x_3^2(1 - 2x_1)$$
$$+ 2A'_{31}x_1x_3(1 - x_1) - 2A'_{23}x_2x_3^2 - 2A'_{32}x_2^2x_3$$
$$+ [\tfrac{1}{2}(A'_{12} + A'_{21} + A'_{13} + A'_{23} + A'_{32}) - Q'](x_2x_3 - 2x_1x_2x_3).$$
$$\tag{6.17-9}$$

Expressions for γ_2 and γ_3 can be obtained from Eq. (6.17-9) by a change of all subscripts on the right-hand side. For γ_2 replace 1 with 2; replace 2 with 3; and 3 with 1. For γ_3 replace 1 with 3; 2 with 1; and 3 with 2.

If no ternary data are available it is possible to estimate a_{123} by a suitable assumption. A reasonable but essentially arbitrary assumption is to set $Q' = 0$. An extensive study of Eq. (6.17-9) has been made by Adler, Friend, and Pigford.[34]

---

†Notice that the constants A'_{12} and A'_{21} are simply related to the constants A' and B' as defined after Eqs. (6.13-6) and (6.13-7) for the case where four-body (and higher) interactions are neglected. The relations are: $A'_{12} = A' + B'$ and $A'_{21} = A' + \tfrac{1}{2}B'$. Similar relations can be written for the 1-3 and 2-3 mixtures.

In principle only one experimental ternary point is required to determine qa_{123}. In practice, however, it is not advisable to base a parameter on one point only. For accurate work it is best to measure vapor-liquid equilibria for several ternary compositions which, in addition to the binary data, can then be used to evaluate a truly representative ternary constant.

The paragraphs above have shown how Wohl's method may be used to derive expressions for activity coefficients in a ternary mixture; exactly the same principles apply for obtaining expressions for activity coefficients containing four, five, or more components. The generalization of Eqs. (6.17-2) and (6.17-6) to mixtures containing any number of components shows that all the constants may be calculated from binary data alone. However, the generalization of Eq. (6.17-8) to solutions containing any number of components shows that the constants appearing in the expressions for the activity coefficients must be found from data on all possible constituent ternaries as well as binaries. For the generalization of Eq. (6.17-8) data on quaternary, quinternary, ... mixtures are not needed.

To illustrate the applicability of the three-suffix Margules equation to ternary systems, we consider three strongly nonideal ternary mixtures at 50°C studied by Severns, et al.[23] They are the systems:

> I. Acetone/methyl acetate/methanol
> II. Acetone/chloroform/methanol
> III. Acetone/carbon tetrachloride/methanol.

Margules constants for the three ternary systems are given in Table 6-6. For System I a good representation of the ternary data was obtained

Table 6-6

THREE-SUFFIX MARGULES CONSTANTS FOR THREE
TERNARY SYSTEMS AT 50°C†

System	Margules Constants	
Acetone (1)/Methyl Acetate (2)/Methanol (3)	$A'_{12} = 0.149$	$A'_{21} = 0.115$
	$A'_{13} = 0.701$	$A'_{31} = 0.519$
	$A'_{23} = 1.07$	$A'_{32} = 1.02$
	$Q' = 0$	
Acetone (1)/Chloroform (2)/Methanol (3)	$A'_{12} = 0.83$	$A'_{21} = -0.69$
	$A'_{13} = 0.701$	$A'_{31} = 0.519$
	$A'_{23} = 0.715$	$A'_{32} = 1.80$
	$Q' = -0.368$	
Acetone (1)/Carbon Tetrachloride (2)/Methanol (3)	$A'_{12} = 0.715$	$A'_{21} = 0.945$
	$A'_{13} = 0.701$	$A'_{31} = 0.519$
	$A'_{23} = 1.76$	$A'_{32} = 2.52$
	$Q' = 1.15$	

†W. H. Severns, A. Sesonske, R. H. Perry, and R. L. Pigford, *A. I. Ch. E. Journal*, **1**, 401 (1955).

by using binary data only and setting $Q' = 0$ in Eq. (6.17-9). In System II the ternary data required a small but significant ternary constant $Q' = -0.368$ which, if it had been neglected, would introduce some, but not serious, error. However, in System III the ternary data required an appreciable ternary constant $Q' = 1.12$ which cannot be neglected since it is of the same order of magnitude as the various A'_{ij} constants for this system.

6.18   Equations of Wilson and Renon
for Multicomponent Mixtures

One of the advantages of both Wilson's equation and Renon's equation for the excess Gibbs energy is that they may be extended to as many components as desired without any additional assumptions and without introducing any constants other than those obtained from binary data.

For a solution of m components Wilson's equation is

$$\frac{g^E}{RT} = -\sum_{i=1}^{m} x_i \ln \left[\sum_{j=1}^{m} x_j \Lambda_{ij} \right], \qquad (6.18\text{-}1)$$

where

$$\Lambda_{ij} \equiv \frac{v_j}{v_i} \exp - \left[\frac{(\lambda_{ij} - \lambda_{ii})}{RT} \right] \qquad (6.18\text{-}2)$$

$$\Lambda_{ji} \equiv \frac{v_i}{v_j} \exp - \left[\frac{(\lambda_{ij} - \lambda_{jj})}{RT} \right]. \qquad (6.18\text{-}3)$$

The activity coefficient for any component k is given by:

$$\ln \gamma_k = - \ln \left[\sum_{j=1}^{m} x_j \Lambda_{kj} \right] + 1 - \sum_{i=1}^{m} \frac{x_i \Lambda_{ik}}{\sum_{j=1}^{m} x_j \Lambda_{ij}}. \qquad (6.18\text{-}4)$$

Equation (6.18-4) requires only parameters which can be obtained from binary data; for each possible binary pair in the multicomponent solution, two parameters are needed.

Orye[27] has tested Eq. (6.18-4) for a variety of ternary mixtures, using only binary data, and finds that for most cases good results are obtained. For example, Fig. 6-24 gives a comparison between calculated and observed vapor compositions for the acetone/methyl acetate/methanol system at 50°C. A similar comparison is also shown for calculations based on the van Laar equation. No ternary data were used in either cal-

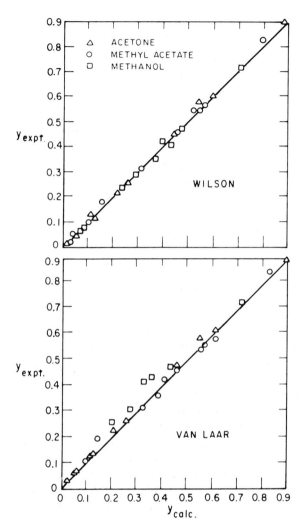

Fig. 6-24   Experimental and calculated vapor compositions for the ternary system acetone/methyl acetate/ methanol at 50°C. Calculations based on binary data only.

culation; the binary constants used are given in Table 6-7. For this ternary system the Wilson equations give a much better prediction than the van Laar equations but, as indicated in Table 6-6, the three-suffix Margules equations (using binary data only) can also give a good prediction since no ternary constant is required.

Similar calculations for the ternary system acetone/methanol/chloro-

Table 6-7

PARAMETERS FOR THE WILSON AND
VAN LAAR EQUATIONS
[Acetone (1)/Methyl Acetate (2)/
Methanol (3) at 50°C][†]

Wilson Equation	van Laar Equation[‡]
$\Lambda_{12} = 0.5781$	$A'_{12} = 0.1839$
$\Lambda_{21} = 1.3654$	$A'_{21} = 0.1106$
$\Lambda_{23} = 0.6370$	$A'_{23} = 0.9446$
$\Lambda_{32} = 0.4871$	$A'_{32} = 1.0560$
$\Lambda_{13} = 0.6917$	$A'_{13} = 0.5965$
$\Lambda_{31} = 0.7681$	$A'_{31} = 0.5677$

[†] W. H. Severns, A. Sesonske, R. H. Perry
and R. L. Pigford, *A. I. Ch. E. Journal*, **1**, 401
(1955).
[‡] Eq. (6.17-7).

form are shown in Fig. 6-25, and again Wilson's equations, based on binary data only, give a better prediction than van Laar's equations. However, for this system the three-suffix Margules equations cannot give as good a prediction if only binary data are used because, as shown in Table 6-6, a significant ternary constant is required.

A final example of the applicability of Wilson's equation is provided by Orye's calculations for the system ethanol/methylcyclopentane/benzene at one atmosphere. Wilson parameters were found from experimental data for the three binary systems;[35] vapor compositions in the ternary system were then calculated for six cases and compared with experimental results as shown in Table 6-8. Wilson's equation again provides a good description for this ternary which has large deviations from ideal behavior.

For a solution of m components, Renon's equation is:

$$\frac{g^E}{RT} = \sum_{i=1}^{m} x_i \frac{\displaystyle\sum_{j=1}^{m} \tau_{ji} G_{ji} x_j}{\displaystyle\sum_{l=1}^{m} G_{li} x_l} \tag{6.18-5}$$

where

$$\tau_{ji} = \frac{(g_{ji} - g_{ii})}{RT} \qquad (g_{ji} = g_{ij}) \tag{6.18-6}$$

$$G_{ji} = \exp\left(-\alpha_{ji}\tau_{ji}\right) \qquad (\alpha_{ji} = \alpha_{ij}) . \tag{6.18-7}$$

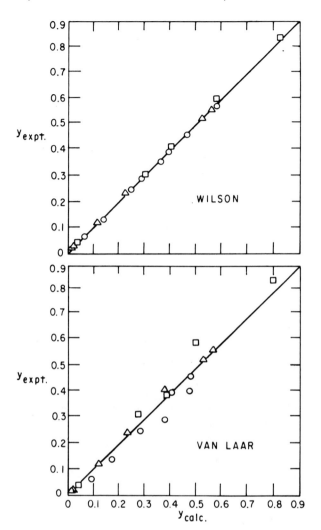

Fig. 6-25 Experimental and calculated vapor compositions for the ternary system acetone/methanol/chloroform at 50° C. Calculations based on binary data only.

The activity coefficient for a component k is given by:

$$\ln \gamma_k = \frac{\sum\limits_{j=1}^{m} \tau_{ji} G_{ji} x_j}{\sum\limits_{l=1}^{m} G_{li} x_l} + \sum\limits_{j=1}^{m} \frac{x_j G_{ij}}{\sum\limits_{l=1}^{m} G_{lj} x_l} \left(\tau_{ij} - \frac{\sum\limits_{r=1}^{m} x_r \tau_{rj} G_{rj}}{\sum\limits_{l=1}^{m} G_{lj} x_l} \right). \tag{6.18-8}$$

Table 6-8

CALCULATED VAPOR COMPOSITIONS FOR A THREE-
COMPONENT SYSTEM USING PARAMETERS OBTAINED
FROM BINARY DATA ONLY
[Ethanol (1)/Methylcyclopentane (2)/Benzene (3) at 1 atm]

$T(^\circ K)$	Component	Experimental†		Calculated y (Wilson)
		x	y	
336.15	1	0.047	0.258	0.258
	2	0.845	0.657	0.661
	3	0.107	0.084	0.081
338.85	1	0.746	0.497	0.502
	2	0.075	0.232	0.223
	3	0.178	0.271	0.275
335.85	1	0.690	0.432	0.434
	2	0.182	0.403	0.401
	3	0.128	0.165	0.165
340.85	1	0.878	0.594	0.603
	2	0.068	0.296	0.283
	3	0.053	0.110	0.114
337.15	1	0.124	0.290	0.300
	2	0.370	0.365	0.365
	3	0.505	0.345	0.335
334.05	1	0.569	0.386	0.383
	2	0.359	0.538	0.542
	3	0.071	0.076	0.075

Wilson Parameters:

$\lambda_{12} - \lambda_{11} = 2204.9 \, \text{cal/g-mol}$ $\lambda_{13} - \lambda_{33} = 125.3 \, \text{cal/g-mol}$
$\lambda_{12} - \lambda_{22} = 245.2$ $\lambda_{23} - \lambda_{22} = 13.3$
$\lambda_{13} - \lambda_{11} = 1389.2$ $\lambda_{23} - \lambda_{33} = 248.7$

†J. E. Sinor and J. H. Weber, *J. Chem. Eng. Data*, **5**, 243 (1960).

Equations (6.18-5) and (6.18-8) contain only parameters obtained from binary data.

For nine ternary systems shown in Table 6-9, Renon[31] predicted ternary vapor equilibria with Eq. (6.18-5) using binary data only. He also calculated ternary equilibria with Wohl's equation [Eq. (6.17-9)] both with and without a ternary constant. Table 6-9 indicates that Eq. (6.18-5) gives a good prediction of multicomponent equilibria from binary equilibrium data alone.

A particularly sensitive test of Renon's equation is provided by ternary liquid-liquid equilibrium calculations. For example, for the system chloroform/acetone/water, Renon extrapolated to lower temperatures parameters obtained from binary vapor-liquid data. He then cal-

Table 6-9

COMPARISON OF RENON'S AND WOHL'S EQUATIONS FOR PREDICTION
OF TERNARY VAPOR-LIQUID EQUILIBRIA

| System | Mean Arithmetic Deviation in Individual Components Vapor Mole Fraction × 1000 | | 95% Confidence Limits in Vapor Mole Fraction × 1000 | |
	Renon (with no ternary constant)	Wohl† (with best ternary constant)	Renon (with no ternary constant)	Wohl† (with best ternary constant)
n-Heptane	3	3	2	8
Toluene	2	− 4	1	5
Methyl ethyl ketone	− 5	1	2	8
n-Heptane	4	0	3	4
Benzene	2	8	4	7
Ethanol (760 mm Hg)	− 6	− 8	6	8
n-Heptane	5	0	7	4
Benzene	3	− 5	4	7
Ethanol (400 mm Hg)	− 7	5	9	8
n-Heptane	− 5	8	4	14
Toluene	− 3	− 2	5	8
Methanol	8	− 6	8	19
Benzene	− 1	− 13	5	22
Carbon tetrachloride	− 3	3	4	20
Methanol (35°C)	4	10	7	39
Benzene	− 3	− 15	3	21
Carbon tetrachloride	− 2	7	4	13
Methanol (55°C)	5	8	7	29
Acetone	− 5	− 11	4	18
Chloroform	− 3	11	4	8
Methanol	8	0	3	12
Acetone	− 4	− 9	3	12
Methanol	1	8	7	15
Methyl acetate	3	1	5	8
Ethanol	− 4	− 6	7	22
Ethyl acetate	5	1	22	57
Water	1	5	17	49

†Eq. (6.17-9).

culated the coexistence curve and tie lines for the ternary system at 25°C with results as shown in Fig. 6-26. The calculations disagree somewhat with experimental observation near the plait point but elsewhere agreement is satisfactory.

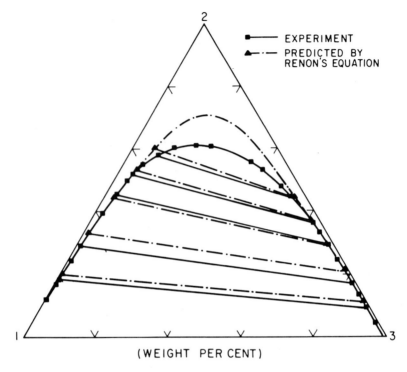

Fig. 6-26   Calculated and observed liquid-liquid equilibria for the system chloroform (1)/acetone (2)/water (3) at 25°C.

6.19   Computer Calculation of Multicomponent
Vapor-Liquid Equilibria

In Chap. 5 we discussed methods for calculating the fugacity of a component in the vapor phase, and in this chapter we have presented techniques for calculating fugacities in liquid mixtures. We continue our discussion of liquid-phase fugacities in the chapters to follow but before we
do so, we want to consider a practical application of the material already
covered. We therefore conclude this chapter with a brief summary of how
multicomponent vapor-liquid equilibria may be estimated with the help
of a computer for rational design of separation equipment such as distillation columns. A more complete discussion is given elsewhere.[36,37,38]

We consider a liquid mixture of m miscible components in equilibrium with its saturated vapor and, as indicated in Chap. 1, we are concerned with relating to each other the variables of interest: P, T, x_1,
$x_2, \ldots x_m$, y_1, $y_2, \ldots y_m$. Various possible combinations can now be constructed of those variables which are given and those which are to be

found. To fix ideas, we will take as given variables the pressure P and the liquid-phase mole fractions $x_1, x_2, \ldots x_m$. We want to find the equilibrium temperature T and the vapor-phase mole fractions $y_1, y_2, \ldots y_m$. We have then, $m + 1$ unknowns, and in order to find them, we require $m + 1$ independent equations. These are, first, m equations of equilibrium, one for each component i:

$$f_i^V = f_i^L \qquad (6.19\text{-}1)$$

where f_i^V is the fugacity of i in the vapor, and f_i^L is the fugacity in the liquid, and second, one stoichiometric relation:

$$\sum_i^m y_i = 1. \qquad (6.19\text{-}2)$$

Introducing the vapor-phase fugacity coefficient φ_i and the liquid-phase fugacity coefficient γ_i, Eq. (6.19-1) becomes:

$$\varphi_i\, y_i\, P = \gamma_i x_i f_i^{(P^r)} \exp \int_{pr}^{P} \frac{\bar{v}_i^L\, dP}{RT} \qquad (6.19\text{-}3)$$

In Eq. (6.19-3), $f_i^{(P^r)}$ is the fugacity of the standard state at the reference pressure P^r; $\bar{v}_i^L$ is the partial molar volume of i in the liquid phase; and γ_i is the activity coefficient corrected to the reference pressure P^r, all at system temperature T. As used here, the activity coefficient γ_i is a function of temperature and composition, but not of total pressure P.

We now simplify the problem by restricting ourselves to low or moderate pressures and to those systems containing only subcritical components, i.e., we consider only those systems where each component in the mixture can exist as a pure liquid at the system temperature.†

The arbitrary reference pressure P^r is conveniently set equal to zero. We assume that the partial molar liquid volume $\bar{v}_i^L$ is equal to v_i^L, the pure liquid molar volume at the same temperature, and that it is independent of the pressure; further, we use the symmetric normalization of activity coefficients (see Sec. 6.4) where for every component i, $\gamma_i \rightarrow 1$ as $x_i \rightarrow 1$. Equation (6.19-3) now becomes

$$\varphi_i\, y_i\, P = \gamma_i x_i f_{\text{pure } i}^{(P0)} \exp \frac{v_i^L\, P}{RT}, \qquad (6.19\text{-}4)$$

where $f_{\text{pure } i}^{(P0)}$ stands for the fugacity of pure liquid i at temperature T and zero pressure.

---

†Vapor-liquid equilibria for systems containing supercritical components are considered in Chaps. 8 and 10.

The quantities $f^{(P0)}_{\text{pure } i}$ and v^L_i are properties of pure i and depend only on the temperature.

We must now decide on techniques for calculating the fugacity coefficient φ_i and the activity coefficient γ_i. At low or moderate pressures, the fugacity coefficient is given by an expression based on the virial equation truncated after the second term, as discussed in Sec. 5.4:

$$\ln \varphi_i = \frac{2}{v} \sum_{j=1}^{m} y_j B_{ij} - \ln z \qquad (6.19\text{-}5)$$

$$z = \frac{Pv}{RT} = 1 + \frac{\displaystyle\sum_{i=1}^{m} \sum_{j=1}^{m} y_i y_j B_{ij}}{v} \qquad (6.19\text{-}6)$$

where v is the molar volume of the vapor mixture, and B_{ij} is the second virial coefficient characteristic of the i-j interaction. B_{ij} is a function only of temperature for a given i-j pair.

For the activity coefficient we must choose one of the (essentially) empirical expressions discussed in the previous sections. The choice we make is somewhat arbitrary; it is dictated by convenience, by the kind of experimental data available, and by the degree of nonideality in the liquid mixture. For example, for a multicomponent mixture of nonpolar liquids where only binary data are at hand it is probably best to use the multicomponent van Laar equations. For a mixture containing some polar components where binary and some ternary data are available, it may be advantageous to use the three-suffix Margules expansion. In the general case, where some components are polar, but where there are only binary data, the most convenient procedure is to use Wilson's or Renon's equations. For our purposes here we leave the choice open and merely write

$$\ln \gamma_i = \mathcal{f}_\gamma (T, x_1, \ldots x_m), \qquad (6.19\text{-}7)$$

where $\mathcal{f}_\gamma$ is some function (van Laar, Margules, Wilson or whatever) which gives the activity coefficient of component i as a function of temperature and liquid mole fractions.

In the problem we are considering, P and x are given and T and y are unknown; this combination of known and unknown variables is frequently encountered in the design of distillation equipment. The equations available to use for finding T and y are highly nonlinear and to obtain a solution we must use an iterative procedure wherein we systematically assume values of T and y until all the equations are satisfied. Such a procedure is tedious and it is usually not practical to attempt a solution without a digital electronic computer.

To compute γ_i from Eq. (6.19-7) we require T, and to compute φ_i from

Eq. (6.19-5), we require y and T. Further, the pure-component properties $f_i^{(P0)}$ and v_i^L all depend on T. At low or moderate pressures φ_i is not far removed from unity and is a weak function of y. Therefore it is clear that the most sensitive unknown is not the vapor mole fraction but the temperature; the equations we are trying to solve are much more sensitive to T than to y.

An iterative technique for solving the equations is shown schematically in Fig. 6-27. The computer is given, as functions of temperature, the quantities $f_i^{(P0)}$ and v_i^L for each component, the function ℓ_γ, and all virial coefficients B_{ij}.

To get started, we first estimate an equilibrium temperature T. The calculations are not sensitive to this first estimate; a reasonable initial guess is to take the saturation temperature (corresponding to the total

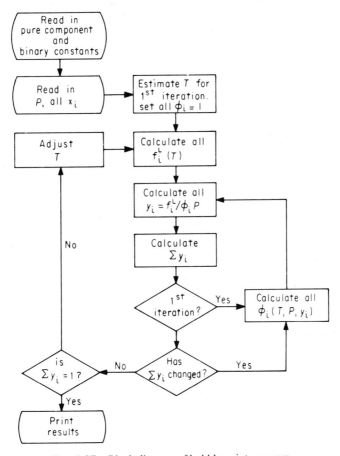

Fig. 6-27 Block diagram of bubble-point program.

pressure P) of one of the components of medium volatility. With this initial temperature and the given pressure P and mole fractions $x_1, x_2, \ldots x_m$, the computer finds all the quantities needed except the φ's, which depend also on the unknown y's. In this first iteration, we let $\varphi_i = 1$ for each i, and we find our first estimate of y by writing for each component:

$$y_i = \frac{f_i^L}{P}, \qquad (6.19\text{-}8)$$

where f_i^L is given by the right-hand side of Eq. (6.19-4). The computer then checks the first estimates of y by calculating $\sum\limits_{i=1}^{m} y_i$. In general, this summation is different from unity and therefore the first estimates are not correct. The computer now uses these first estimates of y (after normalization) to compute the fugacity coefficients φ_i according to Eq. (6.19-5); to perform this calculation the computer first finds the vapor molar volume v from Eq. (6.19-6). With these φ's, we can now obtain our second estimate of y by writing instead of Eq. (6.19-8):

$$y_i = \frac{f_i^L}{\varphi_i P}, \qquad (6.19\text{-}9)$$

where f_i^L is the same as before. The computer again finds $\sum\limits_{i=1}^{m} y_i$ and once more calculates φ's using the second estimates of y; this procedure is repeated until $\sum\limits_{i=1}^{m} y_i$ no longer changes although, in general, $\sum\limits_{i=1}^{m} y_i$ approaches a constant other than unity. If this constant exceeds unity, then it is likely that our initial guess of the temperature was too large; on the other hand, if the constant is less than unity then the assumed temperature was probably too low. The computer now repeats the entire set of calculations just described with a new estimate of T, and this iterative procedure continues until $\sum\limits_{i=1}^{m} y_i$ is not only constant but also equal to unity. When this condition is reached, all the equations are satisfied and the computer prints out the results.

The calculation we have just discussed (P-x known, T-y unknown) is commonly called a bubble-point calculation; similar procedures can be used to solve problems corresponding to other combinations of known and unknown variables. For example, if P-y are given and T-x are to be found (dew-point calculation), the computer is again given an estimate of T and proceeds to find values of x until $\sum\limits_{i=1}^{m} x_i$ is constant and equal to unity. (In this case the first estimate of x is obtained by assuming $\gamma_i = 1$ for each i.)

The method of calculation outlined here is in no way restricted to any particular choice of expressions for φ_i, γ_i, or for any particular choice of the standard state for the activity coefficient. With suitable modifications, either the symmetric or the unsymmetric convention may be used for the normalization of γ and, provided the necessary information is available, the iterative calculations can also take into consideration the variation of partial molar volumes in the liquid with composition and pressure. We shall not go into details here. Our purpose in this section is merely to indicate briefly how the material in Chaps. 5 and 6 may be brought together to solve practical problems. We return now to our discussion of liquid-phase fugacities and in the next chapter we consider to what extent molecular considerations can help us in interpreting and correlating thermodynamic behavior of liquid mixtures.

PROBLEMS

1. Consider a solution of two liquids which are not very dissimilar and which are miscible in all proportions over a wide range of temperature. The excess Gibbs energy of this solution is adequately represented by the equation

$$g^E = A x_1 x_2$$

where A is a constant depending only on temperature.

Over a wide range of temperature the ratio of the vapor pressures of the pure components is constant and equal to 1.649. Over this same range of temperature the vapor phase may be considered ideal.

We want to find out whether or not this solution has an azeotrope. Find the range of values A may have for azeotropy to occur.

2. For a certain equimolar solution, $\partial h^E / \partial T = -2$ cal/g-mol, $^\circ$K. Compute the heat capacity of the solution whose pure components each have a heat capacity of 10 cal/g-mol, $^\circ$K.

3. Consider a liquid mixture of components 1, 2, 3, and 4. The excess Gibbs energies of all the binaries formed by these components obey relations of the form

$$g_{ij}^E = A_{ij} x_i x_j$$

where A_{ij} is the constant characteristic of the i-j binary.

Derive an expression for the activity coefficient of component 1 in the quaternary solution.

4. Limited vapor-liquid equilibrium data have been obtained for a solution of two slightly dissimilar liquids A and B over the temperature range 20–100°C. From these data it is found that the variation of the limiting activity coefficients (symmetric convention) with temperature can be represented by the empirical equation

$$\ln \gamma_A^\infty = \ln \gamma_B^\infty = 0.15 + \frac{10.0}{t}$$

where t is in $°C$ and γ^∞ is the activity coefficient at infinite dilution. Estimate as best you can the heat of mixing of an equimolar mixture of A and B at $60°C$.

5. Total-pressure data are available for the entire concentration range of a binary solution at constant temperature. At the composition $x_1 = a$, the total pressure is a maximum. Show that at the composition $x_1 = a$, this solution has an azeotrope, i.e., that the relative volatility at this composition is unity. Assume that the vapor phase is ideal.

6. Experimental studies have been made on the isothermal vapor-liquid equilibrium of a ternary mixture. The measured quantities are liquid mole fractions x_1, x_2, x_3; vapor mole fractions y_1, y_2, y_3; total pressure P; and absolute temperature T. From these measurements indicate how to calculate the activity coefficient of component 1. All components are liquids at temperature T; the saturation pressure of component 1 is designated by P_1^s. The molar liquid volume of component 1 is v_1^L.

For the vapor phase, assume that the volume-explicit, truncated virial equation of state holds:

$$z = 1 + \frac{B_{\text{mixture}} \, P}{RT}.$$

All necessary virial coefficients B_{ij} $(i = 1, 2, 3,; j = 1, 2, 3)$ are available.

7. From the total-pressure data below compute the y-x diagram for ethyl alcohol/chloroform at $45°C$. Assume ideal gas behavior.

TOTAL PRESSURE DATA FOR THE SYSTEM
ETHYL ALCOHOL (1)/CHLOROFORM (2)
AT $45°C$

x_2	P(mm Hg)	x_1	P (mm Hg)
0	172.8	0	433.5
0.05	200	0.01	438.5
0.10	233	0.02	442.0
0.15	266	0.03	444.9
0.20	298	0.04	447.4
0.25	326	0.05	449.3
0.30	351	0.06	450.9
0.35	372	0.07	452.3
0.40	391	0.08	453.5
0.45	403.5	0.09	454.5
0.50	416	0.10	455.2
0.55	426.5		

Compare your computed results with the experimental data of Scatchard and Raymond [*J. Am. Chem. Soc.*, **60**, 1275 (1938)].

8. Suppose you had isothermal x-y data at $25°C$ for saturated solutions of sodium chloride in mixtures of water and ethanol going from pure water to pure ethanol. Show with suitable equations how you would test these data for thermodynamic consistency.

9. You want to estimate the *y-x* diagram for the liquid mixture carbon tetrabromide/nitroethane at $\frac{1}{2}$ atm total pressure. You have available boiling-point (*T-x*) data for this system at $\frac{1}{2}$ atm as well as pure-component vapor pressure data as a function of temperature. Explain how you would use the available information to construct the desired diagram. Assume that an electronic computer is available. Set up the necessary equations and define all symbols used. State all assumptions made. Are there any advantages in using the Wilson equation for the solution of this problem? Explain.

10. An alcohol is distributed between two immiscible fluids, hexane and dimethyl sulfoxide. Calculate the distribution coefficient of the alcohol between the two liquid phases (when the alcohol concentration is very small) at 30°C and 100 atm. The following binary data are available, all at 0°C and 1 atm pressure:

Alcohol/hexane	Alcohol/dimethyl sulfoxide
$g^E = 600\, x'_A\, x'_H$ cal/g-mol	$g^E = 80\, x''_A\, x''_D$ cal/g-mol
$h^E = 1200\, x'_A\, x'_H$ cal/g-mol	$h^E = 150\, x''_A\, x''_D$ cal/g-mol
$v^E = 16\, x'_A\, x'_H$ cc/g-mol	$v^E = -10\, x''_A\, x''_D$ cc/g-mol

In your notation use ' for the hexane phase and " for the dimethyl sulfoxide phase. State all assumptions made.

The distribution coefficient *K* is defined as

$$K \equiv \lim_{\substack{x'_A \to 0 \\ x''_A \to 0}} \left(\frac{x'_A}{x''_A} \right).$$

REFERENCES

1. Pool, R. A. H., G. Saville, T. M. Herrington, B. D. C. Shields, and L. A. K. Staveley, *Trans. Faraday Soc.*, **58**, 1692 (1962).

2. Scatchard, G., S. E. Wood, and J. M. Mochel, *J. Phys. Chem.*, **43**, 119 (1939).

3. Redlich, O., A. T. Kister, and C. E. Turnquist, *Chem. Eng. Progr. Symp. Ser.*, **48**, No. 2, 49 (1952).

4. Barker, J. A., *Australian J. Chem.*, **6**, 207 (1953).

5. Hermsen, R. W., and J. M. Prausnitz, *Chem. Eng. Sci.*, **18**, 485 (1963).

6. Orye, R. V., and J. M. Prausnitz, *Trans. Faraday Soc.*, **61**, 1338 (1965).

7. Miller, H. C., and H. Bliss, *Ind. Eng. Chem.*, **32**, 123 (1940).

8. Carlson, H. C., and A. P. Colburn, *Ind. Eng. Chem.*, **34**, 581 (1942).

9. Van Ness, H. C., *Classical Thermodynamics of Nonelectrolyte Solutions*, Chap. 6. New York: The Macmillan Company, 1964.

10. Christian, S. D., E. Neparko, and H. E. Affsprung, *J. Phys. Chem.*, **64**, 442 (1960).

11. Bellemans, A., *Bull. Soc. Chim. Belg.*, **68**, 355 (1959).

12. McDermott, C., and S. R. M. Ellis, *Chem. Eng. Sci.*, **20**, 545 (1965).

13. Redlich, O., and A. T. Kister, *Ind. Eng. Chem.*, **40**, 345 (1948).

14. Herington, E. F. G., *Nature*, **160**, 610 (1947).

15. Herington, E. F. G., *J. Appl. Chem.*, **2**, 14 (1952).

16. Snider, G. H., and J. M. Prausnitz, *A. I. Ch. E. Journal*, **5**, 7S (1959).

17. Herington, E. F. G., *J. Inst. Petrol.*, **37**, 457 (1951).

18. Van Ness, H. C., and R. V. Mrazek, *A. I. Ch. E. Journal*, **5**, 209 (1959).

19. Van Ness, H. C., *Chem. Eng. Sci.*, **11**, 118 (1959).

20. Wohl, K., *Trans. A. I. Ch. E.*, **42**, 215 (1946).

21. Weissman, S., and S. E. Wood, *J. Chem. Phys.*, **32**, 1153 (1960).

22. Gadwa, T. A., Dissertation, Massachusetts Institute of Technology, 1936, quoted by H. C. Carlson and A. P. Colburn, *Ind. Eng. Chem.*, **34**, 581 (1942).

23. Severns, W. H., A. Sesonske, R. H. Perry and R. L. Pigford, *A. I. Ch. E. Journal*, **1**, 401 (1955).

24. Scatchard, G., and W. J. Hamer, *J. Am. Chem. Soc.*, **57**, 1805 (1935).

25. Van Ness, H. C. *Classical Thermodynamics of Nonelectrolyte Solutions*, page 129. New York: The Macmillan Company, 1964.

26. Wilson, G. M., *J. Am. Chem. Soc.*, **86**, 127 (1964).

27. Orye, R. V., and J. M. Prausnitz, *Ind. Eng. Chem.*, **57**, No. 5, 19 (1965).

28. Black, C., *A. I. Ch. E. Journal*, **5**, 249 (1959).

29. Brown, I., and F. Smith, *Australian J. Chem.*, **10**, 423 (1957).

30. Kretschmer, C. B., *J. Am. Chem. Soc.*, **70**, 1785 (1948).

31. Renon, H., and J. M. Prausnitz, *A. I. Ch. E. Journal*, **14**, 135 (1968).

32. Butler, J. A. V., *Trans. Faraday Soc.*, **33**, 229 (1937).

33. Shain, S. A., and J. M. Prausnitz, *Chem. Eng. Sci.*, **18**, 244 (1963).

34. Adler, S. B., Leo Friend, and R. L. Pigford, *A. I. Ch. E. Journal*, **12**, 629 (1966).

35. Myers, H. S., *Ind. Eng. Chem.* **48**, 1104 (1956); J. E. Sinor and J. H. Weber, *J. Chem. Eng. Data*, **5**, 243 (1960); A. H. Wehe and J. Coates, *A. I. Ch. E. Journal*, **1**, 241 (1955).

36. Prausnitz, J. M., C. A. Eckert, R. V. Orye, and J. P. O'Connell, *Computer Calculations for Multicomponent Vapor-Liquid Equilibria.* Englewood Cliffs, N.J.: Prentice-Hall, Inc., 1967.

37. Eckert, C. A., J. M. Prausnitz, R. V. Orye, and J. P. O'Connell, "Advances in Separation Techniques," *A. I. Ch. E.-I. Chem. E. Symposium Series* (London), **1**, 1965.

38. Eckert, C. A., and J. M. Prausnitz, *J. Eng. Educ.*, **58**, 1110 (1968).

Fugacities in Liquid Mixtures: Theories of Solutions

7

For the case of two or more pure liquids which are mixed to form a liquid solution, the aim of solution theory is to express the properties of the liquid mixture in terms of the intermolecular forces which determine those properties. To minimize the amount of experimental information required to describe a solution, it is desirable to express the properties of a solution in terms which can be calculated completely from the properties of the pure components. Present theoretical knowledge has not yet reached a stage of development where this can be done with any degree of generality, although some results of limited utility have been obtained. Moreover, since this subject is under active investigation all over the world, results offering greater usefulness are likely to be forthcoming in the near future.† In this chapter we introduce some of the theoretical concepts which have been used to describe and to interpret solution

---

†Most current work in the theory of solutions utilizes the powerful methods of statistical mechanics, which relate macroscopic (bulk) properties to microscopic (molecular) phenomena. These methods are beyond the scope of this chapter but anyone who seriously wishes to obtain an understanding of solution theory must sooner or later familiarize himself with the tools and concepts which statistical mechanics provides.

properties. In one chapter we cannot possibly give a complete treatment; we attempt, however, to give a brief survey of those theoretical ideas which bear promise for practical applications.

The simplest theory of liquid solutions is that due to Raoult, who set the partial pressure of any component equal to the product of its own vapor pressure and its mole fraction in the liquid phase; at modest pressures, this simple relation often provides a reasonable approximation for those liquid solutions whose components are chemically similar. However, Raoult's relation becomes exact only as the components of the mixture become identical, and its failure to represent the behavior of actual solutions is due to differences in molecular size and intermolecular forces of the pure components. It appears logical, therefore, to use Raoult's relation as a reference and to express observed behavior of real solutions as deviations from the behavior calculated by Raoult's law. This treatment of solution properties was formalized by G. N. Lewis in the early twentieth century, and since then it has become customary to express the behavior of real solutions in terms of activity coefficients. Another way of stating the aim of solution theory, then, is to say that it aims to predict numerical values of activity coefficients in terms of properties (or constants) which have molecular significance and which, hopefully, may be calculated from the properties of the pure components.

One of the first systematic attempts to describe quantitatively the properties of fluid mixtures was made by van der Waals and his coworkers very early in the twentieth century, shortly before the work of Lewis. As a result, most of van der Waals' work on fluid mixtures appears in a form which today strikes us as awkward. However, no one can deny that he and his colleagues at Amsterdam were the first great pioneers in a field which, since about 1890, has attracted the serious attention of a large number of outstanding physical scientists, including many Nobel-prize winners. One of van der Waals' students and later collaborators was J. J. van Laar, and it was primarily through van Laar's work that the basic ideas of the Amsterdam school became well-known. It is most convenient, therefore, to begin by discussing van Laar's theory of solutions and then to show how this simple but inadequate theory led to the more useful theory of regular solutions advanced by Scatchard and Hildebrand.

7.1 The Theory of van Laar

One of the essential requirements for a successful theory in physical science is judicious simplification. If one wishes to do justice to all the aspects of a problem, one very soon finds oneself in a hopelessly complicated situation and thus, in order to make progress, it is necessary to ignore certain aspects of a physical situation and to retain others; the wise execution of this choice often makes the difference between a result which is realistic and one which is merely academic.   Van Laar's essential contribution was that he chose good simplifying assumptions which made the problem tractable and yet did not greatly violate physical reality.

Van Laar considered a mixture of two liquids: x_1 moles of liquid 1 and x_2 moles of liquid 2. He assumed that the two liquids mix at constant temperature and pressure in such a manner that: (1) there is no volume change, i.e., $v^E = 0$; and (2) the entropy of mixing is given by that corresponding to an ideal solution, i.e., $s^E = 0$. Since, at constant pressure,

$$g^E = u^E + Pv^E - Ts^E, \qquad (7.1\text{-}1)$$

it follows from van Laar's simplifying assumptions that

$$g^E = u^E. \qquad (7.1\text{-}2)$$

To calculate the energy change of mixing, van Laar constructed a three-step, isothermal, thermodynamic cycle wherein the pure liquids are first vaporized to some arbitrarily low pressure, mixed at this low pressure, and then recompressed to the original pressure, as illustrated in Fig. 7-1. The energy change is calculated for each step and, since energy

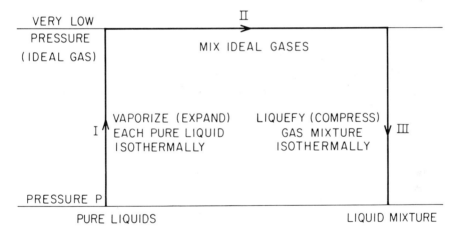

Fig. 7-1   Thermodynamic cycle for forming a liquid mixture from the pure liquids at constant temperature.

is a state function independent of path, the energy change of mixing, Δu, is given by the sum of the three energy changes.   That is,

$$\Delta u = u^E = \Delta u_{\text{I}} + \Delta u_{\text{II}} + \Delta u_{\text{III}}. \qquad (7.1\text{-}3)$$

Step I

The two pure liquids are vaporized isothermally to the ideal-gas state. The energy change accompanying this process is calculated by the rigorous thermodynamic equation

$$\left(\frac{\partial u}{\partial v}\right)_T = T\left(\frac{\partial P}{\partial T}\right)_v - P. \tag{7.1-4}$$

Van Laar then (unfortunately) assumed that the volumetric properties of the pure fluids are given by the van der Waals equation. In that case,

$$\left(\frac{\partial u}{\partial v}\right)_T = \frac{a}{v^2} \tag{7.1-5}$$

where a is the constant appearing in the van der Waals equation. With x_1 moles of liquid 1 and x_2 moles of liquid 2 we obtain exactly one mole of mixture. Then

$$x_1(u_{\text{ideal}} - u)_1 = \int_{v_1^L}^{\infty} \frac{a_1 x_1}{v^2}\, dv = \frac{a_1 x_1}{v_1^L}, \tag{7.1-6}$$

and

$$x_2(u_{\text{ideal}} - u)_2 = \int_{v_2^L}^{\infty} \frac{a_2 x_2}{v^2}\, dv = \frac{a_2 x_2}{v_2^L}, \tag{7.1-7}$$

where u_{ideal} is the energy of the ideal gas and v^L is the molar volume of the pure liquid. Now, according to van der Waals' theory the molar volume of a liquid well below its critical temperature can be replaced approximately by the constant b. Thus

$$\Delta u_{\text{I}} = \frac{a_1 x_1}{b_1} + \frac{a_2 x_2}{b_2}. \tag{7.1-8}$$

Step II

Isothermal mixing of gases at very low pressure (i.e., ideal gases) proceeds with no change in energy. Thus

$$\Delta u_{\text{II}} = 0. \tag{7.1-9}$$

Step III

The ideal-gas mixture is now compressed isothermally and condensed at the original pressure. The thermodynamic equation (7.1-4) also holds for a mixture, and van Laar assumed that the volumetric properties of the mixture are also given by the van der Waals equation. Thus

$$\Delta u_{III} = -\frac{a_{mixture}}{b_{mixture}}.$$ (7.1-10)

It is now necessary to express the constants a and b for the mixture in terms of the constants for the pure components. Van Laar used the expressions

$$\sqrt{a_{mixture}} = x_1 \sqrt{a_1} + x_2 \sqrt{a_2},$$ (7.1-11)

$$b_{mixture} = x_1 b_1 + x_2 b_2.$$ (7.1-12)

Eq. (7.1-11) follows from the assumption that only interactions between two molecules are important and that a_{12}, the constant characteristic of the interaction between two dissimilar molecules, is given by the geometric-mean law. Equation (7.1-12) follows from the assumption that there is no volume change upon mixing the two liquids.

Equations (7.1-8) to (7.1-12) are now substituted in Eq. (7.1-3). Algebraic rearrangement gives

$$g^E = \frac{x_1 x_2 b_1 b_2}{x_1 b_1 + x_2 b_2} \left(\frac{\sqrt{a_1}}{b_1} - \frac{\sqrt{a_2}}{b_2} \right)^2.$$ (7.1-13)

The activity coefficients are obtained by differentiation as discussed in Sec. 6.3 and we obtain:

$$\ln \gamma_1 = \frac{A'}{\left[1 + \dfrac{A'}{B'} \dfrac{x_1}{x_2} \right]^2},$$ (7.1-14)

and

$$\ln \gamma_2 = \frac{B'}{\left[1 + \dfrac{B'}{A'} \dfrac{x_2}{x_1} \right]^2},$$ (7.1-15)

where

$$A' \equiv \frac{b_1}{RT} \left(\frac{\sqrt{a_1}}{b_1} - \frac{\sqrt{a_2}}{b_2} \right)^2,$$ (7.1-16)

and

$$B' \equiv \frac{b_2}{RT}\left(\frac{\sqrt{a_1}}{b_1} - \frac{\sqrt{a_2}}{b_2}\right)^2. \qquad (7.1\text{-}17)$$

Equations (7.1-14) and (7.1-15) are the well-known van Laar equations which relate the activity coefficients to the temperature, to the composition, and to the properties of the pure components, i.e., (a_1, b_1) and (a_2, b_2).

Two important features of the van Laar equations should be noted. One is that the logarithms of the activity coefficients are inversely proportional to the absolute temperature. This result, however, is independent of van Laar's thermodynamic cycle and follows directly from the assumption that $s^E = 0$. The other important feature is that according to van Laar's theory the activity coefficients of both components are never less than unity; hence this theory always predicts positive deviations from Raoult's law. This result follows from Eq. (7.1-11) which says that

$$a_{\text{mixture}} < x_1 a_1 + x_2 a_2 \qquad (7.1\text{-}18)$$

whenever $a_1 \neq a_2$.

Since the constant a is proportional to the forces of attraction between the molecules, Eq. (7.1-11) [or (7.1-18)] implies that the forces of attraction between the molecules in the mixture *are less* than what they would be if they were additive on a molar basis. If van Laar had assumed a combining rule where

$$a_{\text{mixture}} > x_1 a_1 + x_2 a_2, \qquad (7.1\text{-}19)$$

then he would have obtained that

$$\gamma_i \leq 1 \qquad \text{for all } x. \qquad (7.1\text{-}20)$$

On the other hand, had he assumed that

$$a_{\text{mixture}} = x_1 a_1 + x_2 a_2, \qquad (7.1\text{-}21)$$

he would have obtained that

$$\gamma_1 = \gamma_2 = 1 \qquad \text{for all } x. \qquad (7.1\text{-}22)$$

Thus we can see that the combining rules which one uses to express the constants for a mixture in terms of the constants for the pure components have a large influence on the predicted results.

As one might expect, quantitative agreement between van Laar's equations and experimental results is not good. However, this poor agreement is not due as much to van Laar's simplifications as it is to his adherence to the van der Waals equation and to the mixing rules used by van der Waals to extend that equation to mixtures.

One of the implications of van Laar's theory is the relation between solution nonideality and the critical pressures of the pure components. According to van der Waals' equation of state, the square root of the critical pressure of a pure fluid is proportional to $\sqrt{a}/b$. Therefore, van Laar's theory predicts that the nonideality of a solution rises with increasing difference in the critical pressures of the components; in fact, for a solution whose components have identical critical pressures, van Laar's theory predicts ideal behavior. These predictions, unfortunately, are contrary to the experimental facts.

If one regards A' and B' as adjustable parameters, then the van Laar equations are useful empirical relations which have been used successfully to correlate experimental activity coefficients for many binary systems, including some which show large deviations from ideal behavior. (See Sec. 6.13.)

7.2 The Scatchard-Hildebrand Theory

Van Laar had recognized that a simple theory of solutions could be constructed if one restricted attention to those cases where the excess entropy and the excess volume of mixing could be neglected. Several years later Hildebrand found that the experimental thermodynamic properties of iodine solutions in various nonpolar solvents appeared to be substantially in agreement with these simplifying assumptions. Hildebrand called these solutions *regular* and later defined a regular solution as one in which the components mix with no excess entropy provided there is no volume change upon mixing.[1] Another way of saying this is to define a regular solution as one which has a vanishing excess entropy of mixing at constant temperature and constant volume.

Both Hildebrand and Scatchard, working independently and a continent apart, realized that van Laar's theory could be greatly improved if it could be freed from the limitations of van der Waals' equation of state. This can be done by defining a parameter c according to

$$c \equiv \frac{\Delta u^v}{v^L} \qquad (7.2\text{-}1)$$

where Δu^v is the energy of complete vaporization, that is, the energy change upon isothermal vaporization of the saturated liquid to the ideal-

from the Clapeyron equation are find that

$$\frac{\Delta u^{vap}}{v^{L}} = \left[T \left(\frac{dR}{dT} \right) - P \right] \left(\frac{v^{v}}{v^{L}} - 1 \right) = constant$$

gas state (infinite volume). The parameter c is called the *cohesive-energy density*.

Having defined c, the key step made by Hildebrand and Scatchard consisted in generalizing Eq. (7.2-1) to a binary liquid mixture by writing, per mole of mixture,

$$- (u_{\text{liquid}} - u_{\text{ideal gas}})_{\text{binary mixture}} = \frac{c_{11} v_1^2 x_1^2 + 2c_{12} v_1 v_2 x_1 x_2 + c_{22} v_2^2 x_2^2}{x_1 v_1 + x_2 v_2}$$

(7.2-2)

where the superscript L has been dropped from the v's. Equation (7.2-2) assumes that the energy of a binary liquid mixture (relative to the ideal gas at the same temperature and composition) can be expressed as a quadratic function of the volume fraction and it also implies that the volume of a binary liquid mixture is given by the mole-fraction average of the pure-component volumes (i.e., $v^E = 0$). The constant c_{11} refers to interactions between molecules of species 1; c_{22} refers to interactions between molecules of species 2, and c_{12} refers to interactions between unlike molecules. For saturated liquids c_{11} and c_{22} are functions only of temperature.

To simplify notation, we introduce symbols Φ_1 and Φ_2 which designate volume fractions of components 1 and 2 as defined by

$$\Phi_1 \equiv \frac{x_1 v_1}{x_1 v_1 + x_2 v_2}$$

(7.2-3)

$$\Phi_2 \equiv \frac{x_2 v_2}{x_1 v_1 + x_2 v_2}.$$

(7.2-4)

Equation (7.2-2) now becomes

$$-(u_{\text{liquid}} - u_{\text{ideal gas}})_{\text{binary mixture}} = (x_1 v_1 + x_2 v_2)[c_{11} \Phi_1^2 + 2c_{12} \Phi_1 \Phi_2 + c_{22} \Phi_2^2].$$

(7.2-5)

The molar energy change of mixing (which is also the excess energy of mixing) is defined by

$$u^E \equiv u_{\text{binary mixture}} - x_1 u_1 - x_2 u_2.$$

(7.2-6)

Equation (7.2-1) (for each component) and Eq. (7.2-5) are now substituted into Eq. (7.2-6); also, we utilize the relation for ideal gases,

$$u^E_{\text{ideal}} = 0.$$

(7.2-7)

Algebraic rearrangement then gives

$$u^E = (c_{11} + c_{22} - 2c_{12}) \Phi_1 \Phi_2 (x_1 v_1 + x_2 v_2).$$

(7.2-8)

Scatchard and Hildebrand now make what is probably the most important assumption in their treatment. They assume that for molecules whose forces of attraction are due primarily to dispersion forces there is a simple relation between c_{11}, c_{22}, and c_{12} as suggested by London's formula (see Sec. 4.4), viz.,

$$c_{12} = (c_{11} c_{22})^{1/2}. \tag{7.2-9}$$

Substituting Eq. (7.2-9) into Eq. (7.2-8) gives

$$u^E = (x_1 v_1 + x_2 v_2) \Phi_1 \Phi_2 [\delta_1 - \delta_2]^2, \tag{7.2-10}$$

where

$$\delta_1 \equiv c_{11}^{1/2} = \left(\frac{\Delta u^v}{v}\right)_1^{1/2} \tag{7.2-11}$$

and

$$\delta_2 \equiv c_{22}^{1/2} = \left(\frac{\Delta u^v}{v}\right)_2^{1/2}. \tag{7.2-12}$$

The positive square root of c is given the symbol δ, which is called the *solubility parameter*.

To complete their theory of solutions, Scatchard and Hildebrand make one additional assumption, viz., that at constant temperature and pressure the excess entropy of mixing vanishes. This assumption is consistent with Hildebrand's definition of regular solutions because in the treatment outlined above we had already assumed that there is no excess volume. With the elimination of excess entropy and excess volume at constant pressure, we have

$$g^E = u^E. \tag{7.2-13}$$

The activity coefficients follow upon using Eq. (6.3-8). They are

$$\boxed{RT \ln \gamma_1 = v_1 \Phi_2^2 [\delta_1 - \delta_2]^2} \tag{7.2-14}\dagger$$

and

$$\boxed{RT \ln \gamma_2 = v_2 \Phi_1^2 [\delta_1 - \delta_2]^2.} \tag{7.2-15}\dagger$$

†These equations can also be derived by a somewhat more fundamental procedure based on assumptions concerning the radial distribution function of liquids. See App. VI (p. 480).

Equations (7.2-14) and (7.2-15) are known as the *regular-solution equations*, and they have much in common with the van Laar relations [Eqs. (7.1-14) and (7.1-15)]. In fact, the regular-solution equations can easily be rearranged into the van Laar form merely by writing for the parameters A' and B'

$$A' = \frac{v_1}{RT} [\delta_1 - \delta_2]^2 \qquad (7.2\text{-}16)$$

and

$$B' = \frac{v_2}{RT} [\delta_1 - \delta_2]^2. \qquad (7.2\text{-}17)$$

The regular-solution equations always predict $\gamma_i \geq 1$, i.e., a regular solution can exhibit only positive deviations from Raoult's law. This result is again a direct consequence of the geometric-mean assumption; it follows from Eq. (7.2-9), wherein the cohesive-energy density corresponding to the interaction between dissimilar molecules is given by the geometric mean of the cohesive-energy densities corresponding to interaction between similar molecules.

The solubility parameters δ_1 and δ_2 are functions of temperature but the difference between these solubility parameters, $\delta_1 - \delta_2$, is often nearly independent of temperature. Since the regular-solution model assumes that the excess entropy is zero, it follows that at constant composition the logarithm of each activity coefficient must be inversely proportional to the absolute temperature. Hence the model, in effect, assumes that, as the temperature is varied at constant composition,

$$v_1 \Phi_2^2 [\delta_1 - \delta_2]^2 = \text{constant} \qquad (7.2\text{-}18)$$

and

$$v_2 \Phi_1^2 [\delta_1 - \delta_2]^2 = \text{constant}. \qquad (7.2\text{-}19)$$

For many solutions of nonpolar liquids Eqs. (7.2-18) and (7.2-19) are reasonable approximations provided the temperature range is not large and provided the solution is remote from critical conditions.

Table 7-1 gives liquid molar volumes and solubility parameters for some typical nonpolar liquids at 25°C and for a few liquefied gases at 90°K. Merely by inspection of the solubility parameters of different liquids it is easily possible to make some qualitative statements about deviations from ideality of certain mixtures. Remembering that the logarithm of the activity coefficient varies directly as the square of the *difference* in solubility parameters, one can see, for instance, that a mixture of carbon disulfide with *n*-hexane exhibits large positive deviations from Raoult's law whereas a mixture of carbon tetrachloride and cyclohexane is nearly ideal. The difference in solubility parameters of mixture components provides a measure of solution nonideality. For example, the solubility parameters shown in Table 7-1 bear out the well-known fact that whereas mixtures of aliphatic hydrocarbons are nearly ideal, mixtures of aliphatic hydrocarbons with aromatics show appreciable nonideality.

Table 7-1

MOLAR LIQUID VOLUMES AND SOLUBILITY
PARAMETERS OF SOME NONPOLAR LIQUIDS

Liquefied Gases at 90°K	v (cc/g-mol)	δ (cal/cc)$^{1/2}$
Nitrogen	38.1	5.3
Carbon monoxide	37.1	5.7
Argon	29.0	6.8
Oxygen	28.0	7.2
Methane	35.3	7.4
Carbon tetrafluoride	46.0	8.3
Ethane	45.7	9.5

Liquid Solvents at 25°C		
Perfluoro-n-heptane	226	6.0
Neopentane	122	6.2
Isopentane	117	6.8
n-Pentane	116	7.1
n-Hexane	132	7.3
1-Hexene	126	7.3
n-Octane	164	7.5
n-Hexadecane	294	8.0
Cyclohexane	109	8.2
Carbon tetrachloride	97	8.6
Ethyl benzene	123	8.8
Toluene	107	8.9
Benzene	89	9.2
Styrene	116	9.3
Tetrachloroethylene	103	9.3
Carbon disulfide	61	10.0
Bromine	51	11.5

The regular-solution equations give a good semiquantitative representation of activity coefficients for many solutions containing nonpolar components. Because of various simplifying assumptions which have been made in the derivation, one cannot expect complete quantitative agreement between calculated and experimental results but, for approximate work, i.e., for reasonable estimates of (nonpolar) equilibria in the absence of any mixture data, the regular solution equations provide useful

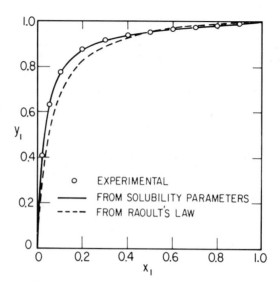

Fig. 7-2 Vapor-liquid equilibria for CO $(1)/CH_4$ (2) mixtures at $90.7°\,K$.

results. Figures 7-2, 7-3 and 7-4 show y-x diagrams for three representative nonpolar systems. Vapor-liquid equilibria were calculated first using Raoult's law and then using the regular-solution equations; experimentally observed equilibria are also shown and it can be seen that for

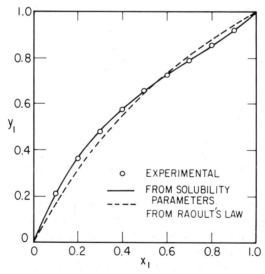

Fig. 7-3 Vapor-liquid equilibria for C_6H_6 $(1)/n$-$C_7H_{16}(2)$ at $70°C$.

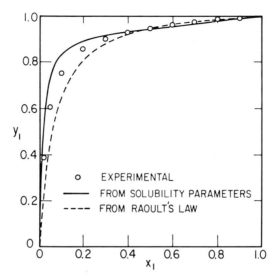

Fig. 7-4   Vapor-liquid equilibria for neo-C_5H_{12} (1)/ CCl_4 (2) at $0°C$.

two of the systems results based on the regular-solution theory are a considerable improvement over those calculated by Raoult's law; for the third system, neopentane/carbon tetrachloride, the regular-solution equations overcorrect. For mixtures of nonpolar liquids it is a fair statement to say that whereas Raoult's law gives a zeroth approximation, the regular-solution equations usually give a first approximation to vapor-liquid equilibria. While regular-solution results are not always good, for non-polar systems they are also hardly ever very bad, and whenever an estimate of phase equilibria is required the theory of regular solutions provides a valuable guide. [It must again be emphasized that Eqs. (7.2-14) and (7.2-15) are not valid for solutions containing polar components.] The only major failure of the theory of regular solutions for nonpolar fluids appears to be when it is applied to certain solutions containing fluorocarbons[2] and the reasons for this failure are only partly understood.

For mixtures which are nearly ideal, the regular-solution equations are often poor in the sense that predicted and observed excess Gibbs energies differ appreciably; however, for nearly ideal mixtures such errors necessarily have only a small effect on calculated vapor-liquid equilibria. For practical applications the regular-solution equations are most useful for mixtures having considerable nonideality. Solubility parameter theory provides a fairly good estimate of the excess Gibbs energies of most mixtures of common nonpolar liquids, especially when the excess Gibbs energy is large. For small deviations from ideality, Eqs. (7.2-14) and (7.2-15) are less reliable because small errors in the geometric-mean assumption and in the solubility parameters become relatively more serious when δ_1 and δ_2 are close to one another.

Scott's review of available data up to 1956[3] has shown that solubility-parameter theory fits excess Gibbs energies of most binary systems of nonpolar liquids to within 10 to 20 percent of the thermal energy RT. (At room temperature RT is nearly 600 cal/g-mol.) McGlashan's review in 1962[4] bears this out as indicated in Fig. 7-5 where a comparison is made between calculated and observed excess Gibbs energies for 21

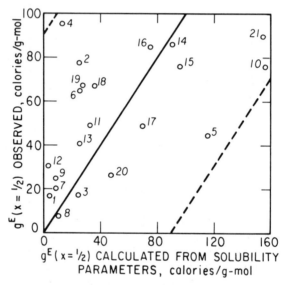

Fig. 7-5   Excess Gibbs energies from the regular-solution equation.†

†Binary systems shown are as follows:

1. cyclo-C_6H_{12}/CCl_4	8. C_6H_6/$C_2H_4Cl_2$	15. C_6H_6/n-C_6H_{14}
2. cyclo-C_6H_{12}/C_6H_6	9. CCl_4/$CHCl_3$	16. $C_6H_5CH_3$/n-C_6H_{14}
3. cyclo-C_6H_{12}/n-C_6H_{14}	10. CCl_4/$C(CH_3)_4$	17. $C_6H_5CH_3$/n-C_7H_{16}
4. cyclo-C_6H_{12}/$C_6H_5CH_3$	11. CCl_4/CH_3I	18. $C_6H_5CT_3$/cyclo-$C_6H_{11}CH_3$
5. cyclo-C_6H_{12}/$C(CH_3)_4$	12. $TiCl_4$/CCl_4	19. CCl_4/CH_2Cl_2
6. C_6H_6/cyclo-C_5H_{10}	13. $SiCl_4$/CCl_4	20. n-C_6H_{14}/CCl_4
7. C_6H_6/CCl_4	14. C_6H_6/n-C_7H_{16}	21. C_6H_6/i-C_8H_{18}

Systems 8, 9, 11, and 19 each contain one component whose polarity is not negligible and, strictly speaking, they should not be included in this list. However, since there are no specific effects (e.g., hydrogen bonding) in these systems, regular-solution theory still gives the right order of magnitude for g^E for these particular mixtures.

Most of the data are at 25°C. The lowest temperature (0°C) is for system 10 and the highest (65°C) for system 18. According to regular-solution theory, the excess Gibbs energy is independent of temperature to a first approximation.

Broken lines indicate $\pm 0.15\ RT$ here taken as ± 90 calories/g-mol.

binary systems near room temperature at the composition midpoint $x_1 = x_2 = \frac{1}{2}$. The broken lines were drawn 90 cal/g-mol ($\sim 0.15\ RT$) above and below the solid line which corresponds to perfect agreement between theory and experiment.

Figure 7-5 suggests that solubility parameters are primarily useful for semiquantitative estimates of activity coefficients in liquid mixtures. This is indeed true. Solubility parameters can tell us readily the magnitude of nonideality which is to be expected in a mixture of two nonpolar liquids. But it is also true that solubility parameters can form a basis for a more quantitative application when modified empirically. One outstanding example of such an application is provided by Chao and Seader[5] who used solubility parameters to correlate phase equilibria for hydrocarbon mixtures over a wide range of conditions. Two other applications, one concerned with gas solubility and the other with solubility of solid carbon dioxide at low temperatures, are discussed in later chapters.

The theory of Scatchard and Hildebrand is essentially the same as that of van Laar but it is liberated from the narrow confines of the van der Waals equation or of any other equation of state. We now know that the assumptions of regularity ($s^E = 0$) and isometric mixing ($v^E = 0$) at constant temperature and pressure are not correct even for simple mixtures, but due to a cancellation of errors, these assumptions frequently do not seriously affect calculations of the excess Gibbs energy. (When regular-solution theory is used to calculate excess enthalpies the results are usually much worse.) However, the most serious defect of the theory is the geometric-mean assumption. This assumption can be relaxed by writing instead of Eq. (7.2-9) the more general relation

$$c_{12} = (1 - l_{12})(c_{11}c_{22})^{1/2}, \qquad (7.2\text{-}20)\dagger$$

where l_{12} is a constant, small compared to unity, characteristic of the 1-2 mixture. From London's theory of dispersion forces an expression can be obtained for l_{12} in terms of molecular parameters but it has been found that such an expression has little quantitative value.

In mixtures of chemically similar components (e.g., cyclohexane/ n-hexane) deviations from the geometric mean are primarily a result of differences in molecular shape and subsequent differences in molecular packing. Our inability to describe properly the geometric arrangement of molecules in the liquid phase is one of the main reasons for the inadequacy of currently existing theories of solution.

When Eq. (7.2-20) is used in place of Eq. (7.2-9) the activity coefficients are given by

$$\ln \gamma_1 = \frac{v_1 \Phi_2^2}{RT} [(\delta_1 - \delta_2)^2 + 2l_{12}\delta_1\delta_2] \qquad (7.2\text{-}21)$$

†The l_{12} used here is related to but different from k_{12} used in Sec. 5.13.

and

$$\ln \gamma_2 = \frac{v_2 \Phi_1^2}{RT} [(\delta_1 - \delta_2)^2 + 2l_{12}\delta_1\delta_2].$$
(7.2-22)

Equations (7.2-21) and (7.2-22) show immediately that if δ_1 and δ_2 are close to one another then even a small value of l_{12} can significantly affect the activity coefficients. For example, suppose $T = 300°K$, $v_1 = 100$ cc/g-mol, and δ_1 and δ_2 are 7.0 and 7.5 (cal/cc)$^{1/2}$ respectively. Then, at infinite dilution, we find that for $l_{12} = 0$, $\gamma_1^\infty = 1.04$. However, if $l_{12} = 0.03$, we obtain $\gamma_1^\infty = 1.77$. Even if l_{12} is as small as 0.01 we obtain $\gamma_1^\infty = 1.24$. These illustrative results show why the solubility-parameter theory is not quantitatively reliable for components whose solubility parameters are very nearly the same. As the difference between δ_1 and δ_2 becomes larger, the effect of deviation from the geometric mean becomes less serious. However, it is apparent that even small deviations from the geometric mean, say one or two percent, can have an appreciable effect on calculated activity coefficients.

7.3 Extension to Multicomponent Systems

One of the main advantages of the regular solution equations is their simplicity, and this simplicity is retained even when the regular-solution model is extended to solutions containing more than two components. The derivation for the multicomponent case is completely analogous to that given for the binary case. The molar energy of a liquid mixture containing m components is written:

$$-(u_{\text{liquid}} - u_{\text{ideal gas}})_{\text{mixture}} = \frac{\displaystyle\sum_i^m \sum_j^m v_i v_j x_i x_j c_{ij}}{\displaystyle\sum_i^m x_i v_i}.$$
(7.3-1)

The volume fraction of component j is now defined by

$$\Phi_j \equiv \frac{x_j v_j}{\displaystyle\sum_i^m x_i v_i},$$
(7.3-2)

and the excess energy of mixing is defined by

$$u^E \equiv u_{\text{mixture}} - \sum_i^m x_i u_i.$$
(7.3-3)

By assumption, the cohesive-energy density c_{ij} is given by the geometric mean,

$$c_{ij} = (c_{ii}c_{jj})^{1/2}. \tag{7.3-4}$$

Again assuming that

$$s^E = v^E = 0, \tag{7.3-5}$$

we again have

$$g^E = u^E. \tag{7.3-6}$$

Substitution and algebraic rearrangement, coupled with Eq. (6.3-8), finally gives a remarkably simple result for the activity coefficient of component j in a multicomponent solution:

$$\boxed{RT \ln \gamma_j = v_j(\delta_j - \bar{\delta})^2,} \tag{7.3-7}$$

where

$$\bar{\delta} \equiv \sum_i^m \Phi_i \delta_i. \tag{7.3-8}$$

The parameter $\bar{\delta}$ is a volume-fraction average of the solubility parameters of *all* the components in the solution; the summation in Eq. (7.3-8) is over all components, including component j.

Equation (7.3-7) has the same advantages and disadvantages as do Eqs. (7.2-14) and (7.2-15). It is useful for providing estimates of equilibria in nonpolar solutions and with empirical modifications, it can serve as a basis for quantitative correlations.†

7.4 The Lattice Theory

Since the liquid state is in some sense intermediate between the crystalline state and the gaseous state, it follows that there are two types of ap-

†Equations (7.2-21) and (7.2-22) can also be generalized for mixtures containing more than two components; the general expression for the activity coefficient, however, is no longer as simple as that given by Eq. (7.3-7). For a mixture of m components, it is:

$$RT \ln \gamma_k = v_k \sum_i^m \sum_j^m \Phi_i \Phi_j [D_{ik} - \tfrac{1}{2} D_{ij}] \tag{7.3-9}$$

where

$$D_{ij} \equiv (\delta_i - \delta_j)^2 + 2l_{ij}\delta_i\delta_j . \tag{7.3-10}$$

For every component i, $l_{ii} = D_{ii} = 0$. Equation (7.3-9) reduces to Eq. (7.3-7) only if $l_{ij} = 0$ for every ij pair.

proaches to a theory of liquids. The first of these considers liquids to be gas-like; a liquid is pictured as a dense and highly nonideal gas whose properties can be described by some equation of state of which van der Waals' is the best-known example. An equation-of-state description of pure liquids can readily be extended to liquid mixtures as was done by van der Waals and by some of his disciples like van Laar and later by Benedict, Webb, and Rubin (see Chap. 3).

The second approach considers a liquid to be solid-like, in a quasi-crystalline state, where the molecules do not translate fully in a chaotic manner as in a gas but where each molecule tends to stay in a small region, a more or less fixed position in space about which it vibrates back and forth. The quasicrystalline picture of the liquid state supposes molecules to sit in a regular array in space, called a *lattice*, and therefore theories of liquids and liquid mixtures based on this simplified picture are called *lattice theories*.[†] These theories are described in detail elsewhere[6, 7] and their proper study requires familiarity with the methods of statistical mechanics. We give here only a brief introduction to the lattice theory of solutions.

Molecular considerations suggest that deviations from ideal behavior in liquid solutions are due primarily to the following effects: First, forces of attraction between unlike molecules are quantitatively different from those between like molecules, giving rise to a nonvanishing heat of mixing; second, if the unlike molecules differ significantly in size or shape, the molecular arrangement in the mixture may be appreciably different from that for the pure liquids, giving rise to a nonideal entropy of mixing; and finally, in a binary mixture, if the forces of attraction between one of three possible pair interactions are very much stronger (or very much weaker) than those of the other two, then there are certain preferred orientations of the molecules in the mixture which, in some cases, may induce thermodynamic instability and demixing (incomplete miscibility).

We consider a mixture of two simple liquids 1 and 2. Molecules of types 1 and 2 are small and spherically symmetric and the ratio of their sizes is close to unity. We suppose that the arrangement of the molecules in each pure liquid is that of a regular array as indicated in Fig. 7-6; all the molecules are situated on lattice points which are equidistant from one another. We suppose further that for a fixed temperature the lattice spacings for the two pure liquids and for the mixture are the same, independent of composition.

[†]An exhaustive attempt has been made by Eyring and coworkers [H. Eyring and T. Ree, *Proc. Nat. Acad. Sci.*, **47**, 526 (1961) and many subsequent papers] to describe liquids as consisting of gas-like and solid-like molecules. While this attempt has had some empirical success its main ideas and assumptions are in direct conflict with many physicochemical investigations of liquid structure [J. H. Hildebrand and R. L. Scott, *Regular Solutions*, Chap. 5 (Englewood Cliffs, N.J.: Prentice-Hall, Inc., 1962)].

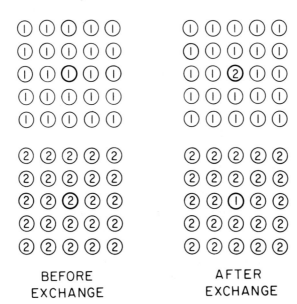

BEFORE
EXCHANGE

AFTER
EXCHANGE

Fig. 7-6   Physical significance of interchange energy. The
energy absorbed in the process above is $2w$. See
Eq. (7.4-9).

In view of the similarity between the two types of molecules, we assume that a mixture of 1 and 2 is perfectly random, i.e., one where all
possible arrangements of the molecules on the lattice are equally probable.
By a simple statistical argument it can be shown that the entropy of
mixing n_1 molecules of 1 with n_2 of 2 is given by

$$\frac{\Delta S_{\text{mixing}}}{R} = -n_1 \ln \frac{n_1}{n_1 + n_2} - n_2 \ln \frac{n_2}{n_1 + n_2}. \tag{7.4-1}$$

This is the entropy of mixing for an ideal solution and we see that for
this type of mixture the excess entropy is zero.

We now proceed to find an expression for the enthalpy of mixing. In
view of the assumptions made about the lattice spacings for the pure
liquids and for the mixture, we assume that the mixing process at constant
pressure and temperature produces no change in volume; since the excess
volume is zero, the enthalpy of mixing is equivalent to the energy of
mixing, or

$$h^E = u^E \qquad (\text{since } v^E = 0). \tag{7.4-2}$$

As indicated in Secs. 7.1 and 7.2 [theories of van Laar and of Scatchard and Hildebrand] a solution for which $s^E = v^E = 0$ is called a *regular
solution*.

To derive an expression for the potential energy of a liquid, pure or mixed, we assume that the potential energy is pairwise additive for all molecular pairs and that only nearest neighbors need be considered in the summation. This means that the potential energy of a large number of molecules sitting on a lattice is given by the sum of the potential energies of all pairs of molecules which are situated immediately next to one another. For uncharged nonpolar molecules, intermolecular forces are short-range and therefore we can neglect contributions to the total potential energy from pairs which are not nearest neighbors.

There are n_1 molecules of type 1 and n_2 molecules of type 2; let each molecule of either type have z nearest (touching) neighbors. (z is called the *coordination number* and may have a value between 4 and 12 depending on the type of packing, i.e., the way in which the molecules are arranged in 3-dimensional space. Empirically, for typical liquids at ordinary conditions z is close to 10.)

In a random mixture, the number of molecules of type 1 which surround a central molecule of type 1 is equal to $zn_1/(n_1 + n_2)$, and the number of molecules of type 2 which surround this same central molecule is $zn_2/(n_1 + n_2)$. Therefore the energy of interaction between this central molecule and all the other molecules in the liquid (the effect of nonnearest neighbors being neglected) is given by:

$$\frac{zn_1}{n_1 + n_2}\, \Gamma_{11} + \frac{zn_2}{n_1 + n_2}\, \Gamma_{12}, \qquad (7.4\text{-}3)$$

where, as in Chap. 4, Γ_{11} is the potential energy of a 1-1 pair and Γ_{12} is the potential energy of a 1-2 pair. Since there are n_1 molecules of type 1, the total energy of interaction between these n_1 molecules with each other and with the remaining molecules is

$$\frac{1}{2}\left[\frac{zn_1^2}{n_1 + n_2}\, \Gamma_{11} + \frac{zn_1 n_2}{n_1 + n_2}\, \Gamma_{12}\right]. \qquad (7.4\text{-}4)$$

The factor $\frac{1}{2}$ is needed to avoid counting each pair twice.

Similarly, for n_2 molecules of type 2, the total energy of interaction between these n_2 molecules with each other and with the remaining molecules is

$$\frac{1}{2}\left[\frac{zn_2^2}{n_1 + n_2}\, \Gamma_{22} + \frac{zn_1 n_2}{n_1 + n_2}\, \Gamma_{12}\right]. \qquad (7.4\text{-}5)$$

The total potential energy of a liquid containing n_1 molecules of 1 and n_2 molecules of 2 is given by the sum of expressions (7.4-4) and (7.4-5):

$$U = \frac{1}{2}\left[\frac{zn_1^2}{n_1 + n_2}\Gamma_{11} + \frac{zn_2^2}{n_1 + n_2}\Gamma_{22} + \frac{2zn_1n_2}{n_1 + n_2}\Gamma_{12}\right]. \qquad (7.4\text{-}6)$$

Equation (7.4-6) gives the potential energy of a binary mixture and also of a pure liquid; in the latter case, either n_1 or n_2 is set equal to zero. The energy of mixing is given by

$$\Delta U_{\text{mixing}} = \underset{\substack{n_1 + n_2 \text{ molecules} \\ \text{in mixture}}}{U} - \underset{\substack{n_1 \text{ molecules} \\ \text{of pure 1}}}{U} - \underset{\substack{n_2 \text{ molecules} \\ \text{of pure 2}}}{U}. \qquad (7.4\text{-}7)$$

Substitution gives

$$\Delta U_{\text{mixing}} = \frac{n_1 n_2 w}{n_1 + n_2}, \qquad (7.4\text{-}8)$$

where w, called the *interchange energy*, is defined by

$$w \equiv z[\Gamma_{12} - \tfrac{1}{2}(\Gamma_{11} + \Gamma_{22})]. \qquad (7.4\text{-}9)$$

The physical significance of w is illustrated in Fig. 7-6; it is the change in potential energy when z dissimilar molecular pairs are formed in a solution from $z/2$ pairs of 1 and $z/2$ pairs of 2. In Fig. 7-6, z pairs of type 1 and z pairs of type 2 are separated to form $2z$ pairs of dissimilar (1-2) molecules. Therefore the change in energy which accompanies the interchange process shown in Fig. 7-6 is equal to $2w$.

For a regular solution the excess Gibbs energy is equal to the excess energy (or energy of mixing) given by Eq. (7.4-8); per mole of mixture,

$$g^E = N_A w x_1 x_2, \qquad (7.4\text{-}10)$$

where N_A is Avogadro's constant. The activity coefficients then follow from Eq. (6.3-8):

$$\boxed{\ln \gamma_1 = \frac{w}{kT}x_2^2} \qquad (7.4\text{-}11)$$

$$\boxed{\ln \gamma_2 = \frac{w}{kT}x_1^2,} \qquad (7.4\text{-}12)$$

where k is Boltzmann's constant.†

† Boltzmann's constant is the gas constant per molecule; $k = R/N_A$.

These results are of the same form as that of the two-suffix Margules equations. However, in Eqs. (7.4-11) and (7.4-12) the parameter w is not just an empirical constant, since it has physical significance.

From Eq. (7.4-9) we see that if the potential energy for a 1-2 pair is equal to the arithmetic mean of the potentials for the 1-1 and 2-2 pairs, then $w = 0$ and $\gamma_1 = \gamma_2 = 1$ for all x and we have an ideal solution. However, as discussed in Chap. 4, for simple, nonpolar molecules Γ_{12} is more nearly equal to the geometric mean than to the arithmetic mean of Γ_{11} and Γ_{22}. Since the magnitude of the geometric mean is always less than that of the arithmetic mean, and since Γ_{12}, Γ_{11}, and Γ_{22} are negative in sign, it follows that for simple nonpolar molecules Eqs. (7.4-11) and (7.4-12) predict positive deviations from ideal-solution behavior, in agreement with experiment.

7.5 Calculation of the Interchange Energy from Molecular Properties

Since the interchange energy w is related to the potential energies, it should be possible to obtain a numerical value for w from information on potential functions. Various attempts to do this have been reported and one of them, due to Kohler,[8] is particularly simple.

The potential function Γ depends on r, the distance between molecules, and Kohler assumes that for the pure liquids,

$$r_{11} = \left(\frac{v_1}{N_A}\right)^{1/3} \tag{7.5-1}$$

and

$$r_{22} = \left(\frac{v_2}{N_A}\right)^{1/3}, \tag{7.5-2}$$

where v stands for the molar liquid volume and N_A is Avogadro's constant.

In the mixture, Kohler assumes

$$r_{12} = \frac{r_{11} + r_{22}}{2}. \tag{7.5-3}\dagger$$

---

†These assumptions are not exactly consistent with the assumptions of the lattice theory, where $r_{11} = r_{22} = r_{12}$. Strictly, the lattice theory requires that $v_1 = v_2$, which very much limits its applicability. A certain degree of inconsistency frequently results when an idealized theory is applied to actual phenomena.

Basing his calculations on London's theory of dispersion forces (see Chap. 4), Kohler then writes:

$$\Gamma_{11} = -\frac{\alpha_1^2}{r_{11}^6}\,\xi_1, \tag{7.5-4}\dagger$$

$$\Gamma_{22} = -\frac{\alpha_2^2}{r_{11}^6}\,\xi_2, \tag{7.5-5}$$

$$\Gamma_{12} = -\frac{2^6\alpha_1\alpha_2}{(r_{11} + r_{22})^6}\cdot\frac{2\xi_1\xi_2}{\xi_1 + \xi_2}, \tag{7.5-6}$$

where α is the polarizability and ξ_i is calculated from $\Delta h_{\text{vap}\ i}$, the molar enthalpy of vaporization, by

$$\xi_i = \frac{2r_{ii}^6}{z\alpha_i^2}\left(\frac{\Delta h_{\text{vap}\ i} - RT}{N_A}\right). \tag{7.5-7}$$

When these expressions are substituted into Eq. (7.4-9) it is possible to obtain the interchange energy w as needed in the calculation of activity coefficients, Eqs. (7.4-11) and (7.4-12). One of the advantages of Kohler's method is that, because of cancellation, no separate estimate of the co-ordination number z is required; further, the three potential energies Γ_{11}, Γ_{22} and Γ_{12} are calculated separately and it is not necessary to assume that Γ_{12} is the geometric mean of the other two.

Using Kohler's method, calculations have been made for the excess Gibbs energies of four simple binary systems, each at the composition midpoint where $x_1 = x_2 = \frac{1}{2}$. The calculated results are compared in Table 7-2 with experimental values and agreement is fairly good. How-

Table 7-2

EXCESS GIBBS ENERGIES FOR EQUIMOLAR, BINARY MIXTURES.
CALCULATIONS BASED ON LATTICE THEORY AND KOHLER'S
METHOD FOR EVALUATING THE INTERCHANGE ENERGY

System	T (°K)	g^E ($x_1 = x_2$), cal/g-mol	
		Lattice Theory	Experimental
Argon/Methane	90.7	16	17
Nitrogen/Methane	90.7	59	32
Benzene/Cyclohexane	298	42	77
Carbon tetrachloride/Cyclohexane	298	13	17

$\dagger\,\xi$ is closely related to the ionization potential. (See Sec. 4.4.)

ever, we must remember that the applicability of this type of calculation is limited to mixtures where the molecules of the two components are not only nonpolar but also essentially spherical and similar in size. As a result, the equations which we have described are useful only for a small class of mixtures; when calculations based on Kohler's method are made for systems outside of this small class, agreement with experimental results is usually poor.

7.6 Nonrandom Mixtures of Simple Molecules

One of the important assumptions made in the previous sections was that when the molecules of two components are mixed, the arrangement of the molecules is completely random; i.e., the molecules have no tendency to segregate either with their own kind or with the other kind of molecule. In a completely random mixture a given molecule shows no preference in the choice of its neighbors.

Since intermolecular forces operate between molecules, a completely random mixture in a two-component system of equisized molecules can only result if these forces are the same for all three possible molecular pairs 1-1, 2-2 and 1-2.† In that event, however, there would also be no energy change upon mixing. Strictly then, only an ideal mixture can be completely random.

In a mixture where the pair energies Γ_{11}, Γ_{22} and Γ_{12} are not the same, a certain amount of ordering (nonrandomness) must result. For example, suppose that the magnitude of the attractive energy between a 1-2 pair is much larger than that between a 1-1 and 2-2 pair; in that case there is a strong tendency to form as many 1-2 pairs as possible. An example of such a situation is provided by the system chloroform/acetone where hydrogen bonds can form between the unlike molecules but not between the like molecules. Or, suppose that the attractive forces between a 1-1 pair are much larger than those between a 1-2 or a 2-2 pair; in that event a molecule of type 1 prefers to surround itself with other molecules of type 1 and more 1-1 pairs exist in the mixture than would exist in a purely random mixture having the same composition. An example of such a situation is the ethyl ether/pentane system; since ethyl ether has a large dipole moment whereas pentane is nonpolar, ether molecules interact by dipole-dipole forces which, on the average, are attractive; but between ether and pentane and between pentane and pentane there are no dipole-dipole forces.

---

†The model based on the lattice theory is actually not quite so restrictive. For molecules of the same size there is no energy of mixing and no departure from randomness when the interchange energy is zero: i.e., when $\Gamma_{12} = \frac{1}{2}(\Gamma_{11} + \Gamma_{12})$.

In the lattice theory (for w independent of T), entropy is a measure of randomness; the entropy of mixing for a completely random mixture (Eq. 7.4-1) is always larger than that of a mixture which is incompletely random regardless of whether nonrandomness is due to preferential formation of 1-2 or 1-1 (or 2-2) pairs. Excess entropy due to ordering (i.e., nonrandomness) is always negative.

Guggenheim[6,7] has constructed a lattice theory for molecules of equal size which form mixtures that are not necessarily random. This theory is not rigorous but utilizes a simplification known as the *quasichemical approximation*. The derivation requires familiarity with statistical mechanics, and need not be reproduced here. The results, in series form, are given by the following expressions for the molar excess Gibbs energy, excess enthalpy, and excess entropy:

$$\frac{g^E}{RT} = \left(\frac{w}{kT}\right) x_1 x_2 \left[1 - \frac{1}{2}\left(\frac{2w}{zkT}\right) x_1 x_2 + \ldots\right] \qquad (7.6\text{-}1)$$

$$\frac{h^E}{RT} = \left(\frac{w}{kT}\right) x_1 x_2 \left[1 - \left(\frac{2w}{zkT}\right) x_1 x_2 + \ldots\right] \qquad (7.6\text{-}2)$$

$$\frac{s^E}{R} = -\left(\frac{w}{kT}\right) x_1 x_2 \left[\frac{1}{2}\left(\frac{2w}{zkT}\right) x_1 x_2 + \ldots\right]. \qquad (7.6\text{-}3)$$

In these equations the excess Gibbs energy is no longer equal to the excess enthalpy and the excess entropy is no longer zero. However, in the limit, as $(2w/zkT) \longrightarrow 0$,

$$g^E \longrightarrow h^E \quad \text{and} \quad s^E \longrightarrow 0,$$

as expected. In other words, the earlier results based on the assumption of complete randomness become a satisfactory approximation as the interchange energy per pair of molecules becomes small relative to the thermal energy kT. For a given mixture, randomness increases as the temperature rises or, at a fixed temperature, randomness increases as the interchange energy falls.

The excess entropy given by Eq. (7.6-3) is never positive; for any nonvanishing value of w, positive or negative, s^E is always negative. For this particular model, therefore, the entropy of mixing is a maximum for the completely random mixture.† However, the contribution of nonrandomness to the excess Gibbs energy and to the excess enthalpy may be positive or negative depending on the sign of the interchange energy.

---

†For many nonpolar mixtures, positive excess entropies have been observed experimentally. These observations are a result of other effects (neglected by the lattice theory) such as changes in volume and changes in excitation of internal degrees of freedom (rotation, vibration) which may result from the mixing process.

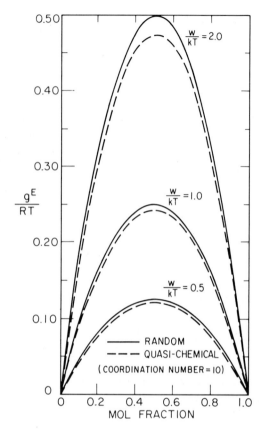

Fig. 7-7   Effect of nonrandomness on excess Gibbs energy of binary mixtures.

The excess Gibbs energy, as given by Eq. (7.6-1) based on the quasi-chemical approximation, is not very different from Eq. (7.4-10) based on the assumption of random mixing. Figure 7-7 compares excess Gibbs energies calculated by the two equations and it is evident that for totally miscible mixtures the correction for nonrandom mixing is not large.

However, deviations from random mixing become significant when w/kT is large enough to induce limited miscibility of the two components. The criteria for incipient demixing (instability) are:†

$$\frac{\partial^2 \Delta g_{mixing}}{\partial x^2} = \frac{\partial^3 \Delta g_{mixing}}{\partial x^3} = 0, \qquad (7.6\text{-}4)$$

†See Sec. 6.15.

where Δg_{mixing} is the change in the total (not excess) molar Gibbs energy upon mixing:

$$\Delta g_{mixing} = RT[x_1 \ln x_1 + x_2 \ln x_2] + g^E. \qquad (7.6\text{-}5)$$

When Eq. (7.4-10) is substituted into Eq. (7.6-4) we find that T^c, the upper consolute temperature, is given by

$$\boxed{T^c = \frac{w}{2k}.} \qquad (7.6\text{-}6)$$

The upper consolute temperature is the maximum temperature at which there is limited miscibility: for $T > T^c$ there is only one stable liquid phase (complete miscibility) whereas for $T < T^c$ there are two stable liquid phases.

In contrast to Eq. (7.6-6), when exact results based on the quasi-chemical approximation are substituted into Eq. (7.6-4), we obtain

$$T^c = \frac{w}{kz[\ln z - \ln(z - 2)]}. \qquad (7.6\text{-}7)$$

When $z = 10$ we find

$$\boxed{T^c = \frac{w}{2.23\, k}.} \qquad (7.6\text{-}8)$$

Equation (7.6-8) shows that the consolute temperature according to the quasichemical approximation, is about ten percent lower than that computed from the assumption of random mixing. This is a significant change although still not large. However, a large effect becomes notice-able when one computes the coexistence curve which is the locus of mutual solubilities of the two components at temperatures below the upper consolute temperature. Figure 7-8 shows calculated results for the change in Gibbs energy due to mixing for four values of w/kT; calculations were performed first assuming random mixing and second, assum-ing the quasichemical approximation. When $w/kT = 1.8$, both theories predict complete miscibility. When $w/kT = 2.0$ the random-mixing theory predicts incipient instability whereas the qasichemical theory pre-dicts complete miscibility. When $w/kT = 2.23$ the random theory indi-cates the existence of two liquid phases whose compositions are given by

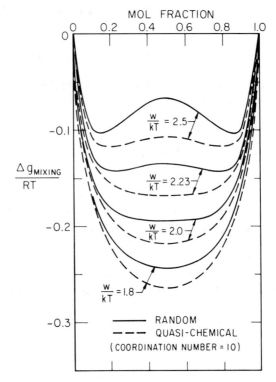

MOL FRACTION

$\dfrac{\Delta g_{MIXING}}{RT}$

$\dfrac{w}{kT} = 2.5$

$\dfrac{w}{kT} = 2.23$

$\dfrac{w}{kT} = 2.0$

$\dfrac{w}{kT} = 1.8$

————— RANDOM
— — — QUASI-CHEMICAL
(COORDINATION NUMBER = 10)

Fig. 7-8   Effect of nonrandomness on Gibbs energy
of mixing.

the two minima in the curves; the more refined theory merely predicts incipient demixing. When $w/kT = 2.5$ both theories indicate the existence of two liquid phases, but the compositions of the two phases as given by one theory are different from those given by the other. These compositions are given by the minima in the curves and we can see that the mutual solubilities predicted by the quasichemical approximation are about twice those predicted by the random-mixing assumption.

These illustrative calculations show that the effect of ordering (that is, nonrandomness) is not important except when the components are near or below their consolute temperature. As a further illustration we show in Figs. 7-9 and 7-10 some calculations of Eckert[9] for the methane/carbon tetrafluoride system. From second-virial-coefficient data for mixtures of the two gases near room temperature, Eckert estimates a value of the interchange energy (Eq. 7.4-9) and then calculates the excess Gibbs energy at 105.5°K for the liquid mixture. The results are shown in Fig. 7-9 along with the experimental data of Thorp and Scott;[10] agreement with experiment is good and there is not much difference between calculations based on random mixing and those based on the quasichemical approximation.

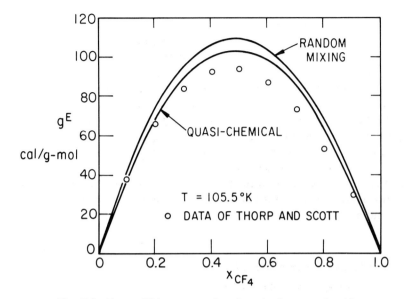

Fig. 7-9   Excess Gibbs energy of methane/carbon tetrafluoride
system.

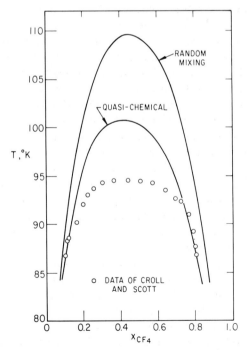

Fig. 7-10   Liquid-liquid coexistence curve for
the methane/carbon tetrafluoride
system.

Croll and Scott[11] have observed that methane and carbon tetrafluoride are not completely miscible below about 94°K; Eckert therefore calculated the coexistence curve and the results are shown in Fig. 7-10. The random-mixing theory predicts a consolute temperature which is too high by about 15°K; the consolute temperature predicted by the quasichemical theory is also too high but considerably less so. It is clear from Fig. 7-10 that for calculation of mutual solubilities in a pair of incompletely miscible liquids, the quasichemical theory provides a significant improvement over the random-mixing theory. However, even in the improved theory there are still many features which are known to be incorrect. While the quasichemical theory is a step in the right direction, it provides no more than an approximation which is still far from a completely satisfactory theory of liquid mixtures.

7.7 Mixtures Whose Molecules Differ
Greatly in Size: Polymer Solutions

One of the most useful applications of the lattice theory of fluids is to liquid mixtures wherein the molecules of one component are very much larger than those of the other component. A common example of such mixtures is provided by solutions of polymers in liquid solvents.

The Gibbs energy of mixing consists of an enthalpy term and an entropy term. In the theory of regular solutions for molecules of similar size it was assumed that the entropy term corresponds to that for an ideal solution and attention was focused on the enthalpy of mixing; however, when considering solutions of molecules of very different size, it has been found advantageous to assume, at least at first, that the enthalpy of mixing is zero and to concentrate on the entropy. Solutions with zero enthalpy of mixing are called *athermal* solutions because their components mix with no liberation or absorption of heat. Athermal behavior is never observed exactly but it is approximated by mixtures of components which are similar in their chemical characteristics even if their sizes are different. An example of such a mixture is a solution of polystyrene in toluene or in ethyl benzene.

Using the concept of a quasicrystalline lattice as a model for a liquid, an expression for the entropy of mixing in an athermal solution was derived independently by Flory[12] and by Huggins.[13] The derivation, which is based on statistical arguments and several well-defined assumptions, is not reproduced here. To understand the derivation some familiarity with statistical mechanics is needed. It is clearly presented in several references[14,15] and therefore we give here only a brief discussion along with the result.

Flory and Huggins assume that a polymer molecule in solution behaves like a chain, that is, as if it consisted of a large number of mobile segments each having the same size as that of a molecule of the solvent. Further, it is assumed that each segment occupies one site in the quasi-lattice and that adjacent segments occupy adjacent sites. Let there be n_1 moles of solvent and n_2 moles of polymer and let there be m segments in a polymer molecule. The total number of sites then is $(n_1 + mn_2)N_A$, where N_A is Avogadro's number. The fractions Φ_1 and Φ_2 of the volume occupied by the solvent and by the polymer are given by

$$\Phi_1 = \frac{n_1}{n_1 + mn_2} \quad \text{and} \quad \Phi_2 = \frac{mn_2}{n_1 + mn_2}. \qquad (7.7\text{-}1)$$

Flory and Huggins have shown that if the amorphous (i.e., non-crystalline) polymer and the solvent mix without any energetic effects (i.e., athermal behavior), the change in Gibbs energy and entropy upon mixing are given by the remarkably simple expression:

$$-\frac{\Delta G_{\text{mixing}}}{RT} = \frac{\Delta S_{\text{mixing}}}{R} = -n_1 \ln \Phi_1 - n_2 \ln \Phi_2. \qquad (7.7\text{-}2)$$

The entropy change in Eq. (7.7-2) is similar in form to that of Eq. (7.4-1) for a regular solution except that volume fractions are used rather than mole fractions. For the special case $m = 1$ the change in entropy given by Eq. (7.7-2) reduces to that of Eq. (7.4-1) as expected.

The expression of Flory and Huggins immediately leads to an equation for the excess entropy which is, per mole of mixture,

$$\frac{s^E}{R} = -x_1 \ln \left[1 - \Phi_2\left(1 - \frac{1}{m}\right)\right] - x_2 \ln \left[m - \Phi_2(m - 1)\right] \qquad (7.7\text{-}3)$$

By algebraic rearrangement of Eq. (7.7-3) and expansion of the resulting logarithmic terms, it can be shown that for all $m > 1$, s^E is positive. Therefore, for an athermal solution of components whose molecules differ in size, the Flory-Huggins theory predicts negative deviations from Raoult's law:

$$\frac{g^E}{RT} = \frac{h^E}{RT} - \frac{s^E}{R} = 0 - \frac{s^E}{R} < 0. \qquad (7.7\text{-}4)$$

The activity coefficient for the solvent follows from differentiation according to Eq. (6.3-8), giving:

$$\ln \gamma_1 = \ln \left[1 - \left(1 - \frac{1}{m}\right)\Phi_2\right] + \left(1 - \frac{1}{m}\right)\Phi_2. \qquad (7.7\text{-}5)$$

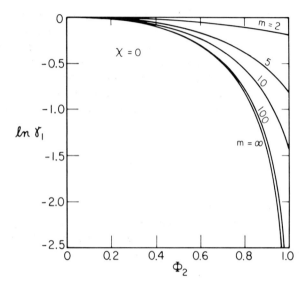

Fig. 7-11   Activity coefficient of solvent in an athermal polymer solution according to equation of Flory and Huggins. The parameter m gives the number of segments in the polymer molecule.

Figure 7-11 shows activity coefficients for the solvent according to Eq. (7.7-5) for several values of the parameter m which measures the disparity in molecular size between the two components. The activity coefficient is a strong function of m for small values of that parameter but for large values ($m \geq 100$) the activity coefficient is essentially independent of m. For solutions of polymers in common solvents m is a very large number and we can see in Fig. 7-11 that large deviations from ideal-solution behavior result merely as a consequence of difference in molecular size even in the absence of any energetic effects.

In order to apply the theoretical result of Flory and Huggins to real polymer solutions, i.e., to solutions which are not athermal, it has become common practice to add to Eq. (7.7-2) a semiempirical term for the enthalpy of mixing. The form of this term is the same as that used in the van Laar-Scatchard-Hildebrand theory of solutions (see Secs. 7.1 and 7.2); the excess enthalpy is set proportional to the volume of the solution and to the product of the volume fractions. The Flory-Huggins equation for real polymer solutions then becomes:

$$\frac{\Delta G_{mixing}}{RT} = n_1 \ln \Phi_1 + n_2 \ln \Phi_2 + \chi \Phi_1 \Phi_2 (\check{n}_1 + mn_2), \quad (7.7\text{-}6)$$

and the activity coefficient of the solvent is

$$\ln \gamma_1 = \ln \left[1 - \left(1 - \frac{1}{m} \right) \Phi_2 \right] + \left(1 - \frac{1}{m} \right) \Phi_2 + \chi \Phi_2^2, \quad (7.7\text{-}7)$$

where χ, called the Flory interaction parameter, is determined by inter-molecular forces between the molecules in the solution. If (as is usually done) we set m equal to the ratio of molar volumes of polymer and solvent, then the definition of volume fraction Φ as given in Eq. (7.7-1) is identical to that used in the Scatchard-Hildebrand theory, Eqs. (7.2-3) and (7.2-4). The Flory-Huggins equation for real polymer solutions is shown in Fig. 7-12.

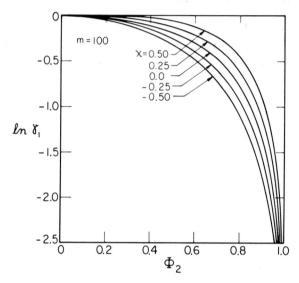

Fig. 7-12   Activity coefficient of solvent in a real polymer solution according to equation of Flory and Huggins. The parameter χ depends on the inter-molecular forces between polymer and solvent.

The dimensionless parameter χ is determined by the energies which characterize the interactions between pairs of polymer segments, between pairs of solvent molecules, and between one polymer segment and one solvent molecule. In terms of the interchange energy [Eq. (7.4-9)], χ is given by:

$$\chi = \frac{w}{kT}. \quad (7.7\text{-}8)$$

In this case the interchange energy refers not to the exchange of solvent and solute molecules but rather to the exchange of solvent molecules and polymer segments. For athermal solutions, χ is zero, and for mixtures of components which are similar chemically, χ is small.

Equations (7.7-6) and (7.7-7) have been used widely to describe thermodynamic properties of polymer solutions. For example, Fig. 7-13 shows activity coefficients for n-heptane in the n-heptane/polyethylene system, and Fig. 7-14 shows how data on rubber solutions may be reduced using Eq. (7.7-7) to obtain values of the Flory-Huggins parameter χ. In Fig. 7-14, $1/m$ has been set equal to zero. For these systems Eq. (7.7-7) gives an excellent representation of the data, but for many other systems the representation is poor because, contrary to the theory, χ varies with the volume fraction of the polymer.

The addition of an enthalpy term to the theoretical result for athermal mixtures is essentially an empirical modification in order to obtain a reasonable expression for the Gibbs energy of mixing. According to the theory, χ should be independent of polymer concentration and of polymer molecular weight but in many systems, especially polar ones, χ changes considerably with both of these variables. Further, the theory erroneously

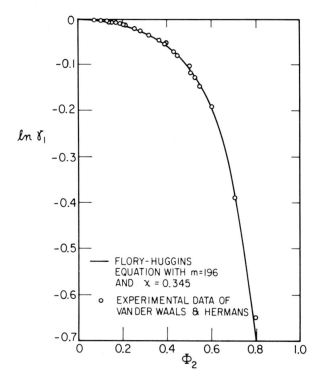

Fig. 7-13   Activity coefficient of heptane in the n-heptane (1)/polyethylene (2) system at 109°C.

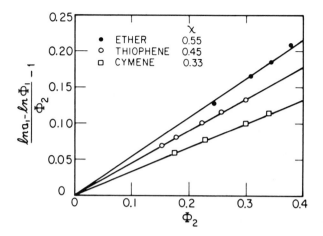

Fig. 7-14   Data reduction using the equation of Flory and Huggins.   Data are for solutions of rubber near room temperature.   The interaction parameter χ is given by the slope of the lines.   (after Huggins)

assumes that the heat of mixing should be given by the last term in Eq. (7.7-6); however, calorimetric data give a value of χ quite different from that which is obtained when experimental Gibbs energies are reduced with Eq. (7.7-6).   The Flory-Huggins equation for real polymer solutions is far from a perfect description of the thermodynamic properties of such solutions but there is little doubt that this relatively simple theory contains most of the essential features which distinguish solutions of very large molecules from those containing molecules of ordinary size.

In terms of solubility parameters, it can be shown that χ for nonpolar systems is given by:

$$\chi = \frac{v_1}{RT}(\delta_1 - \delta_2)^2, \qquad (7.7-9)$$

where v_1 is the molar volume of the solvent and δ_1 and δ_2 are, respectively, the solubility parameters of solvent and polymer.   Equation (7.7-9) is not useful for an accurate quantitative description of polymer solutions but it provides a good guide for a qualitative consideration of polymer solubility.   For good solubility χ should be small or negative.†   According to the Scatchard-Hildebrand theory for nonpolar components, χ cannot be negative (however, in many polar systems negative values have been

†Stability analysis shows that for complete miscibility of components 1 and 2,

$$\chi \le \frac{1}{2}\left(1 + \frac{1}{\sqrt{m}}\right)^2.$$

In polymer solutions $m \ggg 1$ and therefore, for complete miscibility, the largest permissible positive value of χ is $\frac{1}{2}$.

observed), and therefore a criterion of a good solvent for a given polymer is

$$\delta_1 \approx \delta_2. \qquad (7.7\text{-}10)$$

Equation (7.7-10) provides a useful practical guide for nonpolar systems[16,17] and for polar systems an approximate generalization of Eq. (7.7-9) has been suggested by Blanks[17] and by Hansen.[18]

Some solubility parameters for noncrystalline polymers are given in Table 7-3. When these are compared with solubility parameters for common liquids (see Table 7-1) some qualitative statements concerning polymer solubility can easily be made. For example, the solubility parameters show at once that polyisobutylene (δ = 8.1) should be readily soluble in cyclohexane (δ = 8.2) but only sparingly soluble in carbon disulfide (δ = 10). However, it is important to keep in mind that all of the relations given in this section are restricted to amorphous polymers; they are not applicable to crystalline or cross-linked polymers.

Table 7-3

SOLUBILITY PARAMETERS OF SOME
AMORPHOUS POLYMERS NEAR
25°C† $(cal/cc)^{1/2}$

Teflon	6.2
Polydimethyl silicone	7.3
Polyethylene	7.9
Polyisobutylene	8.1
Polybutadiene	8.6
Polystyrene	9.1
Polymethyl methacrylate	9.5
Polyvinyl chloride	9.7
Cellulose diacetate	10.9
Polyvinylidene chloride	12.2
Polyacrylonitrile	15.4

†H. Burrell, *Official Digest*, Federation of Paint and Varnish Production Clubs, October 1955, p. 745.

7.8 Wilson's Extension of the Flory-Huggins Equation

For mixtures of molecules which are chemically similar (athermal solutions) and which differ only in size, Flory and Huggins derived a simple expression (Eq. 7.7-2) for the excess Gibbs energy. According to Eq. (7.7-2), ideal behavior is expected whenever the components in the mixture have the same liquid molar volumes which are taken as measures of the molecular size.

About twenty years after the work of Flory and Huggins, Wilson considered the case where the components in a mixture differ not only in molecular size but also in their intermolecular forces.[19,20] Wilson's point of departure was Eq. (7.7-2), which he used as a basis for an ad hoc extension; his modification was based on loose semitheoretical arguments which lack the theoretical rigor of the lattice theory of Guggenheim.

To derive Wilson's equation we first consider a binary solution of components 1 and 2. If we focus attention on a central molecule of type 1, the probability of finding a molecule of type 2 relative to finding a molecule of type 1 about this central molecule is expressed in terms of the overall mole fractions and two Boltzmann factors:

$$\frac{x_{21}}{x_{11}} = \frac{x_2 \exp\left(-\lambda_{12}/RT\right)}{x_1 \exp\left(-\lambda_{11}/RT\right)}. \qquad (7.8\text{-}1)\dagger$$

Equation (7.8-1) says that the ratio of the number of molecules of type 2 to the number of molecules of type 1 about a central 1 molecule is equal to the ratio of the overall mole fractions of 2 and 1 weighted by the Boltzmann factors $\exp\left(-\lambda_{12}/RT\right)$ and $\exp\left(-\lambda_{11}/RT\right)$. The parameters λ_{12} and λ_{11} are, respectively, related to the potential energies of a 1-2 and 1-1 pair of molecules. An equation analogous to Eq. (7.8-1) can be written for the probability of finding a molecule of type 1 relative to finding a molecule of type 2 about a central molecule 2. This equation is:

$$\frac{x_{12}}{x_{22}} = \frac{x_1 \exp\left(-\lambda_{12}/RT\right)}{x_2 \exp\left(-\lambda_{22}/RT\right)}. \qquad (7.8\text{-}2)$$

Wilson now defines *local* volume fractions using Eqs. (7.8-1) and (7.8-2). The local volume fraction of component 1, designated by ξ_1, is defined by

$$\xi_1 \equiv \frac{v_1 x_{11}}{v_1 x_{11} + v_2 x_{21}}, \qquad (7.8\text{-}3)$$

where v_1 and v_2 are the molar liquid volumes of components 1 and 2. Substitution gives:

$$\xi_1 = \frac{v_1 x_1 \exp\left(-\lambda_{11}/RT\right)}{v_1 x_1 \exp\left(-\lambda_{11}/RT\right) + v_2 x_2 \exp\left(-\lambda_{12}/RT\right)}. \qquad (7.8\text{-}4)$$

Similarly the local volume fraction of component 2 is:

$$\xi_2 = \frac{v_2 x_2 \exp\left(-\lambda_{22}/RT\right)}{v_2 x_2 \exp\left(-\lambda_{22}/RT\right) + v_1 x_1 \exp\left(-\lambda_{12}/RT\right)}. \qquad (7.8\text{-}5)$$

†This notation is slightly different from that used in Ref. 20.

Wilson proposes to use the local volume fractions ξ_1 and ξ_2 rather than the overall volume fractions Φ_1 and Φ_2 in the expression of Flory and Huggins [Eq. (7.7-2)]. That is, he writes for the molar excess Gibbs energy of a binary system:

$$\frac{g^E}{RT} = x_1 \ln \frac{\xi_1}{x_1} + x_2 \ln \frac{\xi_2}{x_2}. \qquad (7.8\text{-}6)\dagger$$

To simplify notation, it is convenient to define two new parameters Λ_{12} and Λ_{21} in terms of the molar volumes v_1 and v_2 and the energies λ_{11}, λ_{22} and λ_{12}. These definitions are:

$$\Lambda_{12} \equiv \frac{v_2}{v_1} \exp \left[-\frac{(\lambda_{12} - \lambda_{11})}{RT} \right] \qquad (7.8\text{-}7)$$

$$\Lambda_{21} \equiv \frac{v_1}{v_2} \exp \left[-\frac{(\lambda_{12} - \lambda_{22})}{RT} \right]. \qquad (7.8\text{-}8)$$

Wilson's modification of the Flory-Huggins equation then becomes:

$$\frac{g^E}{RT} = -x_1 \ln (x_1 + \Lambda_{12} x_2) - x_2 \ln (\Lambda_{21} x_1 + x_2). \qquad (7.8\text{-}9)$$

Equation (7.8-9) has no rigorous basis but is an intuitive extension of the theoretical equation of Flory and Huggins. As discussed in Secs. 6.14 and 6.18, Wilson's equation is suitable for correlating experimental vapor-liquid equilibrium data for a variety of binary systems and, since it can be extended readily to multicomponent systems without any additional assumptions, it is useful for predicting multicomponent vapor-liquid equilibria using only parameters obtained from binary data.

The energy parameters λ_{12}, λ_{11}, and λ_{22} have no precise significance although they are determined by the intermolecular forces which exist in solution. If the cohesive force between two dissimilar molecules 1 and 2 is larger than that between two similar molecules 1 and 1 then $|\lambda_{12}| > |\lambda_{11}|$.‡ On the other hand, if component 1 is a strongly hydrogen-bonded liquid (e.g., ethanol) and component 2 is a nonpolar solvent (e.g., heptane) then $|\lambda_{11}| > |\lambda_{22}|$ and also $|\lambda_{11}| > |\lambda_{12}|$. In this particular case it would be difficult to say anything about the relative values

---

†Notice that according to Wilson's model $x_{21} + x_{11} = 1$ and $x_{12} + x_{22} = 1$. The sum of ξ_1 and ξ_2 is *not* unity except in the limiting case where $\lambda_{12} = \lambda_{11} = \lambda_{22}$.

‡Absolute values must be used because all three energy parameters λ_{12}, λ_{11}, and λ_{22} are relative to the ideal gas and therefore they are always negative.

of $|\lambda_{12}|$ and $|\lambda_{22}|$; however, if the nonpolar solvent were an aromatic (rather than a paraffin) and therefore capable of loose bonding with the alcohol, then it is likely that $|\lambda_{12}| > |\lambda_{22}|$.

To illustrate the variation of Wilson parameters with the chemical nature of the components, Orye[20] has reduced experimental vapor-liquid equilibrium data with Wilson's equation for five binary systems containing acetone as a common component in the region 45–50°C. Let 1 stand for acetone and 2 for the other component; for $(\lambda_{12} - \lambda_{11})$ Orye finds:

Acetone (1) and Component (2)	$\lambda_{12} - \lambda_{11}$ (cal/g-mol)
Carbon tetrachloride	880
Methyl acetate	576
Benzene	535
Chloroform	44
Methanol	−128

These results are shown schematically in Fig. 7-15. The Wilson parameters vary qualitatively as one might expect from the chemical nature of component 2. The 1-2 interaction increases as we proceed from relatively inert, nonpolar carbon tetrachloride through polar methyl

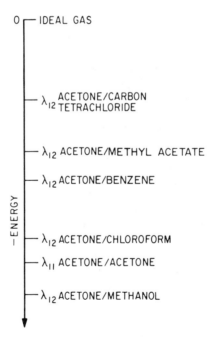

Fig. 7-15   Variation of Wilson energy parameters in acetone solutions (schematic).

acetate to benzene (where interaction of acetone with π-electrons is likely to occur) and finally to components which can form hydrogen bonds with acetone.

An extension of Wilson's equation has been proposed by Heil[21] for describing thermodynamic properties of strongly nonideal polymer solutions.

7.9 Corresponding-States Theory of Liquid Mixtures

The theory of corresponding states asserts that when the residual properties of two different fluids are suitably nondimensionalized (that is, expressed in reduced form) they should coincide when plotted upon the same nondimensional coordinates. Residual properties are thermodynamic properties which depend on the positions of the molecules relative to one another. In the ideal-gas state all molecules are effectively infinitely far apart, and therefore a *residual property* is a property relative to what it would be in the ideal-gas state at the same temperature, composition, and pressure.† All residual properties become zero for ideal gases.

If the residual properties of fluids 1 and 2 can be made to coincide on the same dimensionless plot, it would then appear reasonable to postulate that the residual properties of any mixture of these two fluids should also coincide on that same plot. It is this postulate which extends corresponding-states theory to mixtures. In order to make use of this extension it is necessary to decide how to nondimensionalize mixture properties; the appropriate reducing parameters must in some way depend on the composition of the mixture, but it is not at all obvious just what this dependence is. Determination of this dependence is the essence of the theoretical problem of corresponding-states theory for mixtures. But in addition to the theoretical problem there is also a practical problem, namely that the properties of very few pure liquids conform to corresponding-states behavior with sufficient accuracy; while the residual properties of several liquids may conform reasonably well to the same reduced plot, with only small deviations, it is an unfortunate fact that the excess residual properties for a mixture are frequently of the same order of magnitude as these small deviations. This practical problem is especially severe when corresponding-states theory uses only two parameters to characterize the the intermolecular potential.

Corresponding-states theory has received much attention from various investigators. We give here only a brief introduction to this subject in

---

†Sometimes it is more convenient to define a residual property as a property relative to what it would be in the ideal-gas state at the same temperature, composition, and volume.

order to show the general ideas. A serious study of solution theory based on corresponding states requires the tools of statistical mechanics.†

A presentation of corresponding-states theory for mixtures without the use of statistical mechanics was given by Scott,[22] and we shall summarize the main ideas as given in the first few sections of his paper. As discussed in Sec. 4.6, we assume that the potential energy between any two molecules is due to central forces only and is of the same form regardless of the identity of the molecules. That is, the intermolecular potential energy for any pair of molecules i and j depends only on the distance between molecular centers according to

$$\Gamma_{ij} = \epsilon_{ij}\, \ell\!\left(\frac{r_{ij}}{\sigma_{ij}}\right) \qquad (i = 1 \text{ or } 2; j = 1 \text{ or } 2) \qquad (7.9\text{-}1)$$

where ℓ is the same function for 1-1, 2-2 and 1-2 interactions and where ϵ_{ij} and σ_{ij} are, respectively, characteristic energy and size factors for the ij pair. We assume further that the potential energy of the entire liquid, pure or mixed, is pairwise additive. It then follows that the properly reduced residual properties of liquids 1 and 2 and any of their mixtures must be identical.

Let $g'(T, P)$ stand for the Gibbs energy of a fluid at temperature T and pressure P minus the Gibbs energy of that same fluid in the ideal-gas state at T and P. The theory of corresponding states says that

$$\frac{g'}{\epsilon}(T, P) = \ell_{g'}\!\left(\frac{kT}{\epsilon}, \frac{P\sigma^3}{\epsilon}\right), \qquad (7.9\text{-}2)$$

where ϵ and σ are characteristic constants (scale factors) for the liquid, k is Boltzmann's constant, and $\ell_{g'}$ is a universal function, valid for pure and mixed liquids. When the liquid is pure, it is clear that ϵ and σ are the potential parameters for that pure liquid; these parameters can, in principle, be determined from second virial coefficients, transport properties, critical data, etc. When the liquid is a mixture, however, ϵ and σ depend on the mole fraction in some manner about which corresponding-states theory by itself tells us nothing.

We consider a binary mixture of liquids 1 and 2; we let subscripts 1 and 2 stand for the pure components and subscript M for the mixture at some composition. At constant temperature and pressure, the residual

†For an advanced discussion see I. Prigogine, *The Molecular Theory of Solutions* (Amsterdam: North Holland Publishing Co., 1957), and A. Bellemans, V. Mathot, and M. Simon in *Advances in Chemical Physics*, Vol. XI, ed. by I. Prigogine (New York: Interscience Publishers, 1967). A short, well-written survey is given by N. G. Parsonage and L. A. K. Staveley in *Quart. Rev.* (London) **13**, 306 (1959). A more application-oriented study is given in three articles by C. A. Eckert, H. Renon, and J. M. Prausnitz, *I&EC Fundamentals*, **6**, 52 (1967); **6**, 58 (1967); and **7**, 335 (1968).

Gibbs energy g'_M can be related to g'_1 (or g'_2) by a Taylor series:

$$g'_M = g'_1 + \left(\frac{\partial g'_1}{\partial \epsilon}\right)_1 (\epsilon_M - \epsilon_1) + \left(\frac{\partial g'_1}{\partial (\sigma^3)}\right)_1 (\sigma^3_M - \sigma^3_1)$$

$$+ \frac{1}{2}\left(\frac{\partial^2 g'_1}{\partial \epsilon^2}\right)_1 (\epsilon_M - \epsilon_1)^2 + \left(\frac{\partial^2 g'_1}{\partial \epsilon \partial (\sigma^3)}\right)_1 (\epsilon_M - \epsilon_1)(\sigma^3_M - \sigma^3_1)$$

$$+ \frac{1}{2}\left(\frac{\partial^2 g'_1}{\partial (\sigma^3)^2}\right)_1 (\sigma^3_M - \sigma^3_1)^2 + \ldots \tag{7.9-3}$$

The important feature of Eq. (7.9-3) is that all the derivatives can be evaluated from the thermodynamic properties of pure liquid 1; thermodynamic properties of the mixture need not be known. Using standard thermodynamic identities we can relate all of these derivatives to measurable thermodynamic quantities. For example,

$$\left(\frac{\partial g'_1}{\partial (\sigma^3)}\right)_1 = \frac{\partial g'_1}{\partial \left(\frac{P\sigma^3}{\epsilon}\right)} \cdot \frac{\partial \left(\frac{P\sigma^3}{\epsilon}\right)}{\partial (\sigma^3)} = \frac{\partial g'_1}{\partial \left(\frac{P\sigma^3}{\epsilon}\right)} \cdot \frac{P}{\epsilon}. \tag{7.9-4}$$

Now, since

$$\frac{\partial g'_1}{\partial P} = (v^L - v^G_{\text{ideal}})_1 = v^L_1 - \frac{RT}{P}, \tag{7.9-5}$$

we obtain

$$\left(\frac{\partial g'_1}{\partial (\sigma^3)}\right)_1 = \frac{P}{\sigma^3_1}\left(v^L_1 - \frac{RT}{P}\right), \tag{7.9-6}$$

which can be found from measurable properties of pure liquid 1. Similarly, for the derivative with respect to the characteristic energy ϵ, we find

$$\left(\frac{\partial g'_1}{\partial \epsilon}\right)_1 = \frac{u'_1}{\epsilon_1}, \tag{7.9-7}\dagger$$

where u'_1 is the residual energy of pure liquid 1.

†To derive Eq. (7.9-7), it is important to remember that the parameter ϵ appears twice on the right-hand side of Eq. (7.9-2), once in the dimensionless temperature and once in the dimensionless pressure.

We can also obtain expressions for the three second derivatives but we will not need these for our purpose here.

Upon substitution we get

$$g'_M - g'_1 = u'_1 \left[\frac{\epsilon_M - \epsilon_1}{\epsilon_1} \right] + P\left(v_1^L - \frac{RT}{P} \right) \left[\frac{\sigma_M^3 - \sigma_1^3}{\sigma_1^3} \right]$$

$$+ \text{ second-order terms } + \ldots \tag{7.9-8}$$

Equation (7.9-8) gives the Gibbs energy of the mixture at any composition in terms of thermodynamic properties of pure liquid 1 provided we can decide on how the mixture parameters ϵ_M and σ_M depend on the mole fraction x; that is, we must specify the two functions

$$\epsilon_M = f_\epsilon(x_1, \epsilon_{11}, \epsilon_{22}, \epsilon_{12}) \tag{7.9-9}$$

$$\sigma_M = f_\sigma(x_1, \sigma_{11}, \sigma_{22}, \sigma_{12}). \tag{7.9-10}$$

There has been a lot of discussion on what these two functions should be; they cannot be derived from any general theory but depend on particular assumptions which one chooses to make about the structure of the solution. In addition, something must be said about the parameters ϵ_{12} and σ_{12}; the usual procedure is to use the common mixing rules: arithmetic mean for σ_{12} and geometric mean for ϵ_{12}. Unfortunately the calculations are quite sensitive to the functions chosen for Eqs. (7.9-9) and (7.9-10) and they are extremely sensitive to even very small deviations from the geometric-mean rule for ϵ_{12}.

Once the residual Gibbs energy of a mixture is known, the excess Gibbs energy of the mixture is given by

$$g^E = g'_M - x_1 g'_1 - x_2 g'_2, \tag{7.9-11}$$

and the activity coefficients are found by differentiation according to Eq. (6.3-8).

One of the favorable features of corresponding-states theory is that it gives all excess functions without any further assumptions beyond those already mentioned. Since the Gibbs energy of the mixture is given by Eq. (7.9-11) as a function of temperature and pressure in addition to composition, differentiation with respect to temperature gives the enthalpy and excess entropy of mixing, and differentiation with respect to pressure yields the excess volume.

In summary then, the central idea of a corresponding-states theory of mixture lies in the assumption that if reduced residual properties of a pure liquid are expressed as functions of reduced variables (reduced temperature, and reduced pressure or reduced volume), then these same functions may be used for a mixture merely by changing the reducing parameters. In the case of a pure fluid, the reducing parameters are constants which characterize molecular size and intermolecular forces in the pure fluid; for a mixture the reducing parameters are composition-dependent because they characterize average molecular size and average intermolecular forces in the mixture.

The first practical application of this idea was made by Kay[23] when he suggested that the gas-phase compressibility factor of a mixture of hydrocarbons could be found from a generalized compressibility factor diagram based on volumetric data for pure hydrocarbon gases.† This procedure has become known as the *pseudocritical method*. It was subsequently shown by Joffe[24] and by Watson and Gamson[25] that by using standard thermodynamic relations, Kay's approach could be used without additional assumptions to compute the fugacity of a component in a mixture. Kay used a particularly simple mixing rule for relating the reducing parameters for a mixture to the composition; more realistic mixing rules have been proposed by various authors[26] and the pseudocritical method is now well known, having won wide popularity for calculating properties of gas-phase mixtures. However, what is not known nearly so well is that the general idea of extending reduced purecomponent correlations of residual properties to mixtures can be applied to liquids as well as to gases.

For example, from pure-component volumetric data, it is possible to prepare a generalized chart for liquids giving f/P, the ratio of fugacity to pressure, as a function of reduced temperature T/T_c and reduced pressure P/P_c. If we assume that this chart also holds for the fugacity of a binary liquid mixture,‡ then we can compute the fugacity of component 1 in that mixture by the exact thermodynamic relation, at constant temperature and pressure:

$$\ln \frac{f_1}{x_1 P} = \ln \left(\frac{f}{P}\right)_M + (1 - x_1) \frac{\partial \ln (f/P)_M}{\partial x_1}, \qquad (7.9\text{-}12)$$

where subscript M stands for the mixture.

---

†The generalized compressibility factor chart is of the form

$$z = \pmb{\mathcal{I}}\left(\frac{P}{P_c}, \frac{T}{T_c}\right),$$

where P_c and T_c are the critical pressure and critical temperature. For a mixture, Kay assumed that the same chart is valid provided that for the reducing parameters P_c and T_c one substitutes $\sum_i y_i P_{c_i}$ and $\sum_i y_i T_{c_i}$ where y_i is the mole fraction of component i in the mixture.

‡We limit discussion to binary mixtures to explain the essentials; however, this technique for calculating partial properties of mixtures is applicable to mixtures of any desired number of components. The extension is straightforward and requires no additional assumptions. See, for example, T. W. Leland, Jr. and R. C. Reid, *A. I. Ch. E. Journal*, **11**, 228 (1965); K. S. Pitzer and R. Hultgren, *J. Am. Chem. Soc.*, **80**, 4793 (1958); and T. W. Leland, Jr. and P. S. Chappelear, *Ind. Eng. Chem.*, **60**, 15 (July 1968).

The derivative in Eq. (7.9-12) is related to the variation of T_c and P_c with x_1; since $\ln (f/P)_M$ is a function of T/T_{c_M} and P/P_{c_M} we have

$$d \ln \left(\frac{f}{P}\right)_M = \left[\frac{\partial \ln (f/P)_M}{\partial T/T_{c_M}}\right]_{P/P_{c_M}} d\left(\frac{T}{T_{c_M}}\right) + \left[\frac{\partial \ln (f/P)_M}{\partial P/P_{c_M}}\right]_{T/T_{c_M}} d\left(\frac{P}{P_{c_M}}\right).$$

$$(7.9\text{-}13)$$

The derivatives in Eq. (7.9-13) can be replaced by the standard thermodynamic equations

$$\left(\frac{\partial \ln (f/P)}{\partial T/T_{c_M}}\right)_{P/P_{c_M}} = \frac{-h'}{RT_{c_M}(T/T_{c_M})^2} \qquad (7.9\text{-}14)$$

and

$$\left(\frac{\partial \ln (f/P)}{\partial P/P_{c_M}}\right)_{T/T_{c_M}} = \frac{z - 1}{P/P_{c_M}}, \qquad (7.9\text{-}15)$$

where h' is the residual enthalpy and z is the compressibility factor at T/T_{c_M} and P/P_{c_M}.

Without commiting ourselves to any particular mixing rules let us just say that

$$T_{c_M} = \ell_{T_c}(x_1, T_{c_1}, T_{c_2}), \qquad (7.9\text{-}16)$$

and

$$P_{c_M} = \ell_{P_c}(x_1, P_{c_1}, P_{c_2}). \qquad (7.9\text{-}17)$$

Then Eq. (7.9-13), at constant temperature and pressure, becomes:

$$\frac{\partial \ln (f/P)_M}{\partial x_1} = \frac{h'}{RTT_{c_M}} \frac{d\ell_{T_c}}{dx_1} - \frac{(z - 1)}{P_{c_M}} \frac{d\ell_{P_c}}{dx_1}. \qquad (7.9\text{-}18)$$

When substituted into Eq. (7.9-12), we obtain a simple expression for the fugacity of component 1 in the liquid mixture. In order to use this expression it is necessary to have three reduced correlations for liquids: one for f/P, one for h', and finally, one for z. In addition, the mixing rules ℓ_{T_c} and ℓ_{P_c} must be specified.

This method for calculating fugacities of components in mixtures has been utilized by various authors, notably by Leland and Chappelear[27] who applied Eqs. (7.9-12) and (7.9-18) to both liquid and vapor phases for calculating phase equilibria in hydrocarbon mixtures at high pressures. As a brief schematic illustration of this type of calculation consider the following typical problem: Given the composition of a binary liquid phase and the pressure, what is the equilibrium (bubble point) temperature and what is the composition of the equilibrium vapor? The calculations which are required in order to solve this problem are of the trial-and-error type;

an acceptable solution must satisfy all of the equilibrium conditions: The temperatures and pressures of both phases must be the same and the fugacity of each component in the liquid phase must equal the fugacity of that same component in the vapor phase.

A solution to the problem can be obtained as illustrated in Fig. 7-16. For each component we calculate the fugacity in the liquid phase (at the

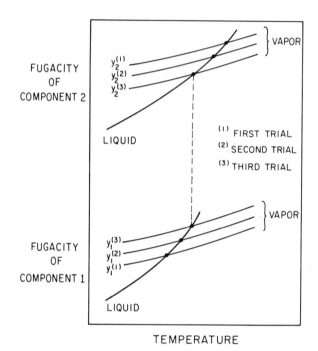

Fig. 7-16   Schematic trial-and-error calculation to determine bubble-point temperature in a binary mixture at fixed liquid composition and pressure. The third trial is successful.

known pressure and liquid composition) as a function of temperature. These calculations are shown by the two lines marked "liquid." Then, we assume a vapor composition (designated by $y^{(1)}$ for the first trial) and, at the known pressure, we calculate the fugacity of each component in the vapor phase as a function of temperature. As shown in Fig. 7-16, the necessary conditions $f_{1\text{(liquid)}} = f_{1\text{(vapor)}}$ and $f_{2\text{(liquid)}} = f_{2\text{(vapor)}}$ cannot be satisfied in the first trial at the same temperature; therefore the composition assumed for the first trial was incorrect and the calculation must be repeated with a new assumed vapor composition, designated by $y^{(2)}$ in Fig. 7-16. Trial 2 is also not successful and thus a third trial is made. As shown in Fig. 7-16, trial 3 is successful because the condition of equality of fugacities is satisfied for both components at the same temperature; this temperature, then, is the desired bubble-point temperature of the mixture.

Calculation of fugacities in mixtures using the theorem of corresponding states is attractive because such calculation is applicable to both liquid and vapor phases, may be used at any conditions of pressure, including the critical region, and no mixture data are required. The calculations are perhaps tedious but, in the era of computers, this is no major obstacle. Unfortunately, however, this method of calculation suffers from two serious disadvantages: First, the method requires that all substances in the mixture follow the *same* law of corresponding states and second, one must know how the reducing parameters for the mixture depend on composition.† It is possible in principle to use a three-parameter theorem of corresponding states (rather than the two-parameter theory used in the previous discussion) but this requires accurate pure-component data and, what is worse, introduces the necessity for still another mixing rule, viz., the variation of the third parameter with composition.

Prediction of fugacities of components in liquid mixtures using corresponding states is at best an approximation and even then it is useful for only a limited class of simple fluids. However, somewhat better results can be obtained if some mixture data‡ are used to serve as a guide in the mixing rules, Eqs. (7.9-16) and (7.9-17). The results are highly sensitive to these mixing rules and we see here, once again, that the key to progress in phase-equilibrium thermodynamics lies in improved understanding of intermolecular forces.

7.10 The Two-Liquid Theory

As discussed in the previous section, extension of corresponding-states theory to mixtures is based on the fundamental idea that a mixture can be considered to be a hypothetical pure fluid whose characteristic molecular size and potential energy are composition averages of the characteristic sizes and energies of the mixture's components or, in macroscopic terms, whose effective critical properties (pseudocriticals) are composition averages of the component critical properties. However, as shown by Scott,[22]

---

†There is no assurance that the functions in Eqs. (7.9-9) and (7.9-10) or (7.9-16) and (7.9-17) are the same for all compositions or, for that matter, that they are independent of temperature and pressure. In the derivations given here such independence has been assumed.

‡Data on second virial coefficients of the mixture are often useful for this purpose because they give some information on the important parameter ϵ_{12} (or $T_{c_{12}}$) which characterizes the energy of interaction between a pair of dissimilar molecules.

this fundamental idea is not limited to one hypothetical pure fluid; we can suppose that the properties of a binary mixture are equivalent to those obtained by averaging the properties of two hypothetical pure fluids, each of which is determined by composition-averaged characteristic constants. While Scott's two-liquid theory was developed in connection with his corresponding-states treatment, its application is more general and is not limited by corresponding-states assumptions.

To fix ideas, consider a binary mixture as illustrated in Fig. 7-17.

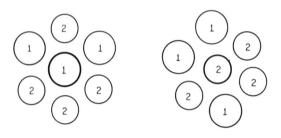

MOLECULE 1 AT CENTER   MOLECULE 2 AT CENTER

Fig. 7-17   Two types of cells according to Scott's two-liquid theory of binary mixtures.

Each molecule is closely surrounded by other molecules; we refer to the immediate region around any central molecule as that molecule's cell. In a binary mixture of components 1 and 2 we have two types of cells: one type contains molecule 1 at its center and the other contains molecule 2 at its center. The chemical nature (1 or 2) of the molecules surrounding a central molecule depends on the mole fractions x_1 and x_2. Let $M^{(1)}$ be some extensive residual property M of a hypothetical fluid consisting only of cells of type 1; similarly, let $M^{(2)}$ be that same residual property of a hypothetical fluid all of whose cells are of type 2. The two-liquid theory says that the extensive residual property M of the mixture is given by

$$M_{\text{mixture}} = x_1 M^{(1)} + x_2 M^{(2)}. \qquad (7.10\text{-}1)$$

For mixtures of nonpolar components, $M^{(1)}$ and $M^{(2)}$ can, perhaps, be calculated from corresponding-states correlations for pure fluids by suitably averaging characteristic molecular (or critical) properties. For example, let M stand for the molar residual enthalpy h'; assume that

$$\frac{h'}{RT} = \mathcal{f}\left(\frac{kT}{\epsilon}, \frac{P\sigma^3}{\epsilon}\right) \qquad (7.10\text{-}2)$$

where σ and ϵ are characteristic size and energy parameters and where $\mathcal{f}$ is a function determined from experimental pure-component properties. To find $h'^{(1)}$ we assume

$$\frac{h'^{(1)}}{RT} = \mathcal{f}\left(\frac{kT}{\epsilon^{(1)}}, \frac{P\sigma^{(1)3}}{\epsilon^{(1)}}\right) \tag{7.10-3}$$

where $\mathcal{f}$ is the same function as in Eq. (7.10-2) and $\epsilon^{(1)}$ and $\sigma^{(1)}$ denote composition averages for cells of type 1. For example, we might assume that

$$\epsilon^{(1)} = x_1 \epsilon_{11} + x_2 \epsilon_{12} \tag{7.10-4}$$

$$\sigma^{(1)} = x_1 \sigma_{11} + x_2 \sigma_{12} \tag{7.10-5}$$

where ϵ_{ij} and σ_{ij} are constants characteristic of the i-j interaction. Similarly,

$$\frac{h'^{(2)}}{RT} = \mathcal{f}\left(\frac{kT}{\epsilon^{(2)}}, \frac{P\sigma^{(2)3}}{\epsilon^{(2)}}\right) \tag{7.10-6}$$

and, using the same assumption,

$$\epsilon^{(2)} = x_2 \epsilon_{22} + x_1 \epsilon_{12} \tag{7.10-7}$$

$$\sigma^{(2)} = x_2 \sigma_{22} + x_1 \sigma_{12}. \tag{7.10-8}$$

The molar residual enthalpy of the mixture is given by Eq. (7.10-1); it is

$$h'_{\text{mixture}} = x_1 h'^{(1)} + x_2 h'^{(2)}. \tag{7.10-9}$$

Application of corresponding-states theory to mixtures as described in Sec. 7.9 represents a special case of the two-liquid theory; it is, in fact, a one-liquid theory where no distinction is made between type-1 cells and type-2 cells.

Scott[22] has also discussed a three-liquid theory for binary mixtures where

$$M_{\text{mixture}} = x_1^2 M^{(11)} + 2x_1 x_2 M^{(12)} + x_2^2 M^{(22)}. \tag{7.10-10}$$

In Eq. (7.10-10), $M^{(11)}$ is the residual property M for pure 1 and $M^{(22)}$ is that for pure 2; $M^{(12)}$ is the residual property of a hypothetical fluid all of whose molecules interact according to Γ_{12}, the intermolecular potential between molecule 1 and molecule 2. Equation (7.10-10) becomes exact for fluids at low density where the volumetric properties are given by the virial equation truncated after the second virial coefficient (see Sec. 5.3).

While Eqs. (7.10-2) to (7.10-8) are based on corresponding-states arguments, the fundamental idea of the two-liquid theory [Eq. (7.10-1)] is independent of corresponding-states theory. The quantities $M^{(1)}$ and $M^{(2)}$ may be, but need not be, determined from corresponding-states calculations.

The two-liquid theory for a binary mixture can immediately be extended to mixtures containing any number of components. If there are m components, then there are m types of cells and the two-fluid binary theory becomes an m-fluid theory for an m-component mixture.

The two-liquid theory provides a convenient point of departure for deriving semiempirical equations to represent thermodynamic excess functions. For example, Renon[28] has derived such equations by applying Wilson's idea of local compositions to the two-liquid concept. Let $g^{(1)}$ and $g^{(2)}$ stand, respectively, for the residual Gibbs energy of one mole of cells of type 1 and one mole of cells of type 2. These Gibbs energies are assumed to be related to the local mole fractions according to

$$g^{(1)} = x_{11}g_{11} + x_{21}g_{21} \tag{7.10-11}$$

$$g^{(2)} = x_{12}g_{12} + x_{22}g_{22} \tag{7.10-12}$$

where g_{11} and g_{22} are parameters characteristic, respectively, of 1-1 and 2-2 interactions, and g_{12} is a parameter characteristic of the 1-2 interaction.† The local mole fraction x_{ij} is the mole fraction of i in the immediate vicinity of a molecule j. The local mole fractions must obey the conservation equations

$$\begin{aligned} x_{21} + x_{11} &= 1 \\ x_{12} + x_{22} &= 1. \end{aligned} \tag{7.10-13}$$

For pure component 1, $x_{21} = 0$ and therefore

$$g^{(1)}_{\text{pure}} = g_{11}. \tag{7.10-14}$$

Similarly, for pure component 2,

$$g^{(2)}_{\text{pure}} = g_{22}. \tag{7.10-15}$$

The molar excess Gibbs energy can now be found by considering first the change in residual Gibbs energy which results when x_1 moles of component 1 are transferred from cells containing only component 1 to their cells in the mixture, and then the similar change for x_2 moles of

†$g_{12} = g_{21}$

component 2:

$$g^E = x_1[g^{(1)} - g_{\text{pure}}^{(1)}] + x_2[g^{(2)} - g_{\text{pure}}^{(2)}]. \tag{7.10-16}$$

Equation (7.10-16) gives the *excess* Gibbs energy (rather than the Gibbs energy of mixing) because a residual property is defined as that property relative to the ideal gas at the same temperature, pressure, *and composition.* As a result the ideal-solution contribution to the Gibbs energy of mixing cancels.

Substituting Eqs. (7.10-11) and (7.10-12) and (7.10-14) and (7.10-15) into (7.10-16), we obtain

$$g^E = x_1 x_{21}(g_{21} - g_{11}) + x_2 x_{12}(g_{12} - g_{22}). \tag{7.10-17}$$

If we assume random mixing, then the local mole fractions are the same as the overall mole fractions; i.e., for random mixing, $x_{21} = x_2$ and $x_{12} = x_1$. In that event, Eq. (7.10-16) simplifies to

$$g^E = x_1 x_2[2g_{12} - g_{11} - g_{22}]. \tag{7.10-18}$$

Equation (7.10-18) is of the same form as that obtained from Guggenheim's theory of regular solutions (see Sec. 7.4). Instead of assuming random mixing, Renon assumes a relation suggested by Guggenheim's quasichemical approximation; it is

$$\frac{x_{21}}{x_{11}} = \frac{x_2 \exp[-\alpha_{12}g_{21}/RT]}{x_1 \exp[-\alpha_{12}g_{11}/RT]} \tag{7.10-19}†$$

and similarly, upon interchanging subscripts,

$$\frac{x_{12}}{x_{22}} = \frac{x_1 \exp[-\alpha_{21}g_{12}/RT]}{x_2 \exp[-\alpha_{21}g_{22}/RT]}. \tag{7.10-20}†$$

Equations (7.10-19) and (7.10-20) are similar to those used by Wilson; they relate the local mole fractions to the overall mole fractions through Boltzmann factors. The parameter α_{12} is a constant characterizing the tendency of the components to mix in a nonrandom fashion; when $\alpha_{12} = 0$, the local mole fractions are equal to the overall mole fractions and mixing is completely random.

Substitution gives the excess Gibbs energy in terms of the overall mole fractions x_1 and x_2 and three binary parameters:

$$\frac{g^E}{RT} = x_1 x_2 \left[\frac{\tau_{21} G_{21}}{x_1 + x_2 G_{21}} + \frac{\tau_{12} G_{12}}{x_2 + x_1 G_{12}} \right] \tag{7.10-21}$$

†$g_{12} = g_{21}$ and $\alpha_{12} = \alpha_{21}$.

where

$$\tau_{12} = \frac{(g_{12} - g_{22})}{RT}$$

$$\tau_{21} = \frac{(g_{21} - g_{11})}{RT}$$

(7.10-22)†

$$G_{12} = \exp\left[-\alpha_{12}\tau_{12}\right]$$

$$G_{21} = \exp\left[-\alpha_{21}\tau_{21}\right].$$

(7.10-23)†

As discussed in Sec. 6.14, Eq. (7.10-21) gives a good representation of excess functions for a variety of liquid mixtures, including those with limited miscibility. Although the parameters $(g_{12} - g_{22})$ and $(g_{21} - g_{11})$ are temperature dependent, it appears that, to a good approximation, α_{12} may be considered independent of temperature and, when data are scarce, α_{12} may be estimated upon considering the chemical nature of the components. For mixtures containing more than two components, Eq. (7.10-21) can readily be generalized[28] without additional assumptions; the general result contains only binary parameters.

7.11   The "Chemical" Theory of Solutions

The theories described in the previous sections attempt to explain solution nonideality in terms of physical intermolecular forces. The regular-solution model, the lattice theory and those theories based on the theorem of corresponding states, relate the activity coefficients to physical quantities which reflect the size of the molecules and the physical forces (primarily London dispersion forces) operating between them. An alternate approach to the study of solution properties is based on a rather different premise, viz., that molecules in a liquid solution interact with each other to form new chemical species and that solution nonideality, therefore, is a consequence of chemical reactions.

We can distinguish between two types of reactions, *association* and *solvation*. Association refers to formation of chemical aggregates‡ or polymers consisting of identical monomers. An association can be represented by reactions of the type

$$n\text{B} \; \rightleftharpoons \; \text{B}_n,$$

where B is the monomer and n is the degree of association (or polymerization).

---

†$g_{12} = g_{21}$ and $\alpha_{12} = \alpha_{21}$.

‡While aggregates are, in effect, loosely bonded chemical compounds, this does not imply that they can necessarily be separated and exist by themselves. In fact, such separations are usually not possible.

A common case of association is dimerization ($n = 2$) and a well-known example is dimerization of acetic acid:

$$2\ CH_3-C\overset{O}{\underset{OH}{\big\langle}} \rightleftharpoons CH_3-C\overset{O---HO}{\underset{OH---O}{\big\langle}}C-CH_3$$

In this case, dimerization is due to hydrogen bonding which is responsible for the most common form of association in liquid solutions.

Solvation refers to formation of chemical aggregates of two or more molecules of which at least two are not identical. It can be represented by the general equation

$$n\,A + m\,B \rightleftharpoons A_n B_m.$$

A well-known example ($n = m = 1$) is solvation of chloroform and diethyl ether:

$$Cl_3-C-H + O\overset{C_2H_5}{\underset{C_2H_5}{\big\langle}} \rightleftharpoons Cl_3-C-H---O\overset{C_2H_5}{\underset{C_2H_5}{\big\langle}}$$

In this case, formation of the new species is again due to hydrogen bonding. Another example of solvation is given by a charge-transfer complex between nitrobenzene and mesitylene:

In this case a weak chemical bond is formed because mesitylene is a good electron donor (Lewis base) and nitrobenzene is a good electron acceptor (Lewis acid).

The chemical theory of solution postulates existence of chemically distinct species in solution which are assumed to be in chemical equilibrium. In its original form, the theory then assumes further that these chemically distinct substances form an ideal solution. According to these assumptions, the observed nonideality of a solution is merely an apparent one due to the fact that it is based on an apparent, rather than a true, account of the solution's composition.

The chemical theory of solution behavior was first developed by Dolezalek[29] at about the same time van Laar was publishing his work on solutions. The different points of view represented by these early workers in solution theory caused them to be hard enemies and some of their publications contain much bitter polemic.[30]

Dolezalek's theory has the advantage that it can readily account for both positive and negative deviations from ideality for molecules of similar size; also, unlike the van Laar-Scatchard-Hildebrand theory of regular solutions, it is applicable to mixtures containing polar and hydrogen-bonded liquids. Its great disadvantage lies in its arbitrariness in deciding what "true" chemical species are present in the solution, and in the fact that we usually cannot assign equilibrium constants to the postulated equilibria without experimental data on the solution under consideration. The chemical theory of solutions, therefore, has little predictive value; it almost never can give quantitative predictions of solution behavior from pure-component data alone. However, for those solutions where chemical forces are dominant, the chemical theory has much qualitative and interpretative value; if some data on such a solution are available, they can often be interpreted along reasonable *chemical* lines and therefore the chemical theory can serve as a tool for interpolation and cautious extrapolation of limited data. A *chemical* rather than a *physical* view is frequently useful for correlation of solution nonidealities in a class of chemically similar mixtures (e.g., alcohols in paraffinic solvents). In the next two sections we discuss the properties of associated solutions. Solvated solutions are discussed in Secs. 7.14 and 7.15.

7.12 Activity Coefficients in Associated Solutions

Suppose we have a liquid mixture of two components, 1 and 2. Component 1 is a nonpolar substance which we designate A, but component 2 is a polar material which we designate B and which, we assume, can dimerize according to

$$2B \; \rightleftharpoons \; B_2 .$$

The equilibrium constant for this dimerization is given by

$$K = \frac{a_{B_2}}{a_B^2} , \qquad (7.12\text{-}1)$$

where a_B is the activity of monomer B molecules and a_{B_2} is the activity of dimer B_2 molecules.

We assume that the species A, B, and B_2 are in equilibrium with one another and we assume further that these three species form an ideal liquid solution. In an ideal solution the activities can be replaced by the mole fractions and Eq. (7.12-1) becomes

$$K = \frac{\mathfrak{z}_{B_2}}{\mathfrak{z}_B^2} ,$$

(7.12-2)

where $\mathfrak{z}$ stands for the "true" mole fraction.

If there are n_1 moles of component 1 and n_2 moles of component 2, then

$$n_1 = n_A$$

(7.12-3)

and

$$n_2 = n_B + 2n_{B_2} ,$$

(7.12-4)

where n_B is the number of moles of monomer B and n_{B_2} is the number of moles of dimer B_2. The "true" total number of moles is $n_A + n_B + n_{B_2}$, which is equal to $n_1 + n_2 - n_{B_2}$. The three "true" mole fractions are:

$$\mathfrak{z}_A = \frac{n_1}{n_1 + n_2 - n_{B_2}} ,$$

(7.12-5)

$$\mathfrak{z}_B = \frac{n_B}{n_1 + n_2 - n_{B_2}} = \frac{n_2 - 2n_{B_2}}{n_1 + n_2 - n_{B_2}} ,$$

(7.12-6)

$$\mathfrak{z}_{B_2} = \frac{n_{B_2}}{n_1 + n_2 - n_{B_2}} .$$

(7.12-7)

If we combine Eq. (7.12-2) with Eqs. (7.12-6) and (7.12-7) and then eliminate n_{B_2} with Eq. (7.12-5), we obtain an expression which we can solve for $\mathfrak{z}_A$ and from this we can obtain the desired result for the activity coefficient of component 1. The algebra is straightforward but involved, and is not reproduced here. Remembering that $x_1 = n_1/(n_1 + n_2)$ and that $x_2 = n_2/(n_1 + n_2)$, we obtain

$$\gamma_1 = \frac{a_1}{x_1} = \frac{2k}{(2k - 1)x_1 + kx_2 + (x_1^2 + 2kx_1 x_2 + kx_2^2)^{1/2}} ,$$

(7.12-8)

where

$$k \equiv 4K + 1.$$

By similar stoichiometric considerations the activity coefficient for the dimerizing component can be shown to be:

$$\gamma_2 = \frac{a_2}{x_2} = \left(\frac{k^{1/2} + 1}{x_2}\right)\left[\frac{- x_1 + (x_1^2 + 2kx_1 x_2 + kx_2^2)^{1/2}}{(2k - 1)x_1 + kx_2 + (x_1^2 + 2kx_1 x_2 + kx_2^2)^{1/2}}\right].$$

(7.12-9)

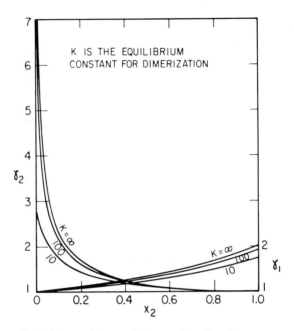

Fig. 7-18   Activity coefficients of a dimerizing component (2) and an "inert" solvent (1).

Figure 7-18 shows plots of Eqs. (7.12-8) and (7.12-9) for three different dimerization equilibrium constants. It is a property of both Eqs. (7.12-8) and (7.12-9) that for all $K > 0$, positive deviations from Raoult's law result; thus $\gamma_1 \geq 1$ and $\gamma_2 \geq 1$, as shown. For $K = \infty$ all molecules of component 2 are dimerized and in this limiting case we have

$$\lim_{K \to \infty} \gamma_1 = \frac{1}{x_1 + (x_2/2)} \quad \text{and} \quad \lim_{K \to \infty} \gamma_2 = \frac{1}{(2x_1 x_2 + x_2^2)^{1/2}}. \quad (7.12\text{-}10)$$

At infinite dilution, then

$$\lim_{\substack{K \to \infty \\ x_1 \to 0}} \gamma_1 = 2 \quad \text{and} \quad \lim_{\substack{K \to \infty \\ x_2 \to 0}} \gamma_2 = \infty . \quad (7.12\text{-}11)$$

For the nondimerizing component the largest possible value of γ_1 is 2.

To test the dimerization model represented by Eqs. (7.12-8) and (7.12-9), we can compare calculated and observed activity coefficients for a binary system containing an organic acid in a nonpolar "inert" solvent since spectroscopic, cryoscopic, and distribution data all indicate that organic acids have a strong tendency to dimerize. Figure 7-19 shows such a comparison for propionic acid dissolved in n-octane. The experi-

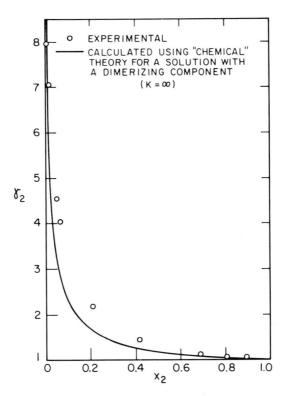

Fig. 7-19 Calculated and experimental activity coefficients for propionic acid (2) dissolved
in n-octane (1) at 1 atm ($T = 121.3 -$
$141.1°C$).

mental data[31] were unfortunately obtained at constant pressure rather
than constant temperature and as a result the comparison is not completely straightforward since the dimerization equilibrium constant depends on temperature.

The experimental data shown in Fig. 7-19 follow the general trend predicted by Eq. (7.12-9), but even with $K = \infty$, the observed deviations
from ideality are larger than those calculated. Since dimerization of an
organic acid is exothermic, dimerization constant K falls with rising
temperature but it is apparent that the variable temperature of the experimental data is not responsible for the lack of agreement between theoretical and experimental results. In this case the chemical theory is evi-
dently able to account qualitatively for the observed activity coefficients,
but since physical effects are neglected by the chemical theory, quantitative agreement is not obtained.

Another comparison between calculated and experimental results is

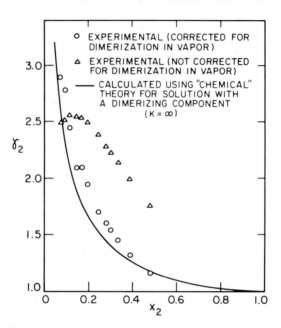

Fig. 7-20   Calculated and experimental activity coefficients for acetic acid (2) dissolved in benzene (1) at 25°C.

given in Fig. 7-20 which shows activity coefficients for acetic acid in the
benzene/acetic acid system at 25°C. Experimental data were reported by
Hovorka and Dreisbach[32] and, again, qualitative agreement is obtained
between experimental and theoretical results.† The activity coefficients
for acetic acid in benzene are somewhat lower than those for propionic
acid in octane; because of the higher polarizability of π electrons in
benzene it is likely that the forces of attraction between benzene and acid
are stronger than those between octane and acid.

The two examples given in Figs. 7-19 and 7-20 show only fair agreement between theory and experiment. But even if the agreement were
good, it should not by itself be considered proof of the validity of the
chemical theory. Good agreement would have shown only that the
assumptions of the chemical theory, with the help of one adjustable

---

†Figure 7-20 presents two sets of results, each based on a particular method of data
reduction. Since the total pressure at 25°C is much less than 1 atmosphere, activity coefficients represented by triangles were calculated from the experimental data without any
vapor-phase correction; neither the fugacity of the acid in the vapor-phase mixture nor the
(standard-state) fugacity of pure acetic acid were corrected for nonideal behavior. However,
as discussed in Sec. 5.9, vapor-phase corrections are important for carboxylic acids even at
pressures of the order of 10 mm Hg. When such corrections are included in data reduction,
the results obtained are those represented by circles.

parameter, are consistent with the experimental facts, but the assumptions of some other theory, again with one adjustable parameter, may be just as consistent with these facts. In other words, when agreement is good, we may say that the chemical theory offers a possible, but by no means unique, explanation of the observed thermodynamic properties. A chemical theory of solution behavior must always be viewed with suspicion unless there is independent evidence to support it. In the case of propionic (or acetic) acid, dissolved in octane or some other relatively "inert" solvent, we believe that the chemical explanation is reasonable because of independent (e.g., spectroscopic and cryoscopic) evidence which has confirmed that organic acids do, in fact, dimerize in nonpolar solvents.

In Eq. (7.12-8) even very large values of the dimerization equilibrium constant K cannot produce high activity coefficients for the "inert" component; yet, such activity coefficients have been observed in some associated solutions, notably for nonpolar solvents dissolved in an excess of alcohol. Furthermore, the activity coefficients of the associating component are often considerably larger than those calculated by Eq. (7.12-9) even with $K = \infty$. In order to explain these very large activity coefficients it has been proposed that the associating component undergoes chain association, i.e., it forms not only dimers but also trimers, tetramers, etc., according to

$$2B \; \rightleftharpoons \; B_2$$
$$B + B_2 \; \rightleftharpoons \; B_3$$
$$B + B_3 \; \rightleftharpoons \; B_4 \quad \text{etc.}$$

For example, phenol is known from a variety of physicochemical data to form a multiple chain according to the structure

$$O{-}H{-}{-}{-}O{-}H{-}{-}{-}O{-}H{-}{-}{-} \quad \text{etc.}$$

Since the tendency of phenol to form chains is a strong function of phenol concentration (especially in the dilute region) it follows that the activity of phenol, when dissolved in some solvent, shows large deviations from ideal behavior.

The thermodynamics of solutions which contain a component capable of multiple association has been considered by many authors; good discussions of this subject are given by Kortüm and Buchholz-Meisenheimer[33] and by Prigogine and Defay.[34] We discuss here only the main concepts and present some typical results.

In order to reduce the number of adjustable parameters it is common to assume that the equilibrium constant for the formation of a chain is independent of the chain length. That is, if we write the general association equilibrium

$$B + B_{n-1} \; \rightleftharpoons \; B_n,$$

then the equilibrium constant K_n is

$$K_n = \frac{a_{B_n}}{a_B \, a_{B_{n-1}}}, \tag{7.12-12}$$

where a is the activity. The simplifying assumption now is that $K_2 = K_3 \ldots K_n = K.$ †

If, as before, we assume that the solution of "true" species is an ideal solution, then we can replace the activity a by the "true" mole fraction $_3$. The mathematical details are tedious and therefore they are not reproduced here, but it can be shown that the activity coefficient of component 2, the associating component, is given by

$$\gamma_2 = \frac{(1 + K)(1 + KZ^2)}{(1 + KZ)^2} \tag{7.12-13}$$

where

$$Z \equiv \frac{[1 + 4Kx_2(1 - x_2)]^{1/2} - 1}{2K(1 - x_2)}. \tag{7.12-14}$$

The activity coefficient of the nonassociating component 1 is given by

$$\gamma_1 = 1 + KZ^2. \tag{7.12-15}$$

Figure 7-21 shows activity coefficients for the multiply-associated component for several values of the equilibrium constant K. At small mole fractions the activity coefficients are now much larger than those shown in Fig. 7-18 which considered only dimerization. The activity coefficients for the associated component are large at the dilute end but fall rapidly as the concentration rises; this characteristic behavior is in excellent agreement with experimental results for solutions of alcohols in nonpolar solvents.

---

†A few studies have been reported where this assumption is not made. However, the algebraic complexity and the amount of data required for meaningful data reduction are much larger when the equilibrium constant is allowed to depend on the degree of association and frequently the extra labor is not justified. One fine example, however, of such a detailed study is that by H. Wolff and A. Höpfner [Z. Elektrochemie, 66, 149 (1962)] for solutions of methylamine and hexane. These authors, using extensive data for the range −55 to 20°C, report equilibrium constants and heats of formation for dimers, trimers and tetramers of methylamine. At a given temperature the authors found that the three equilibrium constants were close to one another but not the same.

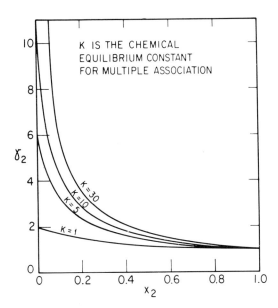

Fig. 7-21   Activity coefficient of a multiply-associated component in an "inert" solvent.

For example, Prigogine, Mathot, and Desmyter[35,36] have shown that observed excess Gibbs energies of solutions of alcohols in carbon tetrachloride can be closely approximated by ascribing the nonideality of these
solutions to multiple association. The small difference between calculated
and observed excess Gibbs energies is probably due to physical forces
which, by assumption, have been neglected in this purely chemical treatment. We shall return to this point in the next section, but first we want to
consider some of the implications of the chemical theory of associated
solutions.

When considering the validity of any theory of solutions it is most important to inquire whether or not it can be supported by evidence based
on several physicochemical properties. The validity of any theory of solution is much enhanced if it can be shown to be in agreement with observed
physical properties other than those used to obtain activity coefficients.
For the theory of associated solutions, where association is due to hydrogen bonding, such support is provided by infrared spectroscopy as shown
by many workers, notably by Hoffmann, Errera, and Sack, and
others.[33,34,37] For example, consider a solution of methanol in a nonpolar
solvent such as carbon tetrachloride. The frequency of vibration of the
OH group is in the infrared spectrum and this frequency is strongly affected by whether or not it is "free" (i.e., attached only to a carbon atom
by a normal covalent bond) or whether it is also attached to another OH
group through hydrogen bonding. Therefore, by measuring the intensity

of absorption at the frequency corresponding to the "free" vibration, it is possible to determine the concentration of alcohol which is in the monomeric, nonassociated state. Spectroscopic measurements thus provide an independent check on the theory of associated solutions. If we let α_{B_1} stand for the fraction of alcohol molecules which are in the monomeric form, then as shown by Prigogine, we can relate α_{B_1} to the unsymmetrically normalized activity coefficients of alcohol (2) and carbon tetrachloride (1) by the remarkably simple equation:

$$\alpha_{B_1} = \frac{\gamma_2^*}{\gamma_1} \tag{7.12-16}$$

where γ_2^* and γ_1 are activity coefficients normalized by the unsymmetric convention (see Sec. 6.4):

$$\gamma_2^* \longrightarrow 1 \quad \text{as} \quad x_2 \longrightarrow 0$$

and

$$\gamma_1 \longrightarrow 1 \quad \text{as} \quad x_1 \longrightarrow 1.$$

Figure 7-22 compares values of α_{B_1} determined spectroscopically with those determined by standard thermodynamic measurements for the system methanol/carbon tetrachloride. The good agreement lends support to the essential ideas of the theory of multiple association for mixtures of alcohols and nonpolar solvents; however, more recent studies by Van Ness[38] indicate that the structure of alcohol solutions is probably considerably more complex than that assumed by Prigogine.

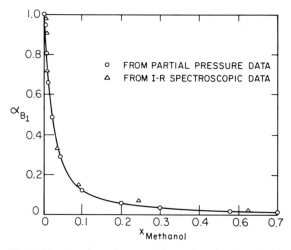

Fig. 7-22 Fraction of monomeric methanol molecules in carbon tetrachloride solution at 20°C.

The simple theory of multiply-associated solutions[33] also establishes a relation between the activity coefficient of the nonassociating solvent and $\bar{n}$, the average chain length of the associating component:

$$\bar{n} = \frac{\gamma_1 x_2}{1 - \gamma_1 x_1}.\qquad (7.12\text{-}17)$$

A comparison can then be made between average chain lengths calculated from activity coefficients and from other physicochemical measurements such as spectroscopy and cryoscopy. Again good agreement is obtained when results based on different methods of measurement are compared with one another. Figure 7-23 shows average chain lengths as a

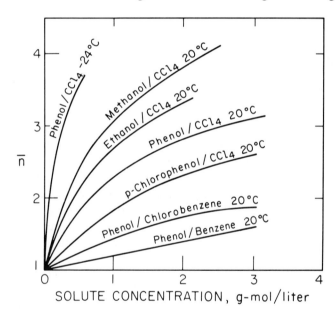

Fig. 7-23   Average chain length of multiply-associated solutes dissolved in various solvents, as determined by I-R spectroscopy. [R. Mecke, *Z. Elektrochemie* **52,** 274 (1948)]

function of alcohol concentration for several systems. The degree of association rises rapidly in the dilute region; this is consistent with the experimental observation that the activity coefficient of an alcohol in a nonpolar solvent is large at infinite dilution but then falls quickly as the mole fraction of alcohol rises. Figure 7-23 shows that for phenol in carbon tetrachloride the extent of association falls with rising temperature, as expected, since the association reaction (hydrogen bonding) is exothermic. However, Fig. 7-23 also shows that for phenol at a constant temperature of $20°C$ the average chain length depends not only on the alcohol concentration but also on the nature of the solvent.† This observation is not consistent with the assumptions of the chemical theory which postulates that the role of any solvent is merely to act as an *inert dispersing agent* for the associated component; in other words, the chemical theory assumes that there are no solvent-solute interactions. According to the theory of multiple association, at a fixed temperature the activity coefficient of a certain associating component should be a function only of its mole fraction regardless of the nature of the "inert" solvent. However, experimental data for many solutions of alcohols in nonpolar solvents show that this is not the case, indicating that the concept of an inert solvent is a convenient, but unreal oversimplification.

7.13   Associated Solutions with
Physical Interactions

The chemical solution theory of Dolezalek assumes that all deviations from ideal behavior are due to formation and destruction of chemical species and that once the mole fractions of the true species are known, the properties of the solution can be calculated without any further consideration being given to interactions between the true species. By contrast, the physical theory states that the true species are the same as the apparent species and that while there are physical (van der Waals) intermolecular forces, there are no chemical reactions in the solution.

The chemical and the physical theories of solutions are extreme, one-sided statements of what we now believe to be the actual situation. In certain limiting cases each theory provides a satisfactory approximation: When forces between molecules are weak, no new stable chemical species are formed and the physical theory applies; on the other hand, when forces between molecules are strong, these forces result in the formation of chemical bonds and, since the energies for chemical bond formation are significantly larger than those corresponding to van der Waals forces,‡ the chemical theory for such cases provides a reasonable description. In general, both physical and chemical forces should be taken into account. A comprehensive theory of solutions should allow for a smooth transition from one limit of a "physical" solution to the other limit of a "chemical" solution.

---

† Notice that $\bar{n}$ for phenol is significantly lower in chlorobenzene and benzene than in carbon tetrachloride. This result is probably due to a certain amount of solvation between the alcohol and the π-electrons of the aromatic solvents.

‡ Weak forces between molecules are those which, roughly speaking, have energies less than RT while strong forces between molecules are those whose energies are considerably larger than RT. At room temperature RT is about 600 cal/g-mol.

It is difficult to formulate a theory which takes account of both physical and chemical effects without thereby introducing involved algebra and, what is worse, a large number of adjustable parameters. Nevertheless, a few attempts have been made and one of the more successful ones is based on the theory of polymer mixtures of Flory.[39] This theory has been applied to mixtures containing aliphatic alcohols and paraffinic hydrocarbons;[40,41] in such mixtures, alcohol polymerizes to form chains but these chains interact with the paraffin only through van der Waals forces. We now briefly outline the essential ideas of the theory of associated solutions following the discussion of Renon.[42]

Let A stand for the hydrocarbon and B for the alcohol. We assume that:

1. The alcohol exists in the solution in the form of linear, hydrogen-bonded polymers B_1, B_2, ..., B_n, ... formed by successive reactions of the type

$$B_1 + B_{n-1} \rightleftharpoons B_n.$$

2. The association constant for the above reaction is independent of n.
3. The molar volume of an n-mer is given by the molar volume of the monomer multiplied by n.
4. There are physical interactions between all of the molecules which can be characterized by expressions of the van Laar type.
5. The temperature dependence of the association constant K is such that the heat of formation of a hydrogen bond is independent of the temperature and the degree of association.

On the basis of his lattice model, Flory[39] derived expressions for the thermodynamic properties of solutions of polymers differing only in molecular weight; these expressions are especially suited for chemical equilibria between linear, polymeric species. We can use Flory's results for the entropy of mixing to obtain the proper expression for the equilibrium constant. It is:

$$K(T) = \frac{\Phi_{B_{n+1}}}{\Phi_{B_n} \Phi_{B_1}} \frac{n}{n+1} \qquad (7.13\text{-}1)$$

where Φ_{B_n} is the volume fraction of species B_n.

The derivation of Eq. (7.13-1) is given elsewhere[39] and we will not repeat it here. Flory's treatment clearly shows that the equilibrium constant should not be expressed in terms of mole fractions as in Sec. 7.12.

As has been shown by several authors,[40,41,42,43] the excess Gibbs energy, taken relative to an ideal solution of alcohol and hydrocarbon, can be separated into two contributions, one chemical and the other physical:

$$g^E = g_c^E + g_p^E. \tag{7.13-2}$$

The chemical contribution g_c^E results from the dependence of the "true" composition of the solution on the chemical equilibria indicated by assumption 1. The excess entropy introduced by the mixing of polymeric species is taken into account in the calculation of the chemical contribution. From Flory's theory we have

$$g_c^E = x_A \ln \frac{\Phi_A}{x_A} + x_B \ln \frac{\Phi_{B_1}}{\Phi_{B_1}^* x_B} + K x_B (\Phi_{B_1} - \Phi_{B_1}^*) \tag{7.13-3}$$

where x_A and x_B are the overall (stoichiometric) mole fractions, Φ_{B_1} is the (true) volume fraction of molecular species B_1, the alcohol monomer, and

$$\Phi_{B_1}^* = \lim_{x_A \to 0} \Phi_{B_1}.$$

We can obtain Φ_{B_1} from the equilibrium constant:

$$\Phi_{B_1} = \frac{1 + 2K\Phi_B - \sqrt{1 + 4K\Phi_B}}{2K^2 \Phi_B}, \tag{7.13-4}$$

where Φ_B is the overall volume fraction of alcohol. The volume fraction of alcohol monomer in pure alcohol then becomes

$$\Phi_{B_1}^* = \frac{1 + 2K - \sqrt{1 + 4K}}{2K^2}. \tag{7.13-5}$$

The physical contribution g_p^E is given by a one-parameter equation as suggested by Scatchard:[40]

$$g_p^E = \beta \Phi_A \Phi_B (x_A v_A + x_B v_B) \tag{7.13-6}$$

where β is a physical interaction parameter related to the hydrocarbon/alcohol monomer interaction, and v_A and v_B are the liquid molar volumes.

The activity coefficients and the enthalpy are found by appropriate differentiation. The results are:

$$\ln \gamma_A = \ln \frac{\Phi_A}{x_A} + \Phi_B \left(1 - \frac{v_A}{v_B}\right) + K \frac{v_A}{v_B} \Phi_B \Phi_{B_1} + \frac{\beta}{RT} v_A \Phi_B^2 \tag{7.13-7}$$

$$\ln \gamma_B = \ln \frac{\Phi_{B_1}}{\Phi_{B_1}^* x_B} + \Phi_A \left(1 - \frac{v_B}{v_A}\right) + K (\Phi_B \Phi_{B_1} - \Phi_{B_1}^*) + \frac{\beta}{RT} v_B \Phi_A^2 \tag{7.13-8}$$

$$h^E = h^E_c + h^E_p \tag{7.13-9}$$

$$h^E_c = -K\Delta h^0 \left[x_B \frac{\partial \ln(\Phi_{B_1}/\Phi^*_{B_1})}{\partial K} + x_B(\Phi_{B_1} - \Phi^*_{B_1}) + Kx_B \left(\frac{\partial \Phi_{B_1}}{\partial K} - \frac{\partial \Phi^*_{B_1}}{\partial K} \right) \right] \tag{7.13-10}$$

$$h^E_p = \beta' \Phi_A \Phi_B (x_A v_A + x_B v_B) \tag{7.13-11}$$

where

$$\beta' = \beta - T\frac{d\beta}{dT} \tag{7.13-12}$$

and Δh^0 is the molar enthalpy of hydrogen-bond formation.

For the Gibbs energy of any alcohol/hydrocarbon system at a fixed temperature, the theory requires only one physical interaction parameter β and one equilibrium constant K. The equilibrium constant, however, depends only on the alcohol and is independent of the hydrocarbon solvent.

Renon's reduction for 11 binary alcohol/hydrocarbon systems is typically represented by Figs. 7-24 and 7-25. Considering the totality of

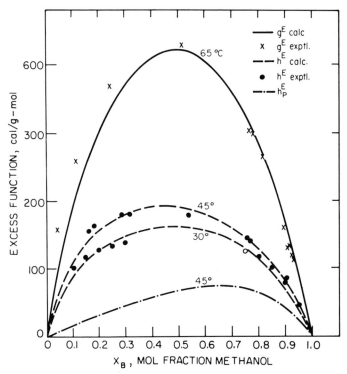

Fig. 7-24   Excess functions for methanol/n-hexane mixtures.

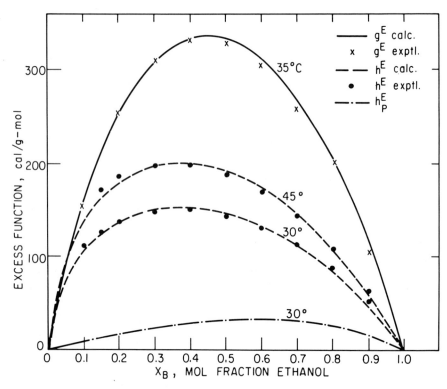

Fig. 7-25   Excess functions for ethanol/n-hexane mixtures.

the data, Renon chose only one value for Δh^0, viz., -6 kcal/g-mol. This value fixes the temperature dependence of K for all alcohols according to

$$\frac{d \ln K}{d(1/T)} = -\frac{\Delta h^0}{R} . \qquad (7.13\text{-}13)$$

Each alcohol is further characterized by the value of K at one temperature. At 50°C, K is 450 for methanol, 190 for ethanol, 60 for isopropanol, and 90 for n-propanol. These constants were obtained upon considering the totality of the data for each alcohol, but giving more weight to the more sensitive data (enthalpies) in the region where the model is physically most reasonable, viz., at low temperatures and at high alcohol concentrations.

Renon's data reduction indicates that the theory described gives a good representation of the experimental data in accordance with the physical meaning of the model. Discrepancies become large only where the degree of alcohol polymerization is small, viz., at low alcohol concentrations and at higher temperatures. A particularly sensitive test of the theory is provided by comparison of calculated and experimental results for excess enthalpies at several temperatures.

In spite of its simplifying assumptions, the Flory-Scatchard model of associated solutions gives a good representation of the properties of concentrated solutions of alcohols in saturated hydrocarbons. It probably takes into account the major effects, but perhaps neglects others, such as formation of cyclic polymers.

The theoretical treatment just described can be extended to include other chemical effects such as, for example, solvation between solute and solvent, or even association of both components, each with itself and with the other. For each assumed chemical equilibrium a characteristic equilibrium constant must be introduced and thus a more general treatment, including various types of chemical equilibria, results in complicated algebraic expressions and, what is worse, requires a large number of empirical parameters.

We must recognize that the distinction between chemical and physical contributions to the excess Gibbs energy leads to an arbitrary and, perhaps, artificial approach based on a simplified picture of solution properties. The designation of molecular interactions as either chemical or physical is merely a convenience which probably cannot be justified by a sophisticated modern theory of intermolecular forces. Nevertheless, a joint chemical and physical description of equilibrium properties of mixtures, as exemplified by Eqs. (7.13-7) and (7.13-8), provides a reasonable and useful treatment for highly nonideal solutions which is a considerable improvement over the ideal chemical theory of Dolezalek on the one hand and the purely physical theory of van Laar on the other.

7.14   Activity Coefficients
in Solvated Solutions

The chemical theory of solutions has frequently been used to describe thermodynamic properties of binary solutions where two components form complexes. There are many experimental studies of such solutions which, if the complex is stable enough, are characterized by negative deviations from Raoult's law. To illustrate, we consider first a simple case, a binary solution where complexes form according to

$$A + B \; \rightleftharpoons \; AB. \qquad (7.14\text{-}1)$$

The equilibrium constant K is related to the activities of the three species by

$$K = \frac{a_{AB}}{a_A a_B} . \qquad (7.14\text{-}2)$$

If the solution is formed from N_1 moles of A and N_2 moles of B, and if at equilibrium N_{AB} moles of complex are formed, then the true mole fractions δ of A, B, and AB are:

$$\delta_A = \frac{N_1 - N_{AB}}{N_1 + N_2 - N_{AB}} \tag{7.14-3}$$

$$\delta_B = \frac{N_2 - N_{AB}}{N_1 + N_2 - N_{AB}} \tag{7.14-4}$$

$$\delta_{AB} = \frac{N_{AB}}{N_1 + N_2 - N_{AB}}. \tag{7.14-5}$$

Following Dolezalek, we assume that the true species form an ideal solution and therefore the activity of each species is equal to its true mole fraction. Equations (7.14-2) to (7.14-5) may then be used to eliminate N_{AB}. The apparent mole fractions of the two components are x_1 (for A) and x_2 (for B). They are given by

$$x_1 = \frac{N_1}{N_1 + N_2} \quad \text{and} \quad x_2 = \frac{N_2}{N_1 + N_2}.$$

Algebraic rearrangement then gives for the activity coefficients:

$$\gamma_1 = \frac{a_A}{x_1} = \frac{kx_1 - 2 + 2(1 - kx_1x_2)^{1/2}}{kx_1^2} \tag{7.14-6}$$

$$\gamma_2 = \frac{a_B}{x_2} = \frac{kx_2 - 2 + 2(1 - kx_1x_2)^{1/2}}{kx_2^2} \tag{7.14-7}$$

where $k \equiv 4K/(K + 1)$. Because of symmetry in Eq. (7.14-1), γ_1 depends on x_1 in exactly the same way as γ_2 depends on x_2.

Figure 7-26 shows γ_1 as a function of x_1 for several values of the equilibrium constant K. When $K = 0$, $\gamma_1 = 1$ for all x_1 as expected, since in that case no complex is formed, and therefore, by assumption, there is no deviation from ideal behavior. At the other extreme, when $K = \infty$, activity coefficients of both components go to zero at the midpoint ($x_1 = x_2 = \frac{1}{2}$) since at this particular composition all the molecules are complexed and no uncomplexed molecules of A or B remain.

Equations (7.14-6) and (7.14-7) predict negative deviations from Raoult's law for $K > 0$ and as a result it has unfortunately become all too common immediately to ascribe observed negative deviations from Raoult's law to solvation effects. It is true that strong negative deviations usually result from complex formation and conversely, if strong complexing is known to occur, negative deviations usually result. However, these conclusions are not always valid because they are based on a strictly

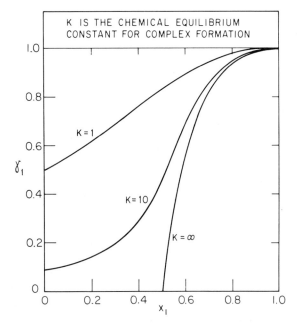

Fig. 7-26   Activity coefficient of a solvating component which forms a bimolecular complex.

chemical theory which neglects all physical effects. In systems where complexing is weak, physical effects are by no means negligible and as a result, weakly solvated solutions sometimes exhibit positive deviations from Raoult's law. Also, small negative deviations from Raoult's law may occur in the absence of complexing as has been observed, for example, in mixtures of normal paraffins.

Equations (7.14-6) and (7.14-7) have been applied to a large number of solutions; one of these is the diethyl ether/chloroform system. Vapor-liquid equilibrium data for this system at several temperatures can be reduced using Eqs. (7.14-6) and (7.14-7) and the following equilibrium constants are then obtained:

Temperature ($^\circ$C)	20	60	80	100
Equilibrium constant K	2.96	1.00	0.80	0.70

Since the formation of a hydrogen bond between ether and chloroform is an exothermic reaction, we expect the equilibrium constant to fall with rising temperature, as indeed it does. From the equilibrium constants we can calculate Δh^0, the enthalpy of hydrogen-bond formation, and we find that for ether/chloroform, Δh^0 is close to -3 kcal/g-mol, a reasonable value, in approximate agreement with those obtained by other experimental measurements.

Equations (7.14-6) and (7.14-7) have been derived for the case where the two components form a 1:1 complex. Similar equations can be derived for cases where the stoichiometry of the complex is 2:1, or 3:1, or 3:2, etc. This flexibility is both a strength and a weakness of the chemical theory of solutions. It is a strength because it can "explain" solution behavior for any sort of chemical interaction and therefore it has, potentially, a wide range of applicability. It is a weakness because, unless other information is available, the stoichiometry of the complex is another adjustable parameter, in addition to the equilibrium constant. Thus, if a particular assumed stoichiometry does not fit the experimental data, one can try another one, and so on, and sooner or later a fit is obtained. Such a fit, however, has no physical significance unless there is independent evidence from the molecular structure of the components to verify the assumed stoichiometry. In the case of diethyl ether/chloroform mixtures, it would be difficult to justify any complex other than one having a 1:1 stoichiometry.

As shown by Harris,[44] we can relax Dolezalek's assumption that the "true" chemical species form an ideal solution. Harris assumed that a mixture of "true" species is described by an equation of the van Laar type. For example, we again consider a mixture of molecules A and B which interact strongly to form complex AB:

$$A + B \; \rightleftharpoons \; AB. \tag{7.14-8}$$

Let K be the equilibrium constant for this chemical equilibrium and let $\mathfrak{z}_A$, $\mathfrak{z}_B$, and $\mathfrak{z}_{AB}$ stand for the true mole fractions. Then

$$K = \frac{\mathfrak{z}_{AB}}{\mathfrak{z}_A \, \mathfrak{z}_B} \cdot \frac{\gamma'_{AB}}{\gamma'_A \, \gamma'_B} \tag{7.14-9}$$

where γ' stands for the true activity coefficient. Dolezalek assumed that all γ' are equal to unity. Harris, however, makes the more reasonable assumption that for any true component k

$$RT \ln \gamma'_k = v_k \left[\sum_j \alpha_{kj} \, \Phi_j - \tfrac{1}{2} \sum_i \sum_j \alpha_{ij} \, \Phi_i \Phi_j \right] \tag{7.14-10}†$$

where Φ is the volume fraction:

$$\Phi_k \equiv \frac{v_k \, \mathfrak{z}_k}{\sum_j v_j \, \mathfrak{z}_j} \, ,$$

---

†For mixtures of components of greatly different size, improved representation of the true activity coefficients may be obtained by adding to the right-hand side of Eq. (7.14-10) the Flory-Huggins term: $RT \left[\ln (\Phi_k / \mathfrak{z}_k) + 1 - \Phi_k / \mathfrak{z}_k \right]$.

and where v_i is the liquid molar volume of i and α_{ij} is a (van Laar) parameter for physical interaction of molecules i and j. The subscripts i, j, and k are understood in this case to range over the three possible species A, B and AB. Equation (7.14-10) contains three physical parameters: α_{A-B}, α_{A-AB}, and α_{B-AB}. In order to limit the number of adjustable parameters to two (one chemical parameter K and one physical parameter α), it is necessary to use plausible physical arguments for relating α_{A-AB} and α_{B-AB} to α_{A-B}, as described by Harris.[44]

To reduce experimental vapor-liquid equilibrium data with Eqs. (7.14-9) and (7.14-10), we use a powerful theorem discussed in detail by Prigogine and Defay.[34] It can be rigorously shown that for a mixture of components 1 (species B) and 2 (species A), the *apparent* activity coefficients and *apparent* mole fractions are related to the *true* activity coefficients and *true* mole fractions by

$$\gamma_2 = \frac{\eth_A\, \gamma_A'}{x_2} \tag{7.14-11}$$

$$\gamma_1 = \frac{\eth_B\, \gamma_B'}{x_1}. \tag{7.14-12}$$

Equations (7.14-11) and (7.14-12) are independent of any physical model. They follow directly from the assumption that the "true" species A, B, and AB are in equilibrium.

For data reduction, Eqs. (7.14-9 to 12) must be combined with material balances relating true mole fractions to apparent mole fractions. This is done most conveniently in terms of the normalized extent of complex formation $\xi(0 \le \xi \le \frac{1}{2})$:

$$\eth_{AB} = \frac{\xi}{1 - \xi} \tag{7.14-13}$$

$$\eth_A = \frac{x_2 - \xi}{1 - \xi} \tag{7.14-14}$$

$$\eth_B = \frac{x_1 - \xi}{1 - \xi}. \tag{7.14-15}$$

To illustrate Harris' extension of Dolezalek's theory, Fig. 7-27 gives results of data reduction for solutions of acetylene in three organic solvents. Acetylene forms hydrogen bonds with butyrolactone and N-methyl pyrrolidone but not with hexane. To represent the experimental data for the two polar solvents, two parameters (K and α) are required whereas the experimental data for hexane are represented with only one parameter, since $K = 0$. In hexane, which is a "physical" solvent, acetylene exhibits positive deviations from Raoult's law over the entire composition range. In chemical solvents, however, acetylene exhibits negative

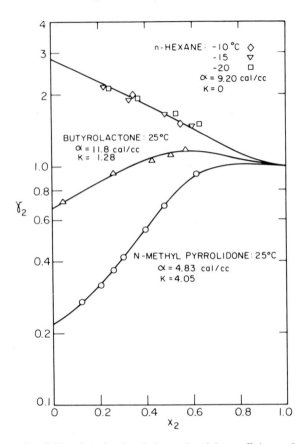

Fig. 7-27 Calculated and observed activity coefficients of acetylene in three organic solvents.

deviations from Raoult's law at the solvent-rich end and positive deviations at the acetylene-rich end. Dolezalek's theory (with one chemical equilibrium constant) cannot account for such behavior. The modification of the chemical theory proposed by Harris combines into one comprehensive treatment the ideas of both Dolezalek and van Laar.

7.15 Solutions Containing Two (or More) Complexes

The chemical theory can be extended in a straightforward way to the case where several complexes of different stoichiometry are formed by the two components. For the same reasons as those stated in the previous section, this possibility of extension carries with it a certain advantage and a certain danger. The advantage lies in the gain in flexibility; the danger arises from the fact that if one postulates enough complexes one necessarily introduces a sufficient number of adjustable parameters to fit data for any system.

An excellent study of a binary solution which contains two types of complexes has been made by McGlashan and Rastogi,[45] who investigated thermodynamic properties of the p-dioxan/chloroform system. P-dioxan is a symmetric cyclic ether containing two oxygen atoms and therefore McGlashan and Rastogi postulated the existence in solution of two complexes having the structures

$$CCl_3-H---O\begin{array}{c} CH_2-CH_2 \\ \diagdown \\ CH_2-CH_2 \end{array}O$$

and

$$CCl_3-H---O\begin{array}{c} CH_2-CH_2 \\ \diagdown \\ CH_2-CH_2 \end{array}O---H-CCl_3.$$

Let A stand for dioxan and B for chloroform. Two equilibria are postulated:

$$A + B \rightleftharpoons AB \tag{7.15-1}$$

$$A + 2B \rightleftharpoons AB_2. \tag{7.15-2}$$

Two equilibrium constants are defined by

$$K_1 = \frac{a_{AB}}{a_A a_B} \tag{7.15-3}$$

$$K_2 = \frac{a_{AB_2}}{a_A a_B^2}. \tag{7.15-4}$$

We again assume that this (apparent) binary, nonideal solution is, in fact, an ideal solution of four components (A, B, AB, and AB_2). The activity of each component is then equal to its true mole fraction $\mathfrak{z}$ and Eqs. (7.15-3) and (7.15-4) become:

$$K_1 = \frac{\mathfrak{z}_{AB}}{\mathfrak{z}_A \mathfrak{z}_B} \tag{7.15-5}$$

$$K_2 = \frac{\mathfrak{z}_{AB_2}}{\mathfrak{z}_A \mathfrak{z}_B^2}. \tag{7.15-6}$$

By material balance,

$$\mathfrak{z}_A + \mathfrak{z}_B + \mathfrak{z}_{AB} + \mathfrak{z}_{AB_2} = 1. \tag{7.15-7}$$

Eliminating $\mathfrak{z}_{AB}$ and $\mathfrak{z}_{AB_2}$ with Eqs. (7.15-5) and (7.15-6), Eq. (7.15-7) becomes, after rearrangement,

$$\frac{1 - \mathfrak{z}_A - \mathfrak{z}_B}{\mathfrak{z}_A \mathfrak{z}_B} = K_1 + K_2 \mathfrak{z}_B \qquad (7.15\text{-}8)$$

or

$$\frac{1 - a_A - a_B}{a_A a_B} = K_1 + K_2 a_B . \qquad (7.15\text{-}9)$$

The activities a_A and a_B are obtained from experimental vapor-liquid equilibrium data.

A plot of the left side of Eq. (7.15-9) versus a_B should give a straight line whose intercept and slope yield the two equilibrium constants K_1 and K_2. Such a plot was constructed by McGlashan and Rastogi who found that at 50°C, $K_1 = 1.11$ and $K_2 = 1.24$.

Once numerical values are given for the two equilibrium constants, activity coefficients can be calculated by somewhat complicated equations which, however, can be derived in a straightforward manner, as shown by McGlashan and Rastogi.[45] The activity coefficient of B (chloroform) is given by

$$\gamma_B = \frac{\mathfrak{z}_B}{x_B}, \qquad (7.15\text{-}10)$$

and the true mole fraction of B is related by material balances to x_B, the apparent mole fraction, by

$$x_B = \frac{(1 + K_1)\mathfrak{z}_B + K_2 \mathfrak{z}_B^2 (2 - \mathfrak{z}_B)}{1 + K_1 \mathfrak{z}_B (2 - \mathfrak{z}_B) + K_2 \mathfrak{z}_B^2 (3 - 2\mathfrak{z}_B)} . \qquad (7.15\text{-}11)$$

The activity coefficient of A (dioxan) is given by

$$\gamma_A = \frac{1 - \mathfrak{z}_B}{(1 + K_1 \mathfrak{z}_B + K_2 \mathfrak{z}_B^2)(1 - x_B)} . \qquad (7.15\text{-}12)$$

Figure 7-28 gives a plot of calculated and observed activity coefficients. The excellent agreement shows that for this system calculations based on the assumption of the existence of two justifiable complexes can account for the system's thermodynamic behavior.

McGlashan and Rastogi also measured calorimetrically the enthalpy of mixing for this system. Using the chemical solution theory just described, the enthalpy of mixing can be related to the enthalpy of complex formation.

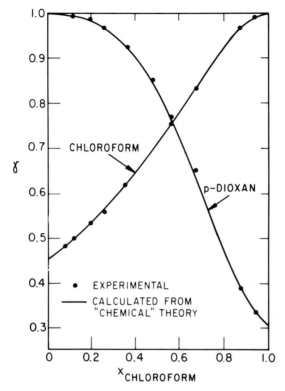

Fig. 7-28 Activity coefficients for a solvated mixture: P-dioxan/chloroform at 50°C (McGlashan and Rastogi).

Since

$$\frac{g^E}{RT} = x_A \ln \gamma_A + x_B \ln \gamma_B, \tag{7.15-13}$$

and since

$$h^E = -RT^2 \left(\frac{\partial (g^E/RT)}{\partial T}\right)_{P,x}, \tag{7.15-14}$$

substitution of Eqs. (7.15-10) and (7.15-11) for γ_A and γ_B gives

$$h^E = \left(\frac{x_A \partial_B}{1 + K_1 \partial_B + K_2 \partial_B^2}\right)(K_1 \Delta h_1^0 + \partial_B K_2 \Delta h_2^0), \tag{7.15-15}$$

where Δh_1^0 is the enthalpy of formation of complex AB and Δh_2^0 is the enthalpy of formation of complex AB_2:

$$\Delta h_1^0 = RT^2 \frac{d \ln K_1}{dT} \qquad (7.15\text{-}16)$$

$$\Delta h_2^0 = RT^2 \frac{d \ln K_2}{dT}. \qquad (7.15\text{-}17)$$

By fitting experimentally determined enthalpies of mixing to Eq. (7.15-15), the enthalpies of complex formation are found to be

$$\Delta h_1^0 = -2.0 \text{ kcal/g-mol}$$
$$\Delta h_2^0 = -3.6 \text{ kcal/g-mol}.$$

These are reasonable values; Δh_1^0 is the enthalpy of formation of one ether-oxygen hydrogen bond and it agrees quite well with results determined by spectroscopic, cryoscopic and other methods. The enthalpy of formation for the complex containing two hydrogen bonds is not quite twice that for the formation of two hydrogen bonds. This is not surprising since the two oxygen atoms in dioxan are only separated by two carbon atoms; therefore the effect of hydrogen bonding on one oxygen atom has an appreciable effect on the other oxygen atom. If the two oxygen atoms were farther apart one might expect that Δh_2^0 would be more nearly equal to $2\Delta h_1^0$.

7.16 Distribution of a Solute Between Two Immiscible Solvents

The "chemical" theory of solution attempts to explain thermodynamic properties in terms of actual (true) chemical species present in solution. An explanation of this sort can sometimes be applied toward interpreting and extending data on partition coefficients for a solute between two immiscible liquid solvents. We end this chapter by giving one example of such an application.[46]

Consider two liquid phases α and β; a solute, designated by subscript 1, is distributed between these two phases. First we consider a simple case. Suppose that the mole fraction of solute in either phase is very small and that we can therefore assume the two solutions to be ideal dilute solutions. We then have

$$f_1^\alpha = H_{1,\alpha} x_1^\alpha \qquad (7.16\text{-}1)$$

$$f_1^\beta = H_{1,\beta} x_1^\beta \qquad (7.16\text{-}2)$$

where $H_{1,\alpha}$ is Henry's constant for solute 1 in phase α and $H_{1,\beta}$ is Henry's constant for solute 1 in phase β. Equating fugacities of component 1 in the two phases, we obtain the partition coefficient K,

$$K = \frac{x_1^\alpha}{x_1^\beta} = \frac{H_{1,\beta}}{H_{1,\alpha}}. \tag{7.16-3}$$

At constant temperature and pressure the partition coefficient given in Eq. (7.16-3) is a true constant independent of composition. Equation (7.16-3) is frequently called the *Nernst distribution law*.

Since the mole fractions x_1^α and x_1^β are very small they are, respectively, proportional to the concentrations of solute 1 in phase α and in phase β; it is therefore customary to use a somewhat different partition coefficient, K', expressed in terms of concentrations c rather than mole fractions x:

$$K' = \frac{c_1^\alpha}{c_1^\beta} = \frac{\rho^\alpha x_1^\alpha}{\rho^\beta x_1^\beta}, \tag{7.16-4}$$

where ρ^α and ρ^β are, respectively, the molar densities of phases α and β. For very small x_1, ρ^α and ρ^β are the densities of the pure solvents.

Now, many cases are known where the Nernst distribution law is not consistent with experiment, even though the solute mole fractions are small; in other words, in these cases the equations for ideal dilute solutions [Eqs. (7.16-1) and (7.16-2)] are not obeyed in either (or both) of the liquid phases at the particular concentrations investigated.[†] In many cases, departure from Nernst's law may be ascribed to chemical effects. We now consider such a case: the distribution of benzoic acid between the two (essentially) immiscible solvents water and benzene near room temperature. The explanation for the failure of Nernst's law can, in this case, be found by taking into account the tendency of organic acids to dimerize in a nonpolar solvent.

We postulate two equilibria as shown in Fig. 7-29:

I. Phase-distribution equilibrium between the two phases:

Acid in water $\rightleftharpoons$ Monomer acid in benzene,

II. Chemical equilibrium in the benzene phase:

Monomer acid $\rightleftharpoons$ Dimer acid.

[†]The ideal dilute solution equation is always approached for any nonelectrolyte when the mole fraction is sufficiently small but just how small it must be depends on the system. For a solute which associates in solution it may be very small indeed, sometimes smaller than can be measured by common analytical methods.

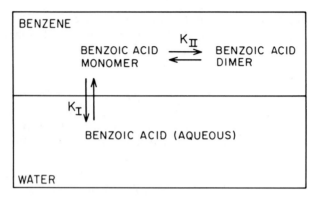

Fig. 7-29   Distribution of benzoic acid between benzene and water.

For each equilibrium there is an equilibrium constant:

$$K_I = \frac{c_M^B}{c^W} \tag{7.16-5}$$

$$K_{II} = \frac{c_D^B}{(c_M^B)^2}. \tag{7.16-6}$$

In Eqs. (7.16-5) and (7.16-6), c stands for concentration of benzoic acid; superscript B stands for the benzene phase and superscript W for the water phase.   Subscript M stands for monomer and subscript D for dimer.†

Let c_T^B stand for the total concentration of benzoic acid in benzene. By material balance,

$$c_T^B = c_M^B + 2c_D^B. \tag{7.16-7}$$

Substitution of Eqs. (7.16-5) and (7.16-6) into Eq. (7.16-7) gives the distribution law

$$\frac{c_T^B}{c^W} = K_I + 2K_I^2 K_{II} c^W. \tag{7.16-8}$$

In this case then, the distribution coefficient (i.e., the ratio of c_T^B to c^W) is not constant, as it would be according to Nernst's law, but varies

---

†In the aqueous phase benzoic acid is probably completely solvated by hydrogen-bonding with water.

linearly with the concentration of benzoic acid in water. Experimental data for this system[47] are plotted in Fig. 7-30 in the form suggested by

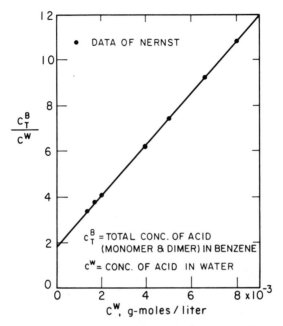

Fig. 7-30   Linearization of distribution data with a "chemical" theory: the system benzoic acid/water/benzene at 20°C.

Eq. (7.16-8). The straight line obtained confirms the prediction based on the chemical theory of solutions; from the slope and intercept $K_I = 1.80$ (dimensionless) and $K_{II} = 176$ liter/g-mol. By plotting the data in this way one can, on the basis of just a few experimental points, interpolate and slightly extrapolate with confidence.

In some systems, where the tendency of the solute to dimerize is strong, K_{II} is very large. If, in the system just discussed, benzoic acid dimerized strongly, we would have:

$$2K_I^2 K_{II} c^W \gg K_I, \qquad (7.16\text{-}9)$$

giving the distribution law

$$\frac{\sqrt{c_T^B}}{c^W} = (2K_I^2 K_{II})^{1/2} = \text{constant}. \qquad (7.16\text{-}10)$$

Various examples of this distribution law have been found and whenever a system behaves according to Eq. (7.16-10) it is considered good evidence that the solute molecules are strongly dimerized in one of the solvents. In the benzoic acid/water/benzene system, dimerization of benzoic acid is moderately strong but the inequality given in Eq. (7.16-9) is not valid until c^W is at least 10^{-2} g-moles per liter.

Numerous other systems have been investigated where deviations from Nernst's distribution law can be reasonably explained by a coupling of chemical and phase-distribution equilibria. For example, data for the distribution of picric acid between water and a nonpolar solvent can be interpreted by taking into account the ionization of picric acid in the aqueous phase. In a similar manner, the distribution (i.e., the solubility) of sulfur dioxide between the gas phase and water can be described quantitatively as shown in Chap. 8.

7.17 Summary

The theory of solutions is an old subject; many of the articles in the first volumes of the *Zeitschrift für Physikalische Chemie* (around 1890) are concerned with the properties of liquid mixtures, and since the early days of physical chemistry thousands of articles have been written in an effort to understand the behavior of mixed fluids. While much progress has been made, we are still far from an adequate theory of liquid mixtures. In this chapter we have indicated a few of the theoretical ideas which have been proposed and it is evident that none of them is sufficiently broad to apply to the general problem; rather, each idea appears to be limited to a particular class of solutions. As a result, we do not have a general theory of liquid mixtures but instead we have several restricted theories each of which is useful for a particular type of mixture.

In order to construct a theory of liquid mixtures, we require essentially two kinds of information: We need to know something about the structure of liquids (i.e., the way in which molecules in a liquid are arranged in space), and we need to know something about intermolecular forces between like and unlike molecules. Unfortunately, information of either kind is inadequate and as a result all of our theories must make simplifying assumptions in order to overcome this disadvantage. Since simplifying assumptions must be made, it follows that we cannot at this time construct a general theory; simplifying assumptions which are reasonable for one type of mixture (e.g., mixtures of hydrocarbons) may be most unreasonable for another type (e.g., aqueous mixtures of organic acids) and since the simplifying assumptions must vary from one type of mixture to another, we inevitably have different theories for different applications: The punishment must fit the crime.

Most theoretical work has been concerned with mixtures of liquids whose molecules are nonpolar and spherical. Some of this work has been indicated here: regular-solution theory, lattice theory, corresponding-states theory, all start out with simple molecules and are then extended, often semiempirically, to more complicated molecules. Recent theoretical work on mixtures of simple molecules (cell theory, conformal solution theory, theories based on the radial distribution function) has not been discussed because an adequate treatment requires familiarity with statistical mechanics. Some of these theories promise to contribute to our understanding of liquid structure, but they do not tell us anything new about intermolecular forces between dissimilar molecules.

For mixtures of nonpolar liquids, the regular-solution theory of Scatchard and Hildebrand frequently provides a good approximation for the excess Gibbs energy. The most serious simplifying assumption in this theory is the geometric-mean rule for the cohesive energy density of the unlike (1-2) interaction. For application, whenever possible, this geometric-mean rule should be modified empirically by utilizing whatever limited experimental binary data may be available. Two important advantages of the Scatchard-Hildebrand regular-solution theory are its simplicity and its ease of extension to systems containing more than two components.

Regular-solution theory, like most theories of solution, is more reliable for excess Gibbs energy than it is for excess enthalpy and excess entropy. All simple theories of solution neglect changes in molecular vibration and rotation which result from the change of molecular environment that is inevitably produced by mixing; these changes, in many cases, affect the excess enthalpy and excess entropy in such a way that they tend to cancel in the excess Gibbs energy.

The lattice theory of solutions, although first developed for monatomic molecules, can be extended to molecules of more complex structure using well-defined assumptions, as shown by Guggenheim, Flory, and others. This extension makes it particularly useful for solutions of molecules which differ appreciably in size, such as polymer solutions. However, the concept of a lattice for liquid structure is a vast over-simplification and as a result, lattice theory becomes increasingly inappropriate as attention is focused on temperatures remote from the melting point. Also, for each binary system, lattice theory requires as an input parameter the interchange energy w, which is difficult to predict and which unfortunately is temperature-dependent.

Corresponding-states theory is attractive because it is not tied to a particular physical model but, essentially, relies on dimensional analysis; by nondimensionalizing the residual properties of pure fluids, one can predict mixture properties by adopting appropriate rules for calculating composition-dependent reducing parameters. For nonpolar fluid mixtures, good results can frequently be obtained but consistently good results cannot be expected because the pure fluids do not with sufficient accuracy obey the assumptions of corresponding-states behavior and because the composition-dependence (mixing rules) of the reducing parameters may vary slightly with temperature and density. Excess functions in the liquid phase are, unfortunately, strongly sensitive to the mixing rules used.

Most simple theories of mixtures assume random mixing of molecules and, for strongly nonideal mixtures, such an assumption can lead to serious error. Since we do not have a rigorous theory of nonrandom mixtures, a semiempirical attempt to introduce nonrandomness is provided by the concept of local concentration leading to the equations of Wilson and Renon. These equations do not have a precise theoretical basis but appear to be of a form which is particularly useful for solutions containing one or more polar components; for such solutions no really satisfactory theory has been developed.

For those who favor a philosophy of idealism, it is attractive to do away with nonideality in solutions by claiming that our observations of nonideality are merely apparent, that all solutions are, in fact, ideal if only we use in our calculations the true, rather than the apparent, molecular concentrations. This idealistic view attributes all observed nonideality to formation of new chemical species in solution; by postulating association or solvation equilibria (or both) and then letting equilibrium constants be adjustable parameters, one can indeed fit experimental data for many liquid mixtures. In fact, the chemical theory of solutions permits one to fit experimental data for any liquid mixture, regardless of complexity, provided there are a sufficient number of adjustable equilibrium constants.

The chemical theory of solutions provides a sensible approximation whenever there is independent evidence that strong chemical forces operate in the liquid mixture; for example, whenever there is appreciable hydrogen bonding between like or unlike components (or both) it is reasonable to assume that the formation (or destruction) of hydrogen-bonded molecules in solution provides the dominant contribution to the solution's thermodynamic properties. If the chemical forces are strong, then physical (van der Waals) forces may often be neglected, at least for a first approximation, but careful study of chemical solution properties has shown that for accurate work both physical and chemical forces must be taken into account. However, the dividing line between physical and chemical forces cannot be determined with rigor and as a result it is necessary to make an essentially arbitrary decision on where that line is to be drawn.

The vagueness of the chemical theory of solutions provides a wide range of applications and, as a result, one must beware of the strong temptation to use it where, in fact, it is inapplicable. Any theory of solutions with a sufficient number of adjustable parameters must always be viewed with suspicion unless supported by independent physico-chemical measurements. Nevertheless, when used judiciously, the chemical theory of solutions provides a useful framework for correlating and extending thermodynamic data for strongly nonideal solutions where currently available physical theories are inappropriate.†

In general, we may say that theories of solutions are mental crutches which enable us to order, interpet and in a vague sense "understand" thermodynamic data for mixtures. These theories provide a framework which enables us to correlate data in a sensible manner; they tell us what to plot against what, the coordinates we must use to get a smooth (and perhaps even straight) line. For engineering work such a framework is extremely useful because it enables us to interpolate and extrapolate limited experimental results and to make reasonable predictions for systems not previously studied, especially for those systems containing more than two components. Finally, however, it is important to remember, as Scatchard has pointed out,‡ that theories of solution are, essentially, working tools; we must not take any theory too seriously since actual liquid mixtures are much more complicated than our oversimplified models. To make progress, we must consistently keep in mind the simplifying assumptions on which our theories are based, for, as Francis Bacon said many years ago, "Truth is more likely to emerge from error than from confusion."

PROBLEMS

1. A liquid hydrocarbon A has a saturation pressure $P_A^s = 100$ mm Hg at 10°C. Its density at 25°C is 0.80 g/cc and its molecular weight is 160. This is all the information available on pure liquid A. An equimolar mixture of A in carbon disulfide at 10°C gives an equilibrium partial pressure of A equal to 60 mm Hg. Estimate the composition of the vapor which at 10°C is in equilibrium with an equimolar liquid solution of A in toluene, using the following data:

	Carbon disulfide	Toluene
Solubility parameters at 25°C $(\text{cal/cc})^{1/2}$	10.0	8.9
Liquid molar volumes at 25°C (cc/g-mol)	61	107
Saturation pressures at 10°C (mm Hg)	13	191

†A major deficiency of the Dolezalek chemical theory follows from its inability to account for phase separation (demixing). This deficiency, however, is removed by the extensions given in Sections 7.13 and 7.14.

‡"The best advice which comes from years of study of liquid mixtures is to use any model insofar as it helps, but not to believe that any moderately simple model corresponds very closely to any real mixture." [*Chem. Rev.*, **44**, 7 (1949)]

2. A liquid mixture containing 40 mol % isobutane and 60 mol % benzene is distilled at a total pressure of 2.5 atmospheres. Compute the K values for both components using the theory of regular solutions.

3. Consider a dilute isothermal solution of acetic acid in benzene. For the dilute region (say up to 5 mol % acid) draw schematically curves for

$$\bar{s}_1^E \text{ vs. } x_1 \quad \text{and} \quad \bar{h}_1^E \text{ vs. } x_1$$

where subscript 1 refers to the acid. Briefly justify your schematic graphs with suitable explanations.

4. Consider a solution of diethyl ether and pentachloroethane. Draw (schematically) a plot of g^E vs. x at constant temperature. Briefly justify your schematic graph with suitable explanations.

5. Liquids A and B when mixed form an azeotrope at 300°K and at a mole fraction $x_A = 0.5$. It is desired to separate a mixture of A and B by distillation and in order to break the azeotrope it is proposed to add a third liquid C into the mixture. Compute the relative volatility of A to B at 300°K when the ternary mixture contains 60 mol % C and equal molar amounts of A and B. Assume ideal gas behavior and assume that A, B, and C are nonreactive nonpolar substances. The data given below are all at 300°K.

	A	B	C
Liquid molar volume (cc/g-mol)	100	100	100
Solubility parameter $(\text{cal/cc})^{1/2}$	7.0	8.0	9.0

6. Chemical engineers are fond of generalized plots. Show how you would prepare a generalized solubility parameter plot for nonpolar liquids on the basis of Pitzer's three-parameter theory of corresponding states.

7. Derive Eq. (7.9-7).

8. At room temperature and atmospheric pressure:
   (a) Give an order-of-magnitude estimate of $\bar{h}^E$ for methanol dissolved in a large excess of isooctane.
   (b) Give an order-of-magnitude estimate of the change in temperature when equal parts of cyclohexane and carbon disulfide are mixed adiabatically. Is ΔT positive or negative?
   (c) Name two polar solvents which are likely to be very good and two others which are likely to be very poor for an extraction separation of hexane and hexene. Explain.

9. A dilute solution of picric acid in water is contacted with n-hexane. Consider the distribution of picric acid between the two solvents; assume that the acid exists as a monomer in both phases but that it ionizes partially in the aqueous phase. Show that the distribution of the acid should be described by an equation of the form

$$\frac{\sqrt{c_H}}{c_W} = a\left(1 - b\,\frac{c_H}{c_W}\right)$$

where c_H is the concentration of picric acid in hexane; c_W is the concentration of picric acid in water; a and b are constants depending only on temperature.

REFERENCES

1. Hildebrand, J. H., *J. Am. Chem. Soc.*, **51**, 66 (1929).

2. Scott, R. L., *J. Phys. Chem.*, **62**, 136 (1958).

3. Scott, R. L., *Ann. Rev. Phys. Chem.*, **7**, 43 (1956).

4. McGlashan, M. L., *Chem. Soc.* (London), *Ann. Rep.*, **59**, 73 (1962).

5. Chao, K. C., and G. D. Seader, *A. I. Ch.E. Journal*, **7**, 598 (1961).

6. Guggenheim, E. A., *Mixtures*. London: Oxford University Press, 1952.

7. Guggenheim, E. A., *Applications of Statistical Mechanics*. London: Oxford University Press, 1966.

8. Kohler, F., *Monats.*, **88**, 857 (1957).

9. Eckert, C. A., Dissertation, University of California, Berkeley, 1964.

10. Thorp, N., and R. L. Scott, *J. Phys. Chem.*, **60**, 670, 1441 (1956).

11. Croll, I. M., and R. L. Scott, *J. Phys. Chem.*, **62**, 954 (1958).

12. Flory, P. J., *J. Chem. Phys.*, **9**, 660 (1941); **10**, 51 (1942).

13. Huggins, M. L., *J. Phys. Chem.*, **9**, 440 (1941); *Ann. N.Y. Acad Sci.*, **43**, 1 (1942).

14. Fast, J. D., *Entropy*, Philips Technical Library. Eindhoven, The Netherlands: Centrex Publishing Co., 1962.

15. Flory, P. J., *Principles of Polymer Chemistry*. Ithaca, N. Y.: Cornell University Press, 1953.

16. Burrell, H., *Interchem. Rev.*, **14**, 3, 31 (1955).

17. Blanks, R. F., and J. M. Prausnitz, *I&EC Fundamentals*, **3**, 1 (1964).

18. Hansen, C. M., *J. Paint Technol.*, **39**, 104 (1967).

19. Wilson, G. M., *J. Am. Chem. Soc.*, **86**, 127 (1964).

20. Orye, R. V., and J. M. Prausnitz, *Ind. Eng. Chem.*, **57**, 18 (May 1965).

21. Heil, J. F., and J. M. Prausnitz, *A. I. Ch. E. Journal*, **12**, 678 (1966).

22. Scott, R. L., *J. Chem. Phys.*, **25**, 193 (1956).

23. Kay, W. B., *Ind. Eng. Chem.*, **28**, 1014 (1936).

24. Joffe, J., *Ind. Eng. Chem.*, **40**, 1738 (1948).

25. Gamson, B. W., and K. M. Watson, *Nat. Petrol. News*, **R623** (August 2, 1944 and Sept. 6, 1944).

26. Reid, R. C., and T. W. Leland, *A. I. Ch. E. Journal*, **11**, 228 (1965).

27. Leland, T. W., P. S. Chappelear and B. W. Gamson, *A. I. Ch. E. Journal*, **8**,

482 (1962); J. W. Leach, P. S. Chappelear, and T. W. Leland, *A. I. Ch. E. Journal*, **14,** 560 (1968).

28. Renon, H., and J. M. Prausnitz, *A. I. Ch. E. Journal*, **14,** 135 (1968).

29. Dolezalek, F., *Z. Physik. Chem.*, **64,** 727 (1908).

30. van Laar, J. J., *Z. Physik. Chem.*, **72,** 723 (1910).

31. Johnson, A. I., W. F. Furter, and T. W. Barry, *Can. J. Technol.*, **32,** 179 (1954).

32. Hovorka, F., and D. Dreisbach, *J. Am. Chem. Soc.*, **56,** 1664 (1934).

33. Kortüm, G., and H. Buchholz-Meisenheimer, *Die Theorie der Destillation und Extraktion von Flüssigkeiten*, Chapter 3. Berlin: Springer Verlag, 1952. See also reference in footnote to Eq. (7.12-12).

34. Prigogine, I., and R. Defay, *Chemical Thermodynamics*, Chapter 26. London: Longmans Green & Co., 1954.

35. Prigogine, I., V. Mathot and A. Desmyter, *Bull. Soc. Chim. Belges*, **58,** 547 (1949).

36. Prigogine, I., and A. Desmyter, *Trans. Faraday Soc.*, **47,** 1137 (1951).

37. Pimentel, G., and A. L. McClellan, *The Hydrogen Bond.* San Francisco: W. H. Freeman & Co., 1960.

38. Van Ness, H. C., J. Van Winkle, H. H. Richtol, and H. B. Hollinger, *J. Phys. Chem.*, **71,** 1483 (1967).

39. Flory, P. J., *J. Chem. Phys.*, **12,** 425 (1944).

40. Scatchard, G., *Chem. Rev.*, **44,** 7 (1949).

41. Kretschmer, C. B., and R. Wiebe, *J. Chem. Phys.*, **22,** 1697 (1954).

42. Renon, H., and J. M. Prausnitz, *Chem. Eng. Sci.*, **22,** 299, Errata, 1891 (1967).

43. Redlich, O., and A. T. Kister, *J. Chem. Phys.*, **15,** 849 (1947).

44. Harris, H. G., and J. M. Prausnitz, *I&EC Fundamentals, Wilhelm Memorial Issue* (May 1969)

45. McGlashan, M. L., and R. P. Rastogi, *Trans. Faraday Soc.*, **54,** 496 (1958).

46. Moelwyn-Hughes, E. A., *J. Chem. Soc.*, 850 (1940).

47. Nernst, W., *Z. Physik. Chem.*, **8,** 110 (1891).

Solubilities of Gases in Liquids

8

Numerous examples in nature illustrate the ability of liquids to dissolve gases; indeed, human life would not be possible if blood could not dissolve oxygen, nor could marine life exist if oxygen did not dissolve in water. Gas mixtures can be separated by absorption because different gases dissolve in a liquid in different amounts and as a result, most gaseous mixtures can be separated by contact with a suitable solvent which dissolves one gaseous component more than another.

The solubility of a gas in a liquid is determined by the equations of phase equilibrium. If a gaseous phase and a liquid phase are in equilibrium, then for any component i the fugacities in both phases must be the same:

$$f_i^{\text{gas}} = f_i^{\text{liquid}}. \tag{1}$$

Equation (1) is of little use unless something can be said about how the fugacity of component i in each phase is related to the temperature, pressure, and composition of that phase. In Chap. 5 we discussed the fugacity of a component in the gaseous phase. In this chapter we consider the fugacity of a component i, normally a gas at the temperature under consideration, when it is dissolved in a liquid solvent.

351

8.1 The Ideal Gas Solubility

The simplest way to reduce Eq. (1) to a more useful form is to rewrite it in a manner suggested by Raoult's law. In doing so, we introduce several drastic but convenient assumptions. Neglecting all gas-phase nonidealities as well as the effect of pressure on the condensed phase (Poynting correction), and also neglecting any nonidealities due to solute-solvent interactions, the equation of equilibrium can be very much simplified by writing

$$p_i = x_i P_i^s \tag{8.1-1}$$

where p_i is the partial pressure of component i in the gas phase,† x_i is the solubility (mole fraction) of i in the liquid, and P_i^s is the saturation (vapor) pressure of pure (possibly hypothetical) liquid i at the temperature of the solution. The solubility x_i, as given in Eq. (8.1-1), is called the *ideal solubility* of the gas.

Aside from the severe simplifying assumptions which were made in obtaining Eq. (8.1-1), an obvious difficulty presents itself in finding a value for P_i^s whenever (as is often the case) the solution temperature is above the critical temperature of pure i. In that case it has been customary to extrapolate the saturation pressure curve of pure liquid i beyond its critical temperature to the solution temperature; the saturation pressure of the hypothetical liquid is usually found from a straight-line extrapolation on a semilogarithmic plot of saturation pressure versus reciprocal absolute temperature as shown in Fig. 8-1. The use of these particular

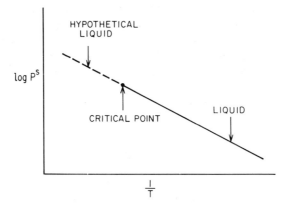

Fig. 8-1   Convenient but arbitrary extrapolation of liquid saturation pressure into hypothetical liquid region.

† The partial pressure p_i is, by definition, equal to the product of the gas-phase mole fraction and the total pressure: $p_i \equiv y_i P$.

coordinates does not have any sound physicochemical basis but is dictated by convenience.

The ideal solubility, as calculated by Eq. (8.1-1) and the extrapolation scheme indicated in Fig. 8-1, usually gives correct order-of-magnitude results provided the partial pressure of the gas is not large and provided the solution temperature is well below the critical temperature of the solvent and not excessively above the critical temperature of the gaseous solute. In some cases where the physical properties of solute and solvent are similar (e.g., chlorine in carbon tetrachloride) the ideal solubility is remarkably close to the experimental value.

Table 8-1 compares ideal and observed solubilities of four gases in a number of solvents at 25°C and 1 atm partial gas pressure. The ideal solubility is significantly different from the observed solubilities but it is of the right order of magnitude.

Table 8-1

SOLUBILITIES OF GASES IN SEVERAL LIQUID SOLVENTS
AT 25°C AND 1 ATM PARTIAL PRESSURE
MOL FRACTION × 10^4

	Ideal	$n\text{-}C_7F_{16}$	$n\text{-}C_7H_{16}$	CCl_4	CS_2	$(CH_3)_2CO$
H_2	8	14.01	6.88	3.19	1.49	2.31
N_2	10	38.7	–	6.29	2.22	5.92
CH_4	35	82.6	–	28.4	13.12	22.3
CO_2	160	208.8	121	107	32.8	–

The ideal solubility given by Eq. (8.1-1) suffers from two serious defects. First, it is independent of the nature of the solvent; Eq. (8.1-1) says that a given gas, at a fixed temperature and partial pressure, has the same solubility in *all* solvents. This conclusion is, of course, completely contrary to observation as illustrated by the data in Table 8-1. Second, Eq. (8.1-1), coupled with the extrapolation scheme shown in Fig. 8-1, predicts that at constant partial pressure the solubility of a gas always decreases with rising temperature. This prediction is frequently correct but not always; near room temperature the solubilities of the light gases helium, hydrogen, and neon increase with rising temperature in most solvents, and at somewhat higher temperatures the solubilities of gases like nitrogen, oxygen, argon, and methane also increase with rising temperature in many common solvents. It is because of these two defects that the ideal-solubility equation is of limited use; it should be employed only whenever no more is desired than a rough estimate of gas solubility.

8.2 Henry's Law and Its
Thermodynamic Significance

It was observed many years ago that the solubility of a gas in a liquid is often proportional to its partial pressure in the gas phase, provided the partial pressure is not large. The equation which describes this observation is commonly known as Henry's law:

$$p_i = y_i P = k x_i \qquad (8.2\text{-}1)$$

where k is a constant of proportionality which, for any given solute and solvent, depends only on the temperature.† Equation (8.2-1) always provides an excellent approximation when the solubility and the partial pressure of the solute are small and when the temperature is well below the critical of the solvent. Just how small the partial pressure and solubility have to be for Eq. (8.2-1) to hold, varies from one system to another, and the reasons for this variation will become apparent a little later.   In general, however, as a *rough* rule for many common systems, the partial pressure should not exceed 5 or 10 atm and the solubility should not exceed about 3 mol %; however, in those systems where solute and solvent are chemically highly dissimilar (e.g., systems containing helium or hydrogen) large deviations from Eq. (8.2-1) are frequently observed at much lower solubilities. On the other hand, in some systems (e.g., carbon dioxide/benzene near room temperature) Eq. (8.2-1) appears to hold up to rather large partial pressures and solubilities, but such cases are rare; the apparent validity of Eq. (8.2-1) at large solubilities is usually fortuitous due to cancellation of two (or more) factors which, taken separately, would cause that equation to fail.

The assumptions on which Eq. (8.2-1) are based can readily be recognized by comparing it with Eq. (1) given in the introduction to this chapter. A comparison of the left-hand sides shows that in Henry's law the gas phase is assumed to be ideal and thus the fugacity is replaced by the partial pressure; this simplification can be avoided by the methods discussed in Chap. 5. A comparison of the right-hand sides show that the fugacity in the liquid phase is assumed to be proportional to the mole fraction and that the constant of proportionality is taken as an empirically determined factor which depends on the natures of solute and solvent and on the temperature. The thermodynamic significance of this constant can be established by comparing the liquid fugacity as given by Henry's law with that obtained in the more conventional manner using

†The constant k has no precise thermodynamic significance although it is very similar to Henry's constant $H_{2,1}$ defined by Eq. (6.3-14). For a given binary system, Henry's constant $H_{2,1}$ depends on temperature and, to a lesser degree, on total pressure, as discussed in Sec. 8.3 and in Chap. 10.

the concept of an activity coefficient γ and some standard-state fugacity f^0:

$$f_2^L = kx_2 = H_{2,1}x_2 = \gamma_2 x_2 f_2^0. \tag{8.2-2}$$

Thus

$$k = H_{2,1} = \gamma_2 f_2^0 \tag{8.2-3}$$

where 1 stands for solvent and 2 stands for solute.

At a given temperature and pressure, the standard-state fugacity is a constant and does not depend on the solute mole fraction in the liquid phase. Since k does not depend on x_2, it follows from Eq. (8.2-3) that the activity coefficient γ_2 must also be independent of x_2; it is this feature, the constancy of the activity coefficient, which contains the essential assumption of Henry's law.

The activity coefficient of a solute is always independent of the solute's mole fraction provided the latter is sufficiently small. This fact can be shown from mathematical as well as molecular considerations. To fix ideas, let us take the case where γ_2 is normalized to approach unity as the mole fraction of 2 goes to unity. As shown in Chap. 6, it is convenient to express $\ln \gamma_2$ as a power series in $(1 - x_2)$:

$$RT \ln \gamma_2 = A(1 - x_2)^2 + B(1 - x_2)^3 + \dots. \tag{8.2-4}$$

where $A, B, \dots$ are constants depending on temperature and on the inter-molecular forces between solute and solvent. Equation (8.2-4) shows at once that if $x_2 \ll 1$, then γ_2 is only weakly dependent on x_2 and Henry's law provides a good approximation.

From the molecular point of view, the activity coefficient of the solute remains constant if the environment in which a solute molecule finds itself remains unchanged. This environment is not altered as long as the mole fraction of solute is small, for under such conditions a given solute molecule hardly every "meets" another solute molecule but "sees" only solvent molecules. As long as $x_2 \ll 1$, each solute molecule does not "know" that there are any other solute molecules in the solution; therefore, as long as $x_2 \ll 1$, the interactions which a solute molecule experiences with its neighbors are insensitive to changes in x_2; thus γ_2 is not affected.

Equation (8.2-4) gives some insight into the well-known observation that Henry's law is a good approximation up to relatively large solubilities for some systems but fails for relatively small solubilities in other systems. Let us consider the case where only the first term in the series is retained, higher terms being neglected. The coefficient A is a measure of nonideality; if A is positive, it indicates the "dislike" between solute and solvent, whereas if it is negative its absolute value may be a measure of the tendency between solute and solvent to form a complex. In any case, it is the absolute value of A/RT which determines the range of validity of Henry's law; in fact, if $A/RT = 0$ (ideal solution) Henry's law holds for the entire range of composition $0 \leq x_2 \leq 1$. If A/RT is small compared to unity, then the activity coefficient γ_2 does not change much even for appreciable x_2, but if it is large, then even a small x_2 can produce a significant change in the activity coefficient with composition. In the limit, as x_2 approaches zero, the logarithm of the activity coefficient approaches the constant value A/RT and therefore Henry's law is always valid as a limiting relation.

As indicated by Eq. (8.2-1), Henry's law assumes that the gas-phase fugacity is equal to the partial pressure. This assumption is not at all necessary and is rather easily removed by including the gas-phase fugacity coefficient φ which has been discussed in detail earlier; more properly, therefore, Henry's law for solute i is:

$$\boxed{f_i = \varphi_i y_i P = H_{i,\,\text{solvent}} x_i.}$$

(8.2-5)

8.3 Effect of Pressure on Gas Solubility

In the previous section we discussed the essential assumption in Henry's law, viz., that at constant temperature the fugacity of solute i is proportional to the mole fraction x_i. The constant of proportionality $H_{i,\,\text{solvent}}$ is not a function of composition but depends on temperature and, to a lesser degree, pressure. The pressure dependence can be neglected as long as the pressure is not large. At high pressures, however, the effect is not negligible and therefore it is necessary to consider how Henry's constant depends on pressure. This dependence is easily obtained by using the rigorous equation

$$\left(\frac{\partial \ln f_i^L}{\partial P} \right)_{T,\,x} = \frac{\bar{v}_i}{RT}$$

(8.3-1)

where $\bar{v}_i$ is the partial molar volume of i in the liquid phase. The thermodynamic definition of Henry's constant is

$$H_{i,\,\text{solvent}} \equiv \lim_{x_i \to 0} \frac{f_i^L}{x_i} \qquad \text{(at constant temperature).}$$

(8.3-2)

Substitution of Eq. (8.3-2) into Eq. (8.3-1) gives

$$\left(\frac{\partial \ln H_{i,\,\text{solvent}}}{\partial P}\right)_T = \frac{\bar{v}_i^\infty}{RT} \qquad (8.3\text{-}3)$$

where $\bar{v}_i^\infty$ is the partial molar volume of solute i in the liquid phase at infinite dilution.† Integrating Eq. (8.3-3) and assuming, as before, that the fugacity of i at constant temperature and pressure, is proportional to x_i, we obtain a more general form of Henry's law:

$$\ln \frac{f_i}{x_i} = \ln H_{i,\,\text{solvent}}^{(P')} + \frac{\displaystyle\int_{P'}^{P} \bar{v}_i^\infty \, dP}{RT} \qquad (8.3\text{-}4)$$

where $H_{i,\,\text{solvent}}^{(P')}$ is Henry's constant evaluated at arbitrary reference pressure P'. As $x_i \to 0$, the total pressure is P_1^s, the saturation (vapor) pressure of the solvent; it is often convenient, therefore, to set $P' = P_1^s$.

If the solution temperature is well below the critical temperature of the solvent, it is reasonable to assume that $\bar{v}_i^\infty$ is independent of pressure. Letting subscript 1 refer to the solvent and subscript 2 to the solute, Eq. (8.3-4) becomes

$$\boxed{\ln \frac{f_2}{x_2} = \ln H_{2,\,1}^{(P_1^s)} + \frac{\bar{v}_2^\infty (P - P_1^s)}{RT}.} \qquad (8.3\text{-}5)$$

Equation (8.3-5) is known as the Krichevsky-Kasarnovsky equation[1] although its first clear derivation was given by Dodge and Newton.[2] This equation is remarkably useful for representing solubilities of sparingly soluble gases up to very high pressures. Figures 8-2 and 8-3 show that solubility data for hydrogen and nitrogen in water up to 1000 atm are accurately reproduced by the Krichevsky-Kasarnovsky equation; in this case the vapor pressure of the solvent is completely negligible in comparison to the total pressure and therefore the abscissa reads P, rather than $P - P_1^s$. The intercepts of these plots give $H_{2,1}^{(P_1^s)}$, and the slopes yield the partial molar volumes of the gaseous solutes in the liquid phase. At 25°C, Figs. 8-2 and 8-3 give a partial molar volume for hydrogen of 19.5 cc/g-mol; that for nitrogen is 32.8 cc/g-mol. These results are in good agreement with partial molar volumes for these gases in water as obtained from dilatometric measurements.

Equation (8.3-5) can be expected to hold for all those cases which conform to the two assumptions on which the equation rests. One of these is

†See Sec. 10.5 for a discussion of liquid-phase partial molar volumes.

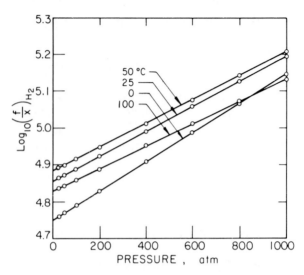

Fig. 8-2   Solubility of hydrogen in water at high pressures.

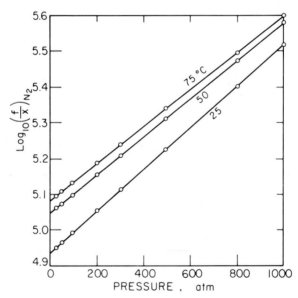

Fig. 8-3   Solubility of nitrogen in water at high pressures.

that the activity coefficient of the solute does not change noticeably over the range of x_2 being considered; in other words, x_2 must be small, as discussed in the previous section. The other assumption states that the infinitely dilute liquid solution must be essentially incompressible, which is approximately true at temperatures far removed from the critical temperature of the solution.

To illustrate the use and limitation of the Krichevsky-Kasarnovsky equation, consider the high-pressure solubility data of Wiebe and Gaddy[3] for nitrogen in liquid ammonia. These are shown in Fig. 8-4 plotted in the

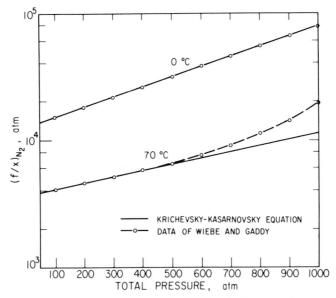

Fig. 8-4   Success and failure of the Krichevsky-Kasarnovsky equation. Solubility of nitrogen in liquid ammonia.

manner indicated by Eq. (8.3-5). At 0°C the Krichevsky-Kasarnovsky equation holds up to 1000 atm but at 70°C it breaks down after about 600 atm. This striking difference is readily explained upon considering the two assumptions just mentioned; at 0°C liquid ammonia is still an unexpanded liquid solvent (the critical temperature of ammonia is 132.3°C) and the solubility of nitrogen is small throughout, being only 2.2 mol% at 1000 atm. At 0°C, therefore, the assumptions of Eq. (8.3-5) are reasonably satisfied. However, at 70°C liquid ammonia is already quite expanded (and compressible) and the solubility of nitrogen is no longer small, being 12.9 mol % at 1000 atm. Under these conditions it is not reasonable to expect that the activity coefficient of nitrogen in the liquid phase is independent of composition, nor is it likely that the partial molar volume is constant. As a result, it is not surprising that the Krichevsky-Kasarnovsky equation gives an excellent representation of the data for the entire pressure range at 0°C but fails at the higher pressures for the data at 70°C.

The variation of the activity coefficient of the solute with mole fraction can be taken into account by one of the methods discussed in Chap. 6. In the simplest case we may assume that the activity coefficient of the *solvent* is given by a two-suffix Margules equation

$$\ln \gamma_1 = \frac{A}{RT} x_2^2 \qquad (8.3\text{-}6)$$

where A is an empirical constant determined by the intermolecular forces in the solution.

The activity coefficient γ_2^* of the solute, normalized according to the unsymmetric convention (see Sec. 6.4), is then found from the Gibbs-Duhem equation: It is given by

$$\ln \gamma_2^* = \frac{A}{RT} (x_1^2 - 1) . \qquad (8.3\text{-}7)$$

The fugacity of component 2 at pressure P_1^s is

$$f_2 = \gamma_2^* H_{2,1}^{(P_1^s)} x_2 \qquad (8.3\text{-}8)$$

and instead of Eq. (8.3-5) we obtain

$$\ln \frac{f_2}{x_2} = \ln H_{2,1}^{(P_1^s)} + \frac{A}{RT} (x_1^2 - 1) + \frac{\bar{v}_2^\infty (P - P_1^s)}{RT} . \qquad (8.3\text{-}9)\dagger$$

Equation (8.5-9) is called the Krichevsky-Ilinskaya equation[4] and because of the additional parameter, it has a wider applicability than does Eq. (8.3-5). It is especially useful for solutions of light gases (such as helium and hydrogen) in those liquid solvents where the solubility is appreciable. For example, Orentlicher[5] found that Eq. (8.3-9) could be successfully used to correlate solubility data for hydrogen in a variety of solvents at low temperatures and at pressures up to about 100 atm. Table 8-2 gives the parameters reported by Orentlicher. In the systems studied, the solubility of hydrogen may be as large as 20 mol % and therefore the data could not be adequately represented by the simpler Krichevsky-Kasarnovsky equation.

---

†Equation (8.3-9) assumes that the partial molar volume of the solute is independent of pressure and composition over the pressure and composition ranges under consideration.

Table 8-2

THERMODYNAMIC PARAMETERS FOR HYDROGEN
SOLUBILITY†

Solvent	T (°K)	$H_{2,1}^{(P_1^s)}$ (atm)	A (l atm/g-mol)	$\bar{v}_2^\infty$ (cm³/g-mol)
Ar	87	830	5.0 ± 0.5	30
	100	660		31
	120	500		35
	140	385		44
CO	68	640	7.0 ± 0.7	31.2
	73	550		31.8
	78	470		32.6
	83	440		33.6
	88	400		34.4
N$_2$	68	540	7.0 ± 0.7	30.4
	79	450		31.5
	86	390		32.0
	90	370		33.0
	95	340		34.4
CH$_4$	90	1824	15 ± 3	29.7
	110	1036		31.0
	116	970		31.6
	127	838		32.6
	144	630		36.0
C$_2$H$_4$	144	3050	19.5 ± 1.5	37.1
	158	2400		37.8
	172	2050		39.4
	200	1400		44.5
C$_2$H$_6$	144	2600	25 ± 2	37.9
	172	2050		40.7
	200	1650		44.2
	228	1210		54.3
C$_3$H$_8$	228	1670	25 ± 2	50
	255	1300		51
	282	1030		63
C$_3$H$_6$	172	3300	25 ± 2	39.6
	200	2600		41.7
	228	1850		48.8
	268	1320		55.6

†M. Orentlicher and J. M. Prausnitz, *Chem. Eng. Sci.*, **19**,
775 (1964).

8.4 Effect of Temperature on Gas Solubility

Many elementary books on chemistry state without qualification that the solubility of a gas falls with rising temperature. Unfortunately this is not always true although it is more often correct than not.

The temperature derivative of the solubility, as calculated from the Gibbs-Helmholtz equation,† is directly related to either the partial molar enthalpy or the partial molar entropy of the gaseous solute in the liquid phase. Therefore, if something can be said about the enthalpy or entropy change of solution, insight can be gained on the effect of temperature on solubility. A general derivation of the thermodynamic relations is given elsewhere;[6,7] we consider here only the relatively simple case where the solvent is essentially nonvolatile and where the solubility is sufficiently small to make the activity coefficient of the solute independent of the mole fraction. With these restrictions it can be shown that

$$\left(\frac{\partial \ln x_2}{\partial 1/T}\right)_P = -\frac{\Delta \bar{h}_2}{R} \tag{8.4-1}$$

and

$$\left(\frac{\partial \ln x_2}{\partial \ln T}\right)_P = \frac{\Delta \bar{s}_2}{R} \tag{8.4-2}$$

where x_2 is the mole fraction of gaseous solute at saturation and

$$\Delta \bar{h}_2 \equiv \bar{h}_2^L - h_2^G, \qquad \Delta \bar{s}_2 \equiv \bar{s}_2^L - s_2^G.$$

First, we consider Eq. (8.4-2); if the partial molar entropy change of the solute is positive, then the solubility increases with rising temperature; otherwise it falls. To understand the significance of the entropy change it is convenient to divide it into two parts:

$$\Delta \bar{s}_2 = (s_2^L - s_2^G) + (\bar{s}_2^L - s_2^L) \tag{8.4-3}$$

where s_2^L is the entropy of the (hypothetical) pure liquid at the temperature of the solution. The first term on the right-hand side of Eq. (8.4-3) is the entropy of condensation of the pure gas and in general we expect this term to be negative since the entropy (disorder) of a liquid is lower than that of a saturated gas at the same temperature. The second term is the partial molar entropy of solution of the condensed solute and, assuming ideal entropy of mixing for the two liquids, we can write

$$\bar{s}_2^L - s_2^L = -R \ln x_2. \tag{8.4-4}$$

†See Eq. (6.2-6).

Since $x_2 < 1$, the second term in Eq. (8.4-3) is positive and the smaller the solubility, the larger this term must be. It therefore follows that $\Delta \bar{s}_2$ should be positive for those gases which have very small solubilities and negative for the others; this result leads to the expectation that gases which are sparingly soluble (very small x_2) show positive temperature coefficients of solubility, whereas gases which are readily soluble (relatively large x_2) show negative temperature coefficients. This expectation is, in fact, observed. This semiquantitative interpretation of the sign of $\Delta \bar{s}_2$ is in good agreement with observed behavior of gas-liquid solutions as shown in Fig. 8-5 which relates the observed partial molar entropy change [Eq. (8.4-2)] to the ideal partial molar entropy of the condensed solute [Eq. (8.4-4)].[8, 9] The plot gives experimental results at 25°C for fifteen gases in six solvents at 1 atm partial pressure. The figure can be divided into two parts: The upper part corresponds to a positive, and the lower to a negative temperature coefficient of solubility. As suggested by the foregoing discussion, Fig. 8-5 shows that as a general

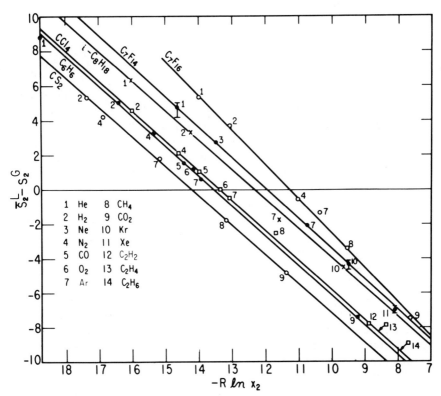

Fig. 8-5   Entropy of solution of gases in liquids as a function of gas solubility x_2 at 25°C and 1 atm (from Hildebrand and Scott).

rule, the solubility of a gas rises with increasing temperature whenever x_2 is small ($-R \ln x_2$ is large), and it falls with increasing temperature whenever x_2 is large ($-R \ln x_2$ is small). This is only a general rule, and there are differences in behavior between different gases and different solvents. For example, consider the solubility which a gas must have in order to have a zero temperature coefficient of solubility (no change in solubility with temperature). In perfluoroheptane $\Delta \bar{s}_2$ is zero when $-R \ln x_2 = 11.3$ cal/g-mol, $°K$; therefore, in this solvent the temperature coefficient of solubility changes sign when $x_2 = 3.43 \times 10^{-3}$. In carbon disulfide, however, the corresponding value is $x_2 = 76 \times 10^{-3}$. Going from left to right, the solvents in Fig. 8-5 are arranged in order of decreasing solubility parameters. At constant solubility (i.e., constant x_2) the temperature coefficient of solubility for a given gas has a tendency to increase (algebraically) as the solubility parameter of the solvent falls.

Some qualitative insight into the effect of temperature on gas solubility can also be obtained from the partial molar enthalpy change [Eq. (8.4-1)]. Again, it is useful to divide this change into two parts:

$$\Delta \bar{h}_2 = (h_2^L - h_2^G) + (\bar{h}_2^L - h_2^L) \tag{8.4-5}$$

where h_2^L is the enthalpy of the (hypothetical) pure liquid at the temperature of the solution.

The first term in Eq. (8.4-5) is the enthalpy of condensation of pure solute and since the enthalpy of a liquid is generally lower than that of a gas at the same temperature, we expect this quantity to be negative.† The second quantity is the partial enthalpy of mixing for the liquid solute; in the absence of solvation between solute and solvent this quantity tends to be positive (endothermic) and the theory of regular solutions (Chap. 7) tells us that the larger the difference between the cohesive energy density of the solute and that of the solvent, the larger the heat of mixing. If this difference is very large (e.g., hydrogen and benzene) the second term in Eq. (8.4-5) dominates; $\Delta \bar{h}_2$ is then positive and the solubility increases with rising temperature. However, if the difference in cohesive energy densities is small (e.g., chlorine in carbon tetrachloride); then the first term in Eq. (8.4-5) dominates; $\Delta \bar{h}_2$ is then negative and the solubility falls as the temperature increases.

If there are specific chemical interactions between solute and solvent (e.g., ammonia and water), then both terms in Eq. (8.4-5) are negative (exothermic) and the solubility decreases rapidly as the temperature becomes larger.

The effect of temperature on solubility is sensitive to the intermolecular forces of the solute-solvent system. When the partial pressure of the

†At $T/T_{c_2} \gg 1$, $(h_2^L - h_2^G)$ may be positive.

solute is small, the solubility typically decreases with temperature, goes through a minimum, and then rises. To illustrate, Fig. 8-6 shows the solubility of methane in *n*-heptane over a wide temperature range. For most systems the temperature corresponding to minimum solubility lies well above room temperature but for light gases, minimum solubility is usually observed at low temperatures.

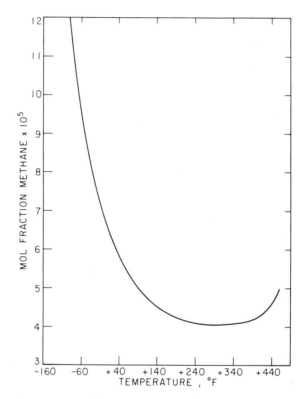

Fig. 8-6   Solubility of methane in *n*-heptane when vapor-phase fugacity of methane is 0.01 atm.

8.5   Estimation of Gas Solubility

Reliable data on the solubility of gases in liquids are not plentiful, especially at temperatures well removed from 25°C. In recent years, however, J. H. Hildebrand and co-workers have obtained a large amount of accurate data in the vicinity of room temperature and therefore we consider first the semiempirical correlations which they have established.

Figure 8-7 shows solubilities of twelve gases (at 25°C and 1 atm partial

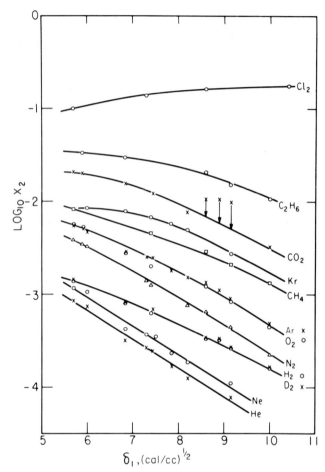

Fig. 8-7   Solubility of gases in liquids at 25°C and at a partial
gas pressure of 1 atm as a function of solvent
solubility parameter δ_1 (from Hildebrand and Scott).

pressure) as a function of the solubility parameter of the solvent, and
Fig. 8-8 shows solubilities of thirteen gases (at the same conditions) in
nine solvents as a function of Lennard-Jones energy parameters (ϵ/k)
which were determined from second virial coefficient data for the solutes.†
These plots, presented by Hildebrand and Scott,[6] indicate that the solubilities of nonpolar gases in nonpolar solvents can be correlated in terms
of two parameters, viz., the solubility parameter of the solvent and the

†The parameters used in Fig. 8-8 were taken from J. O. Hirschfelder, C. F. Curtiss, and
R. B. Bird, *Molecular Theory of Gases and Liquids* (New York: John Wiley & Sons, Inc.,
1954).

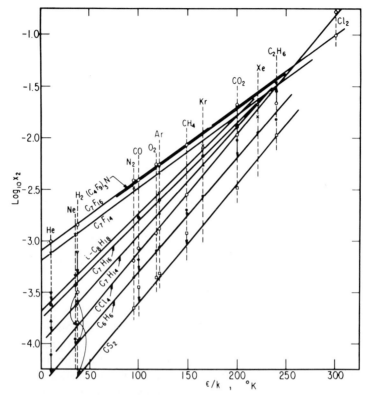

Fig. 8-8   Solubility of gases in liquids at 25°C and a partial pressure of 1 atm as a function of solute characteristic energy ϵ/k (Lennard-Jones) taken from Hirschfelder, Curtiss and Bird (from Hildebrand and Scott).

Lennard-Jones energy parameter for the solute. Figures 8-7 and 8-8, therefore, may be used with some confidence to predict solubilities in non-polar systems where no experimental data are available; these predictions are necessarily limited to systems at 25°C but, with the help of Eq. (8.4-2) and the entropy data shown in Fig. 8-5, it is possible to predict solubilities at other temperatures not far removed from 25°C.

Figures 8-7 and 8-8 are useful, but in addition to their utility we must recognize that even for nonpolar systems at one temperature there are significant deviations from "regular" behavior; the straight lines correlate most of the experimental results but there are notable exceptions. In Fig. 8-7 the solubilities of carbon dioxide in benzene, toluene and carbon tetrachloride are somewhat larger than those predicted from results obtained in other solvents. The high solubility in aromatic hydrocarbons can probably be attributed to a Lewis acid-base interaction between acidic carbon dioxide and basic aromatic; apparently there is also some specific chemical interaction between carbon dioxide and carbon tetrachloride about which, however, little is known. In Fig. 8-8 the solubilities of the quantum gases, helium, hydrogen, and neon, are a little higher than expected, the discrepancies becoming larger as the solubility parameter of the solvent increases. These anomalies are not yet fully understood.

Hildebrand's correlations provide a good basis for estimating gas solubilities in nonpolar systems at temperatures near 25°C but since the entropy of solution is temperature-dependent, these correlations are not helpful at temperatures well removed from 25°C. Unfortunately, solubility data at temperatures much larger or smaller than room temperature are scarce and therefore strictly empirical correlations cannot be used; rather, it is necessary to resort to whatever theoretical methods might be available and appropriate. A rigorous method for the prediction of gas solubilities requires a valid theory of solutions. Such a theory does not now exist, but for a semitheoretical description of nonpolar systems, the theory of regular solutions and the theorem of corresponding states can serve as the basis of a correlating scheme,[10] which we now describe.

Consider a gaseous component at fugacity f_2^G dissolved isothermally in a liquid not near its critical temperature. The solution process is accompanied by a change in enthalpy and in entropy, just as occurs when two liquids are mixed. However, in addition, the solution process for the gas is accompanied by a large decrease in volume; since the partial molar volume of the solute in the condensed phase is much smaller than that in the gas phase. It is this large decrease in volume which distinguishes solution of a gas in a liquid from solution of another liquid or of a solid. Therefore, in order to apply regular solution theory (which assumes no volume change) it is necessary first to "condense" the gas to a volume close to the partial molar volume which it has as a solute in a liquid solvent. The isothermal solution process is then considered in two steps, I and II:

$$\Delta g = \Delta g_I + \Delta g_{II} \qquad (8.5\text{-}1)$$

$$\Delta g_I = RT \ln \frac{f_{pure\ 2}^L}{f_2^G} \qquad (8.5\text{-}2)$$

$$\Delta g_{II} = RT \ln \gamma_2 x_2 \qquad (8.5\text{-}3)$$

where $f_{pure\ 2}^L$ is the fugacity of (hypothetical) pure liquid solute and γ_2 is the symmetrically normalized activity coefficient of the solute referred to the (hypothetical) pure liquid ($\gamma_2 \rightarrow 1$ as $x_2 \rightarrow 1$).

In the first step, the gas isothermally "condenses" to a hypothetical state having a liquid-like volume. In the second step, the hypothetical, liquid-like fluid dissolves in the solvent. Since the solute in the liquid solution is in equilibrium with the gas which is at the fugacity f_2^G, the equation of equilibrium is

$$\Delta g = 0. \tag{8.5-4}$$

We assume that the regular-solution equation gives the activity coefficient for the gaseous solute:†

$$RT \ln \gamma_2 = v_2^l (\delta_1 - \delta_2)^2 \Phi_1^2 \tag{8.5-5}$$

where δ_1 is the solubility parameter of solvent, δ_2 is the solubility parameter of solute, v_2^l is the molar "liquid" volume of solute, and Φ_1 is the volume fraction of solvent.

Substitution of Eqs. (8.5-1), (2), (3), and (5) into Eq. (8.5-4) gives the solubility:

$$\frac{1}{x_2} = \frac{f_{\text{pure 2}}^L}{f_2^G} \exp \frac{v_2^l (\delta_1 - \delta_2)^2 \Phi_1^2}{RT}. \tag{8.5-6}$$

Equation (8.5-6) forms the basis of the correlating scheme; it involves three parameters for the gaseous component as a hypothetical liquid: the pure liquid fugacity, the liquid volume, and the solubility parameter. These parameters are all temperature dependent; however, the theory of regular solutions assumes that at constant composition

$$\ln \gamma_2 \propto \frac{1}{T} \tag{8.5-7}$$

and therefore, the quantity $v_2^l (\delta_1 - \delta_2)^2 \Phi_1^2$ is not temperature-dependent. As a result, any convenient temperature may be used for v_2^l and δ_2 provided the same temperature is also used for δ_1 and v_1^l. (The most convenient temperature—that used here—is 25°C.) The fugacity of the hypothetical liquid, however, must be treated as a function of temperature.

The three correlating parameters for the gaseous solutes were calculated by Shair[10] from experimental solubility data. The molar volume v_2^l and the solubility parameter δ_2, both at 25°C, are given for eleven gases in Table 8-3. However, as explained above, Eq. (8.5-6) is not restricted to 25°C; in principle, it may be used at any temperature provided it is well removed from the critical temperature of the solvent. A correlation similar to that of Shair has been given by Yen and McKetta.[11]

†See Sec. 7.2 for a discussion of the regular-solution equation. A list of liquid molar volumes and solubility parameters for some common solvents is given in Table 7-1.

Table 8-3

"LIQUID" VOLUMES AND SOLUBILITY
PARAMETERS FOR GASEOUS SOLUTES
AT 25°C†

Gas	v^L (cc/g-mol)	δ (cal/cc)$^{1/2}$
N_2	32.4	2.58
CO	32.1	3.13
O_2	33.0	4.0
Ar	57.1	5.33
CH_4	52	5.68
CO_2	55	6.0
Kr	65	6.4
C_2H_4	65	6.6
C_2H_6	70	6.6
Rn	70	8.83
Cl_2	74	8.7

†J. M. Prausnitz and F. H. Shair,
A. I. Ch. E. Journal, 7, 682 (1961).

The fugacity of the hypothetical pure liquid has been correlated in a corresponding-states plot as shown in Fig. 8-9; the fugacity of the solute, divided by its critical pressure is shown as a function of the ratio of the solution temperature to the solute's critical temperature. The fugacities in Fig. 8-9 are for a total pressure of one atmosphere. If the solution under consideration is at a considerably higher pressure, the Poynting correction should be applied to the fugacity as read from Fig. 8-9; thus

$$\underset{\substack{\text{(at total pressure } P)}}{f^L_{\text{pure 2}}} = \underset{\substack{\text{(from Fig. 8-9)}}}{f^L_{\text{pure 2}}} \; \exp \frac{v^L_2 (P-1)}{RT} \qquad (8.5\text{-}8)$$

where P is in units of atmospheres.

Equation (8.5-6) contains Φ_1, the volume fraction of the solvent and, therefore, solving for x_2 requires a trial-and-error calculation; however the calculation converges rapidly. If the partial pressure of the gas is not large, x_2 is usually very small, and thus setting Φ_1 equal to unity in Eq. (8.5-6) provides an excellent approximation.

Shair's technique for correlating gas solubilities on the basis of regular-solution theory can readily be extended to mixed solvents. To estimate the solubility of a gas in a mixture of two or more solvents, Eq. (8.5-6) should be replaced by

$$\frac{1}{x_2} = \frac{f^L_{\text{pure 2}}}{f^G_2} \exp \frac{v^L_2 (\delta_2 - \bar{\delta})^2}{RT} \qquad (8.5\text{-}9)$$

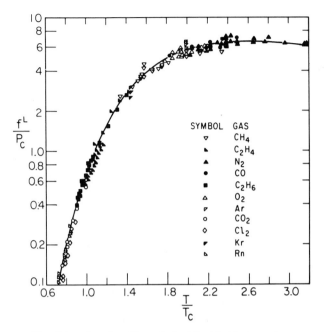

Fig. 8-9   Fugacity of hypothetical liquid at 1 atmosphere.

where $\bar{\delta}$ is an average solubility parameter for the entire solution:

$$\bar{\delta} \equiv \sum_i \Phi_i \delta_i \quad (i = 1, 2, \ldots).$$

The summation in the above definition is over *all* components, including the solute.

Table 8-3 and Fig. 8-9 do not include information for the light gases, helium, hydrogen, and neon. For these gases, Hildebrand's correlations are useful near room temperature and for low temperatures, a separate correlation for hydrogen is available.† Good solubility data for hydrogen, helium, and neon are rare at higher temperatures.

The correlation given by Eq. (8.5-6), Table 8-3, and Fig. 8-9 gives fair estimates of gas solubilities over a wide temperature range for nonpolar gases and liquids. It is not to be expected that a simple correlation such as this should give highly accurate results, but in view of the poor accuracy of many of the experimental solubility data which have been reported in the literature, the estimated solubilities may, in some cases, be more reliable than the experimental ones. Whenever really good experimental data are available they should of course be given priority, but

†See Table 8-2.

whenever there is serious disagreement between observed results and those calculated from a pertinent correlation one must not, without further study, immediately give preference to the experimental value.

8.6   Gas Solubility in Mixed Solvents

Good solubility data for gases in pure liquids are not plentiful, while solubility data in mixed solvents are extremely scarce. With the aid of a simple thermodynamic model, however, it is possible to make a fair estimate of the solubility of a gas in a simple solvent mixture, provided the solubility of the gas is known in each of the pure solvents which comprise the mixture. The procedure for making such an estimate is, essentially, based on Wohl's expansion discussed in Secs. 6.13 and 6.17. We consider here only the simplest case and shall follow the procedure given by O'Connell.[12]   A more detailed discussion is given in App. VII.

Let subscript 2 stand for the gas as before, and let subscripts 1 and 3 stand for the two (miscible) solvents. To simplify matters, we confine ourselves to low or moderate pressures where the effect of pressure on liquid-phase properties can be neglected. For the ternary liquid phase we write the simplest (two-suffix Margules) expansion of Wohl for the excess Gibbs energy at constant temperature:

$$\frac{g^E}{RT}\,\text{(ternary)} = a_{12}\,x_1 x_2 + a_{13}\,x_1 x_3 + a_{23}\,x_2 x_3 \qquad (8.6\text{-}1)$$

where each a_{ij} is a constant characteristic of the ij binary pair.

From Eq. (8.6-1) we can compute the symmetrically normalized activity coefficient γ_2 of the gaseous solute using Eq. (6.3-8). As shown by O'Connell[12] the unsymmetrically normalized activity coefficient γ_2^* can then be found by

$$\gamma_2^* = \gamma_2 \exp\left(-a_{12}\right) \qquad (8.6\text{-}2)$$

where the definition of γ_2^* is

$$\gamma_2^* \equiv \frac{f_2}{x_2\,H_{2,1}}. \qquad (8.6\text{-}3)$$

As in previous sections, $H_{2,1}$ stands for Henry's constant of component 2 in pure solvent 1 at the system temperature. O'Connell also has shown that for this simple model, the parameters a_{23} and a_{21} are related to the two Henry's constants:

$$a_{23} = a_{12} + \ln \frac{H_{2,3}}{H_{2,1}} \qquad (8.6\text{-}4)$$

where $H_{2,3}$ is Henry's constant for the solute in pure solvent 3 at the system temperature.

From Eqs. (8.6-1) and (8.6-2), utilizing Eq. (6.3-8), we obtain

$$\ln \gamma_2^* = a_{12}[x_1(1 - x_2) - 1] + a_{23}x_3(1 - x_2) - a_{13}x_1x_3. \qquad (8.6-5)$$

We now use Eqs. (8.6-4) and (8.6-5) to obtain an expression for $H_{2,M}$ which is Henry's constant for the solute in the mixed solvent. For some fixed ratio of solvents 1 and 3,

$$H_{2,M} \equiv \underset{x_2 \to 0}{\text{limit}} \frac{f_2}{x_2} = \underset{x_2 \to 0}{\text{limit}} \gamma_2^* H_{2,1}. \qquad (8.6-6)$$

From Eq. (8.6-5)

$$\underset{x_2 \to 0}{\text{limit}} \ln \gamma_2^* = (a_{23} - a_{12})x_3 - a_{13}x_1x_3. \qquad (8.6-7)$$

Substitution of Eqs. (8.6-4) and (8.6-7) into Eq. (8.6-6) gives the desired result:

$$\boxed{\ln H_{2,M} = x_1 \ln H_{2,1} + x_3 \ln H_{2,3} - a_{13}x_1x_3.} \qquad (8.6-8)$$

Equation (8.6-8) says that the logarithm of Henry's constant in a binary solvent mixture is a linear function of the solvent composition whenever the two solvents (without solute) form an ideal mixture ($a_{13} = 0$). If the solute-free solvent mixture exhibits positive deviations from Raoult's law ($a_{13} > 0$), Henry's constant in the mixture is smaller (or solubility is larger) than that corresponding to a perfect solvent mixture of the same composition. Similarly, if $a_{13} < 0$, the solubility of the gas is lower than what it would be if the solvents formed an ideal mixture. The constant a_{13} must be estimated from vapor-liquid equilibrium data on the solvent mixture.†

The conclusions just reached follow from the important assumption that Eq. (8.6-1) gives a valid description of the excess Gibbs energy of the ternary mixture. This assumption provides a reasonable approximation for some solutions of simple fluids‡ but for mixtures containing polar or

---

†For nonpolar solvents an estimate can sometimes be made using the theory of regular solutions:

$$a_{13} \approx \frac{(\delta_1 - \delta_3)^2(v_1^L + v_3^L)}{2RT}$$

where δ is the solubility parameter and v^L the liquid molar volume. See Sec. 7.2.

‡For accurate work, a two-suffix Margules expansion may not be sufficient. See App. VII.

hydrogen-bonded liquids a more complicated form of Wohl's expansion is required.

According to Eq. (8.6-8) the effect of nonideal mixing of the solvents is not large. To illustrate, two calculated examples at 25°C are shown in Fig. 8-10. The first one considers Henry's constants for hydrogen in mixtures of toluene and heptane. Since toluene and heptane show only modest positive deviations from Raoult's law (g^E = 48 cal/g-mol for the equimolar mixture), a_{13} is small and, as shown in Fig. 8-10, Henry's con-

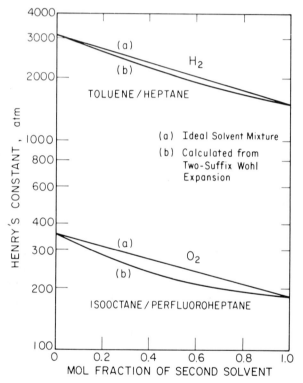

Fig. 8-10   Calculated Henry's constants in solvent mixtures at 25°C.

stant as calculated from Eq. (8.6-8) differs little from that calculated assuming an ideal solvent mixture. The second example considers Henry's constants for oxygen in mixtures of isooctane and perfluoroheptane, which exhibit large deviations from Raoult's law; for the equimolar mixture of these solvents g^E = 330 cal/g-mol and at temperatures only slightly below 25°C, these two liquids are no longer completely miscible. In this case there is a more significant difference between Henry's constants calculated with and without nonideality of the solvent mixture. For an equimolar mixture of the two solvents, the solubility of oxygen is enhanced by nearly 20% as a result of the solvent mixture's nonideality.

Equation (8.6-8) is readily generalized to solvent mixtures containing any desired number of solvents. For an m-component system in which the gas is designated by subscript 2, Henry's constant for the gaseous solute is given by

$$\ln H_{2,M} = \sum_{\substack{j=1 \\ j\neq 2}}^{m} x_j \ln H_{2,j} - \sum_{\substack{j=1 \\ j\neq 2}}^{m-1} \sum_{\substack{k>j \\ k\neq 2}}^{m} a_{jk} x_j x_k. \qquad (8.6\text{-}9)$$

Equation (8.6-9), like Eq. (8.6-8), follows directly from the simplest form of Wohl's expansion as given by Eq. (8.6-1) for a ternary mixture.

8.7   Chemical Effects on Gas Solubility

The gas solubility correlations of Hildebrand and Scott and of Shair discussed in Sec. 8.5 are based on a consideration of physical forces between solute and solvent; they are not useful for those cases where chemical forces are present. Sometimes these chemical effects are not large and therefore a physical theory is a good approximation. But frequently the chemical forces are dominant; in that case correlations are hard to establish since specific chemical forces are not subject to simple generalizations. An extreme example of a chemical effect in a gas-liquid solution is provided by the interaction between ammonia and water (to form ammonium hydroxide); somewhat milder examples are the interaction between acetylene and acetone (hydrogen bonding) or that between ethylene and an aqueous solution of silver nitrate (Lewis acid-base complex formation); a still weaker case of the effect of chemical forces was mentioned earlier in connection with Fig. 8-7 where the solubility of (acidic) carbon dioxide in (basic) benzene and toluene is larger than that predicted by the physical correlation. In each of these cases the solubility of the gas is enhanced as a result of a specific affinity between solute and solvent.

Systematic studies of the effect of chemical forces on gas solubility are rare, primarily because it is difficult to characterize chemical forces in a quantitative way; the chemical affinity of a solvent (unlike its solubility parameter) depends on the nature of the solute, and therefore a measure of chemical effects in solution is necessarily relative rather than absolute. Three examples, given below, illustrate how the effect of chemical forces on gas solubility can be studied in at least a semiquantitative way.

The first example concerns the solubility of dichloromonofluoromethane (Freon-21) at 32.2°C in a variety of solvents. Solubility data in fifteen solvents were obtained by Zellhoefer, Copley and Marvel[13] and are shown in Fig. 8-11 as a function of pressure. Since the critical temperature of Freon-21 (178.5°C) is well above 32.2°C, it is possible to calculate an ideal solubility [Eq. (8.1-1)] without extrapolation of the vapor-pressure data, and this ideal solubility is shown in Fig. 8-11 by a dotted line. Figure 8-11 indicates that the solvents fall roughly into three groups: in the first group the solubility is less than ideal (positive deviations from Raoult's law); in the second group deviations from Raoult's law are very small; and in the third group the solubility is larger than ideal (negative deviations from Raoult's law). These data can be explained qualitatively by the concept of hydrogen bonding; Freon-21 has an active hydrogen

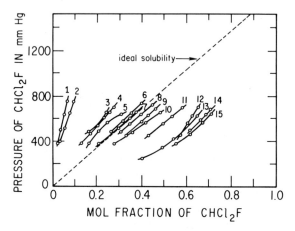

Fig. 8-11   Solubility of Freon-21 in liquid solvents at 32.2°C (data of Zellhoefer, Copley, and Marvel).   Liquid solvents shown are as follows:

1. Ethylene glycol
2. Trimethylene glycol
3. Decalin
4. Aniline
5. Benzotrifluoride
6. Nitrobenzene
7. Tetralin
8. Bis-β-methylthiolethyl sulfide
9. Dimethylaniline
10. Dioxan
11. Diethyl oxalate
12. Diethylene glycol diethoxy
13. Tetrahydrofurfuryl laurate
14. Tetraethyl oxamide
15. Dimethyl ether of tetraethylene glycol

atom and whenever the solvent can act as a proton acceptor it is a relatively good solvent. In fact, the stronger the proton affinity (Brönsted basicity) of the solvent, the better that solvent is, provided the proton-accepting atom of the solvent molecule is not already "taken" by a proton from a neighboring solvent molecule. In other words, if the solvent molecules are strongly self-associated, they will not be available to form hydrogen bonds with the solute. As a result, strongly associated substances are poor solvents for solutes such as Freon-21, which can form only weak hydrogen bonds and thus cannot compete successfully for proton acceptors. It is for this reason that the glycols are poor solvents for Freon-21; while no data for alcohols are given in Fig. 8-11, it is likely that methanol and ethanol would also be poor solvents for Freon-21.

The aromatic solvents aniline, benzotrifluoride and nitrobenzene are weak proton acceptors and it appears that for these solvents chemical forces (causing negative deviations from Raoult's law) are just strong enough to overcome physical forces (which usually cause positive deviations from Raoult's law), with the result that the observed solubilities are close to ideal. Finally, those solvents which are powerful proton acceptors —and whose molecules are free to accept protons—are excellent solvents for Freon-21. Dimethylaniline, for example, is a much stronger base and hence a better solvent than aniline. The oxygen atoms in ethers and the nitrogen atoms in amides are also good proton acceptors for solute molecules since ethers and amides do not self-associate to an appreciable extent.

The second example concerns the solubility of hydrogen chloride gas in heptane solutions of aromatic hydrocarbons at a low temperature. The solubility data of Brown and Brady[14] are shown in Fig. 8-12. Since the pressure of the hydrogen chloride gas was kept very low, Henry's law holds; Henry's constants (the slopes of the lines in Fig. 8-12) are shown in Table 8-4.

The solubility data show that the addition of small amounts of aromatics to heptane increases the solubility of hydrogen chloride. This increase is a result of the electron-donating properties of aromatic molecules which, because of their π-electrons, can act as Lewis bases. The solubility data, therefore, can be explained by postulating the formation of a complex between hydrogen chloride and aromatics; the stability of the complex depends on the ease with which the aromatic can donate electrons and this, in turn, is determined by the nature of the substituent on the benzene ring. Halide groups such as $-CF_3$ and $-Cl$ withdraw electrons and decrease the electron density on the benzene ring, while methyl groups donate electrons and increase the electron density on the ring. Therefore, relative to benzene, toluene is a better Lewis base while chlorobenzene is weaker. The abilities of substituents on benzene to donate or withdraw electrons have also been determined from theoretical calcula-

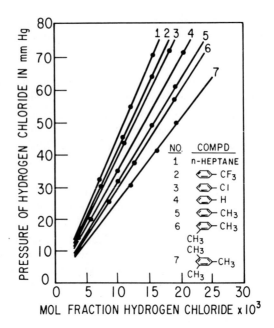

Fig. 8-12   Solubility of hydrogen chloride at
$-78.51°C$ in n-heptane and in 5 mol
% solutions of aromatics in n-hep-
tane (data of Brown and Brady).

tions, from ionization potentials, and from chemical kinetic data[15] and the results are in excellent agreement with those obtained from solubility studies. Further evidence that hydrogen chloride and aromatics form stable complexes at low temperatures is given by the freezing-point data of Cook, Lupien and Schneider[16] for mixtures of hydrogen chloride and mesitylene shown in Fig. 8-13. The maximum at a mole fraction of one-half shows that the stoichiometric ratio of hydrogen chloride to aromatic in the complex is one-to-one.

Brown and Brady reduced their solubility data by calculating dissociation equilibrium constants for the complexes; the dissociation constant is defined by

$$K \equiv \frac{x_{\text{free aromatic}} \cdot p_{\text{HCl}}}{x_{\text{complex}}} \tag{8.7-1}$$

where x stands for mole fraction and p for partial pressure. According to this definition, the stability of a complex falls as the dissociation constant rises. By simple stoichiometry and by assuming that the complex is non-volatile, Brown and Brady were able to calculate K from the change in

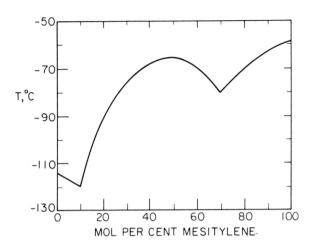

Fig. 8-13   Evidence for complex formation: freezing points
for hydrochloric acid/mesitylene mixtures.

Henry's constant in heptane which results when aromatic molecules are
added to the solvent. Their results are shown in Table 8-4. The aromatic

Table 8-4

HENRY'S CONSTANTS FOR SOLUBILITY OF HCl IN
5 MOL % SOLUTIONS OF AROMATICS IN
n-HEPTANE, AND DISSOCIATION CONSTANTS FOR
COMPLEX FORMATION (ALL AT $-78.51°$C)†

Aromatic Solute	H (mm Hg)	K (mm Hg)
None (pure heptane)	4520	–
Benzotrifluoride	4220	3200
Chlorobenzene	4000	1500
Benzene	3500	720
Toluene	3170	460
m-Xylene	2980	360
Mesitylene	2550	240

†H. C. Brown and J. D. Brady, *J. Am. Chem. Soc.*,
74, 3570 (1952).

components are listed in order of rising basicity and as the basicity increases, the solubility also increases and Henry's constant falls. Although
the concentration of the aromatic solvent is only 5 mol % in an excess of
"inert" heptane, the effect of the aromatic component on the solubility
of HCl is pronounced; the presence of 5 mol % mesitylene almost doubles
the solubility.

In the two examples just discussed we showed the effect of chemical forces on Henry's constant. In the final example we show how chemical forces can account for deviations from Henry's law; as an illustration we consider the solubility of a gas which dissociates in solution.

Deviations from Henry's law may result from chemical effects even at very low solute concentrations. In Secs. 8.2 and 8.3 we discussed how departures from Henry's law may be due to advanced pressures, to large dissimilarity between solvent and solute, or both. However, whenever the gaseous solute experiences a chemical change such as association or dissociation in the solvent, Henry's law fails because the equilibrium between the vapor phase and the liquid phase is then coupled with an additional (chemical) equilibrium in the liquid phase. We briefly discuss such a case here by analyzing data for solubility of sulfur dioxide in water at 25°C, shown in Table 8-5.

Table 8-5

SOLUBILITY OF SULFUR DIOXIDE IN WATER AT 25°C†

P_{SO_2} Partial Pressure (atm)	m Molality (g-mol SO_2 / 1000 g H_2O)	α Fraction Ionized	m_M Molality of Molecular SO_2 (g-mol SO_2 / 1000 g H_2O)
0.0104	0.0271	0.524	0.0129
0.0450	0.0854	0.363	0.0544
0.0971	0.1663	0.285	0.1189
0.1790	0.2873	0.230	0.2212
0.3330	0.5014	0.184	0.4092
0.5260	0.7643	0.154	0.6470
0.7230	1.0273	0.134	0.8897
0.9190	1.290	0.120	1.134
1.0680	1.496	0.116	1.329

†H. F. Johnstone and P. W. Leppla, *J. Am. Chem. Soc.*, **56**, 2233 (1934); W. B. Campbell and O. Maass, *Can. J. Research*, **2**, 42 (1930); O. M. Morgan and O. Maass, *Can. J. Research*, **5**, 162 (1931).

The first column in Table 8-5 gives the partial pressure of sulfur dioxide, which in this case is essentially equal to the fugacity. The second column gives the molality of sulfur dioxide in aqueous solution, which at these very low concentrations is proportional to the solute mole fraction. However, when plotted, these data do not yield a straight line; despite the very low concentrations of sulfur dioxide, Henry's law is not valid.

The reason for the failure of Henry's law becomes apparent when we consider that sulfur dioxide plus water produces hydrogen ions and bisulfite ions. When sulfur dioxide gas is in contact with liquid water, we must consider two equilibria:

Gas phase SO_2
$\overline{\phantom{Gas phase \qquad SO_2 }}$
Liquid phase SO_2 (aqueous) $\rightleftharpoons$ $H^+ + HSO_3^-$

Henry's law governs only the (vertical) equilibrium between the two phases; in this case, Henry's law must be written

$$p = Hm_M \qquad (8.7\text{-}2)$$

where p is the partial pressure of sulfur dioxide, H is a "true" Henry's constant and m_M is the molality of molecular (nonionized) sulfur dioxide in aqueous solution.

Johnstone and Leppla[17] have calculated α, the fractional ionization of sulfur dioxide in water, from electrical conductivity measurements; these are given in column 3 of Table 8-5. Column 4 gives m_M which is the product of m, the total molality, and $(1 - \alpha)$. When the partial pressure of sulfur dioxide is plotted against m_M (rather than m) a straight line is obtained.

This case is a particularly fortunate one because independent conductivity measurements are available and thus it was easily possible quantitatively to reconcile the solubility data with Henry's law. In a more typical case, independent data on the liquid solution (other than the solubility data themselves) would not be available; however, even then it is possible to linearize the solubility data by construction of a simple but reasonable model, similar to that given in Sec. 7.16.

For equilibrium between sulfur dioxide in the gas phase and molecular sulfur dioxide in the liquid phase we write

$$p = Hm_M = Hm(1 - \alpha). \qquad (8.7\text{-}3)$$

For the ionization equilibrium in the liquid phase we write

$$K = \frac{m_{H^+} \cdot m_{HSO_3^-}}{m_M} = \frac{\alpha^2 m^2}{m_M} \qquad (8.7\text{-}4)$$

where K is the ionization equilibrium constant. Substituting Eq. (8.7-3) we have

$$K = \frac{\alpha^2 m^2}{(p/H)} \qquad (8.7\text{-}5)$$

from which we obtain α:

$$\alpha = \frac{\sqrt{p}}{m}\left(\frac{K}{H}\right)^{1/2}. \qquad (8.7\text{-}6)$$

Further substitution and rearrangement finally gives

$$\frac{m}{\sqrt{p}} = \frac{\sqrt{p}}{H} + \left(\frac{K}{H}\right)^{1/2} . \tag{8.7-7}$$

Equation (8.7-7) shows the effect of ionization on Henry's law; if there were no ionization, $K = 0$ and Henry's law is recovered. The ability of a solute to ionize in solution increases its solubility; however, as the concentration of solute in the solvent rises, the fraction ionized falls. Therefore the "effective" Henry's constant p/m rises with increasing pressure and a plot of p vs. m is not linear but convex toward the horizontal axis.

Figure 8-14 presents solubility data for the sulfur dioxide/water system plotted according to Eq. (8.7-7). A straight line is obtained. A

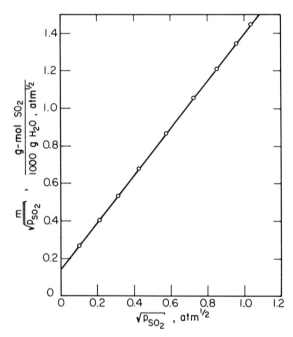

Fig. 8-14   Effect of ionization on solubility. Linearization of solubility data for sulfur dioxide in
water at 25°C.

plot such as this is useful for smoothing, interpolating and cautiously extrapolating limited gas-solubility data for any system where the gaseous solute tends to ionize (or dissociate) in the solvent.

These three examples illustrate how chemical effects may have a large influence on solubility behavior. It must be remembered that a solvent is never an inert material which merely acts as a cage for the solute, although it is frequently tempting to think of it that way. Solvent and solute always interact to a certain extent and, historically, when the interaction is strong enough to result in a new species, we call it *chemical.* But, as discussed previously in Secs. 7.11–7.16, this is only a convenience, useful for classification. There is no sharp boundary between physical and chemical forces; all molecules have forces acting between them and it is these forces which determine solubility regardless of whether they fall into the physical or the chemical types. For interpreting solubility behavior, however, it appears worthwhile to utilize these classifications by first taking physical forces into account and then correcting for specific chemical effects. This is essentially what Brown and Brady did since they assumed that the physical forces between hydrogen chloride and hydrocarbons were the same regardless of the nature of the hydrocarbon and that any excess solubility was due to chemical-complex formation. Such a procedure is both simple and practical but its molecular significance must not be taken too seriously unless there is additional independent evidence for the existence of a complex. Similarly, the solubility behavior of sulfur dioxide in water could be explained in terms of a physical Henry's law coupled with a chemical equilibrium for a dissociation reaction whose existence is supported by physicochemical data other than the solubility data themselves.

A thorough review of gas-solubility data is given by Battino and Clever.[18]

PROBLEMS

1. Compute the K value for methane at 200 psia, 200°F in a hydrocarbon mixture whose composition in mole percent is 20% benzene, 60% xylene and 20% n-hexane. Use Shair's correlation.

2. At 20°C, the solubility of helium in argyle acetate is $x_2 = 1.00 \times 10^{-4}$ at 25 atmospheres. At 75 atmospheres the solubility $x_2 = 2.86 \times 10^{-4}$. Estimate as best you can the solubility of helium in this solvent at 20°C and 150 atmospheres. State all assumptions made.

3. Estimate as best you can the solubility of hydrogen gas (at 1 atm partial pressure) in liquid air at 90°K. The following experimental information is available, all at 90°K: x_{H_2} (1 atm partial pressure) in liquid methane is 0.0549×10^{-2}; x_{H_2} (1 atm partial pressure) in liquid carbon monoxide is 0.263×10^{-2}.

Pure-component data (at 90°K):

	Liquid Molar Volume (cc/g-mol)	Heat of Vaporization (cal/g-mol)
CH_4	35.6	2090
CO	37.0	1320
N_2	37.5	1220
O_2	27.9	1565

4. Compute the solubility x_2 of hydrogen in nitrogen at 77° K and 100 atm. At these conditions the vapor-phase fugacity coefficient for pure hydrogen is 0.88 and the saturation pressure of pure liquid nitrogen is 1 atm. Use Orentlicher's correlation.

REFERENCES

1. Krichevsky, I. R., and J. S. Kasarnovsky, *J. Am. Chem. Soc.*, **57**, 2168 (1935).

2. Dodge, B. F., and R. H. Newton, *Ind. Eng. Chem.*, **29**, 718 (1937).

3. Wiebe, R., and V. L. Gaddy, *J. Am. Chem. Soc.*, **59**, 1984 (1937).

4. Krichevsky, I. R., and A. A. Ilinskaya, *Zh. fiz. khim. USSR*, **19**, 621 (1945).

5. Orentlicher, M., and J. M. Prausnitz, *Chem. Eng. Sci.*, **19**, 775 (1964).

6. Hildebrand, J. H., and R. L. Scott, *Regular Solutions.* Englewood Cliffs, N.J.: Prentice-Hall, Inc., 1962.

7. Sherwood, A. E., and J. M. Prausnitz, *A. I. Ch. E. Journal*, **8**, 519 (1962); Errata, **9**, 246 (1963).

8. Jolley, J. E., and J. H. Hildebrand, *J. Am. Chem. Soc.*, **80**, 1050 (1958).

9. Kobatake, Y., and J. H. Hildebrand, *J. Phys. Chem.*, **65**, 331 (1961).

10. Prausnitz, J. M., and F. H. Shair, *A. I. Ch. E. Journal*, **7**, 682 (1961).

11. Yen, L., and J. McKetta, *A. I. Ch. E. Journal*, **8**, 501 (1962).

12. O'Connell, J. P., and J. M. Prausnitz, *I&EC Fundamentals*, **3**, 347 (1964).

13. Zellhoefer, G. F., M. J. Copley, and C. S. Marvel, *J. Am. Chem. Soc.*, **60**, 1337 (1938).

14. Brown, H. C., and J. O. Brady, **74**, 3570 (1952).

15. Hine, J., *Physical Organic Chemistry*, 2nd ed. New York: McGraw-Hill Book Company, 1962.

16. Cook, D., Y. Lupien and W. G. Schneider, *Can. J. Chem.*, **34**, 964 (1956).

17. Johnstone, H. F., and P. W. Leppla, *J. Am. Chem. Soc.*, **56**, 2233 (1934).

18. Battino, R., and H. L. Clever, *Chem. Rev.*, **66**, 395 (1966).

Solubility of Solids in Liquids   9

The extent to which solids can dissolve in liquids varies enormously; in some cases a solid solute may form a highly concentrated solution in a solvent (e.g., calcium chloride in water) and in other cases the solubility may be so small as to be barely detectable (e.g., paraffin wax in mercury). In this chapter we consider some of the thermodynamic principles which govern equilibrium between a solid phase and a liquid phase.

9.1 Thermodynamic Framework

Solubility is a strong function of the intermolecular forces between solute and solvent, and the well-known guide "like dissolves like" is merely an empirical statement of the fact that, in the absence of specific chemical effects, the intermolecular forces between chemically similar species lead to a smaller endothermic heat of solution than those between dissimilar species. Since dissolution must be accompanied by a decrease in the Gibbs energy, a low endothermic heat is more favorable than a large one. However, there are factors other than intermolecular forces between solvent and solute which also play a large role in determining the solubility of a solid. To illustrate, consider the solubilities of two isomers, phenanthrene and anthracene, in benzene at 25°C, as given below:

Solute	Structure	Solubility in Benzene (mol %)
Phenanthrene		20.7
Anthracene		0.81

The solubility of phenanthrene is about 25 times greater than that of anthracene even though both solids are chemically similar to each other and to the solvent.   The reason for this large difference in solubility follows from something that is all too often overlooked, viz., that solubility depends not only on the activity coefficient of the solute (which is a function of the intermolecular forces between solute and solvent) but also on the fugacity of the standard state to which that activity coefficient refers and on the fugacity of the pure solid.   Let the solute be designated by the subscript 2.   Then the equation of equilibrium is:

$$f_{2 \text{(pure solid)}} = f_{2 \text{(solute in liquid solution)}}, \tag{9.1-1}\dagger$$

or

$$f_{2 \text{(pure solid)}} = \gamma_2 x_2 f_2^0, \tag{9.1-2}$$

where x_2 is the solubility (mole fraction) of the solute in the solvent, γ_2 is the liquid-phase activity coefficient, and f_2^0 is the standard-state fugacity to which γ_2 refers.

From Eq. (9.1-2) the solubility is

$$x_2 = \frac{f_{2 \text{(pure solid)}}}{\gamma_2 f_2^0}. \tag{9.1-3}$$

Thus the solubility depends not only on the activity coefficient but also on the ratio of two fugacities as indicated by Eq. (9.1-3).

The standard-state fugacity f_2^0 is arbitrary, the only thermodynamic requirement being that it must be at the same temperature as that of the solution.   Although other standard states can be used, it is most convenient to define the standard-state fugacity as the fugacity of pure, subcooled liquid at the temperature of the solution and at some specified pressure.   This is a hypothetical standard state but it is one whose prop-

†Equation (9.1-1) is based on the assumption that there is no appreciable solubility of the liquid solvent in the solid phase.

erties can be calculated with fair accuracy provided the solution temperature is not very far removed from the triple point of the solute.

To show the utility of Eq. (9.1-3) we consider first a very simple case. Assume that the vapor pressures of the pure solid and of the subcooled liquid are not large; in that case we can substitute vapor pressures for fugacities without serious error. This simplifying assumption is an excellent one in the majority of typical cases. Next, let us assume that the chemical natures of the solvent and of the solute (as a subcooled liquid) are similar. In that case we can assume $\gamma_2 = 1$ and Eq. (9.1-3) becomes

$$x_2 = \frac{P^s_{2(\text{pure solid})}}{P^s_{2(\text{pure subcooled liquid})}}. \tag{9.1-4}$$

The solubility x_2 given by Eq. (9.1-4) is called the *ideal solubility*. The significance of Eq. (9.1-4) can best be seen by referring to a typical pressure-temperature diagram for a pure substance as shown in Fig. 9-1.

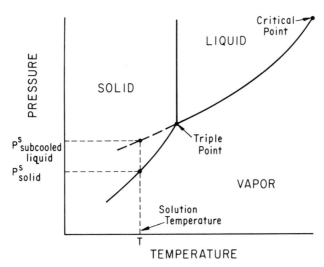

Fig. 9-1   Extrapolation of liquid vapor pressure on pressure-temperature diagram for a pure material (schematic).

If the solute is a solid, then the solution temperature is necessarily below the triple-point temperature. The vapor pressure of the solid at T is easily found from the solid-vapor pressure curve but the vapor pressure of the subcooled liquid must be found by extrapolating the liquid-vapor pressure curve from the triple-point temperature to the solution temperature T. Since the slope of the solid-vapor pressure curve is always larger than that of the extrapolated liquid-vapor pressure curve, it follows from Eq. (9.1-4) that the solubility of a solid in an ideal solution must always be less than unity except at the triple-point temperature, where it is equal to unity.

Equation (9.1-4) easily explains why phenanthrene and anthracene have very different solubilities in benzene: Because of structural differences, the triple-point temperatures of the two solids are significantly different.† As a result, the pure-component fugacity ratios at the same temperature T also differ for the two solutes.

The extrapolation indicated in Fig. 9-1 is easy to make when the solution temperature T is not far removed from the triple-point temperature. However, any essentially arbitrary extrapolation involves uncertainty, and when the extrapolation is made over a wide temperature range, the uncertainty may be prohibitively large. It is therefore important to establish a systematic method for performing the desired extrapolation; this method should be substituted for the arbitrary graphical construction shown in Fig. 9-1. Fortunately, a systematic extrapolation can readily be derived by using a thermodynamic cycle as indicated in Sec. 9.2. This extrapolation does not require the assumption of low pressures but yields an expression which gives the fugacity, rather than the pressure, of the saturated, subcooled liquid.

9.2 Calculation of the Pure-Solute Fugacity Ratio

We define the standard state as the pure, subcooled liquid at the temperature T under its own saturation pressure. Assuming negligible solubility of the solvent in the solid phase, the equilibrium equation is

$$x_2 = \frac{f_{2(\text{pure solid})}}{\gamma_2 f_{2(\text{pure, subcooled liquid})}}. \tag{9.2-1}$$

To simplify notation, let

$$f_{2(\text{pure solid})} = f_2^\diamond,$$

and let

$$f_{2(\text{pure, subcooled liquid})} = f_2^L.$$

These two fugacities depend only on the properties of the solute (component 2); they are independent of the nature of the solvent. The ratio of these two fugacities can readily be calculated by the thermodynamic cycle indicated in Fig. 9-2. The molar Gibbs energy change for

---

†The normal melting temperatures for phenanthrene and anthracene are, respectively, 100 and 217°C. These are very close to the respective triple-point temperatures.

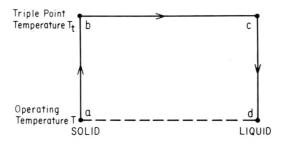

Fig. 9-2   Thermodynamic cycle for calculating the fugacity of a pure subcooled liquid.

component 2 in going from a to d is related to the fugacities of solid and subcooled liquid by:

$$\Delta g_{a \to d} = RT \ln \left(\frac{f^L}{f^\delta} \right) \tag{9.2-2}$$

where, for simplicity, subscript 2 has been omitted. This Gibbs energy change is also related to the corresponding enthalpy and entropy changes by

$$\Delta g_{a \to d} = \Delta h_{a \to d} - T \Delta s_{a \to d}. \tag{9.2-3}$$

The thermodynamic cycle shown in Fig. 9-2 provides a method to evaluate the enthalpy and entropy changes given in Eq. (9.2-3); since both enthalpy and entropy are state functions independent of the path, it is permissible to substitute for the path $a \to d$ the alternate path $a \to b \to c \to d$. For the enthalpy change in going from a to d we have

$$\Delta h_{a \to d} = \Delta h_{a \to b} + \Delta h_{b \to c} + \Delta h_{c \to d}. \tag{9.2-4}$$

Equation (9.2-4) can be rewritten in terms of heat capacity c_p and latent heat of fusion Δh^f:

$$\Delta h_{a \to d} = \Delta h^f_{\text{at } T_t} + \int_{T_t}^{T} \Delta c_p \, dT \tag{9.2-5}\dagger$$

where $\Delta c_p \equiv c_{p(\text{liquid})} - c_{p(\text{solid})}$ and T_t is the triple-point temperature. Similarly, for the entropy change in going from a to d,

$$\Delta s_{a \to d} = \Delta s_{a \to b} + \Delta s_{b \to c} + \Delta s_{c \to d} \tag{9.2-6}$$

---

†Equations (9.2-5) and (9.2-7) neglect the effect of pressure on the properties of solid and subcooled liquid. Unless the pressure is large, this effect is negligible.

which becomes

$$\Delta s \atop a \to d = \Delta s^f_{\text{at } T_t} + \int_{T_t}^{T} \frac{\Delta c_p}{T} dT \ . \qquad (9.2\text{-}7)\dagger$$

At the triple point, the entropy of fusion Δs^f is

$$\Delta s^f = \Delta h^f / T_t. \qquad (9.2\text{-}8)$$

Substituting Eqs. (9.2-3), (9.2-5), (9.2-7), and (9.2-8) into Eq. (9.2-2), and assuming that Δc_p is constant over the temperature range $T \longrightarrow T_t$, we obtain:

$$\ln \frac{f^L}{f^\Delta} = \frac{\Delta h^f}{RT} \left[1 - \frac{T}{T_t} \right] - \frac{\Delta c_p}{R} \left(\frac{T_t - T}{T} \right) + \frac{\Delta c_p}{R} \ln \frac{T_t}{T}. \qquad (9.2\text{-}9)$$

Equation (9.2-9) gives the desired result; it expresses the fugacity of the subcooled liquid at temperature T in terms of measurable thermo-dynamic properties. The fugacity ratio for solid and subcooled liquid carbon dioxide is shown in Fig. 9-3.

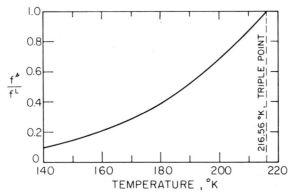

Fig. 9-3   Fugacity ratio for solid and subcooled liquid carbon dioxide.

Two simplifications in Eq. (9.2-9) are frequently made, but these usually introduce only a slight error. First, for most substances there is little difference between the triple-point temperature and the normal melting temperature; also, the difference in the heats of fusion at these two

† See footnote on previous page.

temperatures is often negligible. Therefore, in practice it is common to substitute the normal melting temperature for T_t and to use for Δh^f the heat of fusion at the melting temperature. Second, the three terms on the right-hand side of Eq. (9.2-9) are not of equal importance; the first term is the dominant one and the remaining two, being of opposite sign, have a tendency approximately to cancel each other, especially if T and T_t are not far apart. Therefore, in many cases it is sufficient to consider only the term which includes Δh^f and to neglect the terms which include Δc_p.

9.3 Ideal Solubility

An expression for the ideal solubility of a solid solute in a liquid solvent has already been given by Eq. (9.1-4), but no clear-cut method was given for finding the saturation pressure of the subcooled liquid. However, this difficulty can be avoided by substituting Eq. (9.2-9) into Eq. (9.2-1); if we assume that the solution is ideal, then $\gamma_2 = 1$ and we obtain for the solubility x_2:

$$\ln \frac{1}{x_2} = \frac{\Delta h^f}{RT}\left[1 - \frac{T}{T_t}\right] - \frac{\Delta c_p}{R}\left(\frac{T_t - T}{T}\right) + \frac{\Delta c_p}{R}\ln \frac{T_t}{T}. \qquad (9.3\text{-}1)$$

Equation (9.3-1) provides a reliable method for estimating solubilities of solids in liquids where the chemical nature of the solute is similar to that of the solvent. For example, in Sec. 9.1 we quoted the experimental solubilities of phenanthrene and anthracene in benzene at 25°C; they are 20.7 and 0.81 mol % respectively. The corresponding solubilities calculated from Eq. (9.3-1) are 22.1 and 1.07 mol %.

Equation (9.3-1) immediately leads to two useful conclusions concerning the solubility of solids in liquids. Strictly, these conclusions apply only to ideal solutions but they are useful guides for other solutions which do not deviate excessively from ideal behavior:

1. For a given solid/solvent system, the solubility increases with rising temperature. The rate of increase is approximately proportional to the enthalpy of fusion and, to a first approximation, does not depend on the melting temperature.

2. For a given solvent and at a fixed temperature, if two solids have similar enthalpies of fusion, the solid with the lower melting temperature has the higher solubility; similarly, if the two solids have nearly the same melting temperature, the one with the lower enthalpy of fusion has the higher solubility.

A typical application of Eq. (9.3-1) is provided by the solubility data of McLaughlin and Zainal[1] for nine aromatic hydrocarbons in benzene in the temperature range 30–70°C. With the help of Eq. (9.3-1) these data can be correlated in a simple manner. Since the melting points of the nine solutes are not more than about 100°C above the temperatures of the solutions, the terms which include Δc_p in Eq. (9.3-1) may be neglected; also, it is permissible to substitute melting temperatures for triple temperatures. Equation (9.3-1) may then be rewritten:

$$\ln x_2 = -\frac{\Delta s^f}{R}\left[\frac{T_m}{T} - 1\right], \qquad (9.3\text{-}2)$$

where T_m is the normal melting temperature. For the nine solutes considered here, the entropies of fusion do not vary much and an average value is 13.0 calories per degree per gram-mole. Therefore, a semilogarithmic plot of $\log_{10} x_2$ vs. T_m/T should give a nearly straight line with a slope approximately equal to $-(13.0/2.303R)$ and with intercept $\log x_2 = 0$ when $T_m/T = 1$. Such a plot is shown by the dashed line in Fig. 9-4 and it is evident that this line gives a good representation of the ex-

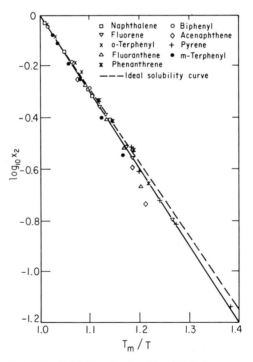

Fig. 9-4   Solubility of aromatic solids in benzene
(data of McLaughlin and Zainal).

perimental data. Thus the assumption of ideal solution is appropriate for these systems. However, for precise work the assumption of ideality is only an approximation. Solutions of aromatic hydrocarbons in benzene, even though the compounds are chemically similar, show slight positive deviations from ideality and therefore the observed solubilities are a little lower than those calculated with Eq. (9.3-2). The continuous line in Fig. 9-4, which was determined empirically as the best fit of the data, has a slope equal to $-(13.8/2.303R)$.

Equation (9.3-1) gives the ideal solubility of solid 2 in solvent 1. By interchanging subscripts, we may use the same equation to calculate the ideal solubility of solid 1 in solvent 2; by repeating such calculations at different temperatures we can then obtain the freezing diagram of the entire binary system as a function of composition as shown in Fig. 9-5

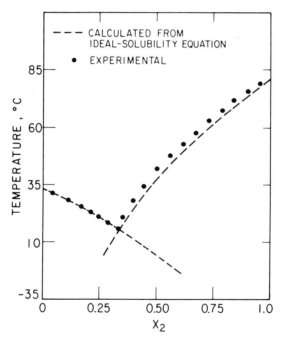

Fig. 9-5   Freezing points for the system o-chloronitro-
benzene (1)/p-chloronitrobenzene (2).

taken from Prigogine and Defay.[2] In these calculations we assume ideal behavior in the liquid phase and complete immiscibility in the solid phase. The left side of the diagram gives the equilibrium between the liquid mixture and solid o-chloronitrobenzene while the right side gives the equilibrium between the liquid mixture and solid p-chloronitrobenzene.

At the point of intersection, called the eutectic point, all three phases are in equilibrium.

9.4 Nonideal Solutions

Equation (9.3-1) assumes ideal behavior but Eqs. (9.2-1) and (9.2-9) are general. Whenever there is a significant difference in the nature and size of the solute and solvent molecules, we may expect that γ_2 is not equal to unity; in nonpolar solutions, where only dispersion forces are important, γ_2 is generally larger than unity (and thus the solubility is less than that corresponding to ideal behavior) but in cases where polar or specific chemical forces are important, the activity coefficients may well be less than unity with correspondingly higher solubilities. Such enhanced solubilities, for example, have been observed for unsaturated hydrocarbons in liquid sulfur dioxide.

In Fig. 9-4 we showed the solubilities of nine aromatic hydrocarbons in benzene and it is apparent that the assumption of ideal-solution behavior is approximately valid. By contrast, Fig. 9-6 shows the solubilities

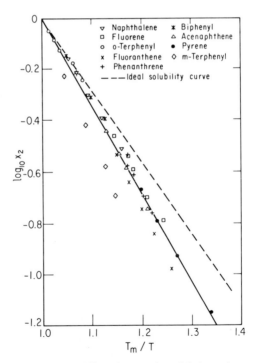

Fig. 9-6   Solubility of aromatic solids in carbon tetrachloride (data of McLaughlin and Zainal).

of the same nine hydrocarbons in carbon tetrachloride; the data are again those of McLaughlin and Zainal.[3]   The dashed line shows the ideal solubility curve using, as before, $\Delta s^f = 13.0$ calories per degree per gram-mole; the continuous line, which best represents the data, has a slope of $-(15.9/2.303R)$.

A comparison of Figs. 9-4 and 9-6 shows that at the same temperature, the solubilities in carbon tetrachloride are lower than those in benzene; in other words, the activity coefficients of the solutes in carbon tetrachloride are larger than those in benzene.   If the activity coefficients of the solutes in carbon tetrachloride are represented by the simple empirical relation

$$\ln \gamma_2 = \frac{A}{RT}(1 - x_2)^2, \tag{9.4-1}$$

then typical values of A are in the range of 100–300 calories per gram-mole.

As in the case of solutions of liquid components, there is no general method for predicting activity coefficients of solid solutes in liquid solvents.   For nonpolar solutes and solvents, however, a reasonable estimate can frequently be made with the Scatchard-Hildebrand relation

$$\ln \gamma_2 = \frac{v_2^L(\delta_1 - \delta_2)^2 \Phi_1^2}{RT} \tag{9.4-2}$$

where v_2^L is the molar volume of the subcooled liquid, δ_2 is the solubility parameter of the subcooled liquid, δ_1 is the solubility parameter of the solvent, and

$$\Phi_1 = \frac{x_1 v_1^L}{x_1 v_1^L + x_2 v_2^L}$$

is the volume fraction of the solvent.

The molar liquid volume and solubility parameter of the solvent can be determined easily from the thermodynamic properties of the solvent, but it is necessary to use a thermodynamic cycle (as illustrated in Fig. 9-2) to calculate these functions for the subcooled liquid solute.

Let Δv^f stand for the volume change of fusion at the triple-point temperature; that is,

$$\Delta v^f \equiv v_t^L - v_t^\diamond, \tag{9.4-3}$$

where the subscript t refers to triple-point temperature.

Let $v^\diamond$ be the molar volume of the solid at the temperature T of the solution.   The molar volume of the subcooled liquid is then given by

$$v^L = v^\diamond + \Delta v^f + (v_t^\diamond \alpha^\diamond - v_t^L \alpha^L)(T_t - T), \tag{9.4-4}$$

where α^s and α^L are the volumetric coefficients of expansion of the solid and liquid respectively.

The energy of vaporization of the subcooled liquid is found in a similar manner. Let Δh^f stand for the enthalpy of fusion of the solid at the triple-point temperature and let Δh^{sb} stand for the enthalpy of sublimation of the solid at the temperature T. The energy of vaporization of the subcooled liquid is then

$$\Delta u = \Delta h^{sb} - \Delta h^f + \Delta c_p(T_t - T) - P^s(v^G - v^L), \qquad (9.4\text{-}5)$$

where P^s is the saturation pressure of the subcooled liquid and v^G is the volume of the saturated vapor in equilibrium with the solid, all at temperature T.

In Eqs. (9.4-4) and (9.4-5) it is often convenient to replace the triple-point temperature with the melting temperature and at moderate pressures this substitution usually introduces insignificant error; in that case, all subscripts t should be replaced by subscript m. Also, if the temperature T is not far removed from the triple-point (or melting) temperature, the last term in Eq. (9.4-4) may be neglected, and finally, if the saturation pressure of the subcooled liquid is small, as is usually the case, $v^G \gg v^L$, and the last term in Eq. (9.4-5) may be replaced by RT.

The square of the solubility parameter is defined as the ratio of the energy of complete vaporization to the liquid volume;† therefore, if the vapor pressure of the subcooled liquid is large it will be necessary to add a vapor-phase correction to the energy of vaporization given by Eq. (9.4-5). Such a correction, however, is rarely required and for most cases of interest the solubility·parameter of the subcooled liquid is given with sufficient accuracy by

$$\delta_2 = \left(\frac{\Delta u_2}{v_2^L}\right)^{1/2}. \qquad (9.4\text{-}6)$$

To illustrate the applicability of Eq. (9.4-2), we consider the solubility of white phosphorus in n-heptane at 25°C. The melting point of white phosphorus is 44.2°C; the heat of fusion and the heat capacities of the solid and liquid have been measured by Young and Hildebrand.[4] From Eq. (9.3-1) the ideal solubility at 25°C is $x_2 = 0.942$. A much better approximation can be obtained from regular-solution theory as given by Eq. (9.4-2). From the extrapolated thermal and volumetric properties, the solubility parameter and molar volume of subcooled liquid phosphorus are, respectively, 13.1 $(\text{cal/cc})^{1/2}$ and 70.4 cc/g-mol at 25°C. The solubility parameter of n-heptane is 7.4 $(\text{cal/cc})^{1/2}$ and therefore one can immediately conclude that subcooled phosphorus and heptane form a

† See Sec. 7.2.

highly nonideal liquid solution. When Eqs. (9.4-2) and (9.2-9) are substituted into the fundamental Eq. (9.2-1), the calculated solubility is $x_2 = 0.022$, a result strikingly different from that obtained by assuming liquid-phase ideality. The experimentally observed solubility is $x_2 = 0.0124$.

An extension of the Scatchard-Hildebrand equation was introduced by Myers[5] in his study of the solubility of solid carbon dioxide in liquefied light hydrocarbons. Since carbon dioxide has a large quadrupole moment, Myers considered separately the contribution of dispersion forces and quadrupole forces to the cohesive-energy density of subcooled liquid carbon dioxide.† Myers divided the energy of vaporization into two parts:

$$\Delta u = \Delta u_{disp} + \Delta u_{quad}. \tag{9.4-7}$$

As a result, two cohesive-energy densities can now be computed, corresponding to the two types of intermolecular forces:

$$c_{disp} \equiv \frac{\Delta u_{disp}}{v} \tag{9.4-8}$$

$$c_{quad} \equiv \frac{\Delta u_{quad}}{v}, \tag{9.4-9}$$

where the superscript L has been omitted.

These cohesive-energy densities for carbon dioxide are shown as a function of temperature in Fig. 9-7.

The activity coefficient of component 2, the solute, dissolved in nonpolar solvents, is now written

$$RT \ln \gamma_2 = v_2 \Phi_1^2 [c_1 + c_{2_{total}} - 2(c_1 c_{2_{disp}})^{1/2}], \tag{9.4-10}$$

where

$$c_{2_{total}} = c_{2_{disp}} + c_{2_{quad}}.$$

The last term in the brackets follows from the geometric-mean assumption for the attractive dispersion forces between solute and solvent. If $c_{2_{quad}} = 0$, Eq. (9.4-10) reduces to Eq. (9.4-2).

Splitting the cohesive-energy density into a dispersion part and a quadrupole part has an important effect on the calculated solubility of solid carbon dioxide. Even though the contribution from quadrupole forces is small, appreciable error is introduced by not separately considering this contribution. As an example, let us consider the solubility

---

†The solubility parameter is the square root of the cohesive-energy density. See Sec. 7.2.

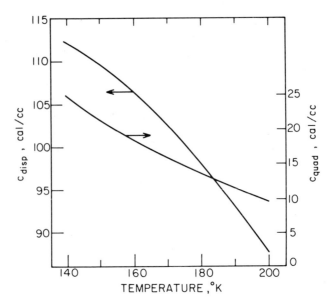

Fig. 9-7   Cohesive-energy density due to dispersion forces and
due to quadrupole forces for subcooled liquid carbon
dioxide.

of solid carbon dioxide in a nonpolar (and nonquadrupolar) solvent
having the (typical) value of c_1 = 60 cal/cc for its cohesive-energy density
at 160°K. If we do not take proper account of the presence of quadrupolar forces we calculate, from Eq. (9.4-2), a solubility x_2 = 0.067. However, if we do account properly for quadrupolar forces [Eq. (9.4-10)] we
obtain the significantly different result x_2 = 0.016. In this illustrative
calculation we have, for simplicity, assumed that Φ_2 = x_2.

To calculate the cohesive-energy density due to quadrupole forces,
Myers derived the relation

$$c_{i\,\text{quad}} = \frac{\beta Q_i^4}{kT\left(\dfrac{v_i}{N_A}\right)^{13/3}}, \qquad (9.4\text{-}11)$$

where Q_i is the quadrupole moment of species i, v_i is the molar liquid
volume, N_A is Avogadro's number, k is Boltzmann's constant, T is the
absolute temperature and β is a numerical, dimensionless constant. If the
solvent, component 1, also has a significant quadrupole moment, then an
additional term must be added to the bracketed quantity in Eq. (9.4-10)
to account for quadrupole forces between the dissimilar components;
further, the geometric-mean term must be modified to include only the
dispersion cohesive-energy density of component 1. The bracketed term

in Eq. (9.4-10) then becomes

$$[c_{1\,total} + c_{2\,total} - 2(c_{1\,disp} c_{2\,disp})^{1/2} - 2c_{12\,quad}].$$

Using the theory of intermolecular forces, Myers showed that

$$c_{12\,quad} = \frac{\beta Q_1^2 Q_2^2}{kT\left(\dfrac{v_{12}}{N_A}\right)^{13/3}}, \qquad (9.4\text{-}12)$$

where β is the same constant as that in Eq. (9.4-11).

For v_{12} Myers used the mixing rule

$$v_{12}^{1/3} = \tfrac{1}{2}(v_1^{1/3} + v_2^{1/3}). \qquad (9.4\text{-}13)$$

The term $c_{12\,quad}$ is frequently negligible but it is important, for example, in carbon dioxide/acetylene mixtures since in this system both solute and solvent have a significant quadrupole moment.

In nonpolar systems, the ideal solubility is generally larger than that observed. This result is correctly predicted by the Scatchard-Hildebrand theory (Eq. 9.4-2) since $\gamma_2 \geq 1$; as a consequence, this equation says that the ideal solubility is the maximum possible since the larger γ_2, the smaller x_2. However, whenever there is a tendency for the solvent to solvate with the solute, i.e., whenever there are strong specific forces between the dissimilar molecules, the actual solubility may well be larger than the ideal solubility. Enhanced solubility is observed whenever there are negative deviations from Raoult's law in the liquid solution; such deviations frequently occur in polar systems, especially in systems where hydrogen bonding between solute and solvent is strong. However, even in nonpolar systems specific solvation forces may, on occasion, be sufficiently strong to result in solubilities above the ideal. For example, Weimer[6] measured the solubility of hexamethylbenzene (m.p. 165.5°C) in carbon tetrachloride at 25°C and found $x_2 = 0.077$. Contrary to predictions using the Scatchard-Hildebrand theory, this solubility is *larger* than the ideal solubility $x_2 = 0.062$. From spectroscopic and other evidence we know that carbon tetrachloride forms weak charge-transfer complexes with aromatic hydrocarbons and we also know that the stability of the complex increases with methyl substitution on the benzene ring. For the solutes listed in Fig. 9-6 the complex stability is small, but in the carbon tetrachloride/hexamethylbenzene system the tendency to complex is sufficiently strong to produce an enhanced solubility; in this system, complex-formation overshadows the "normal" physical forces which would tend to give a solubility lower than the ideal. Weimer found that in carbon tetrachloride solution the activity coefficient of hexamethylbenzene at saturation (referred to pure subcooled liquid) is $\gamma_2 = 0.79$, indicating negative deviation from Raoult's law.

9.5 Distribution of a Solid Solute

Between Two Immiscible Liquids

If the solubility of a solid is known in each of two immiscible liquids it is possible to make good predictions of how the solid solute distributes itself between the two liquids when they are brought into contact.

Consider the following situation: A solute, designated by subscript 2, distributes itself between two immiscible solvents which are designated by prime (') and double prime ("). For a given concentration of solute in one solvent, we desire to find the equilibrium concentration in the other solvent. The equation of equilibrium is

$$f_2' = f_2''. \tag{9.5-1}$$

Introducing the activity coefficient γ_2 whose standard state is sub-cooled liquid 2 at the temperature of the solution, Eq. (9.5-1) becomes

$$\gamma_2' x_2' = \gamma_2'' x_2''. \tag{9.5-2}$$

Since we are here dealing with two liquid solutions of the same solute, the fugacity of the pure subcooled liquid (i.e., the standard-state fugacity) appears on both sides of the equation of equilibrium and cancels. We see from Eq. (9.5-2) that if the solute forms two liquid solutions such that $\gamma_2' = \gamma_2''$, the distribution of the solute shows no preference for one solvent over the other since, from Eq. (9.5-2), $x_2' = x_2''$.

Activity coefficients γ_2' and γ_2'' are, respectively, functions of the mole fractions x_2' and x_2''; if the solutions are only moderately nonideal, these functions can be expressed by

$$\ln \gamma_2' = \alpha'(1 - x_2')^2 \tag{9.5-3}$$
$$\ln \gamma_2'' = \alpha''(1 - x_2'')^2 \tag{9.5-4}$$

where α' and α'' are interaction constants which, for a given solute-solvent pair, depend only on the temperature.

The distribution coefficient K_2 is defined by

$$K_2 \equiv \frac{x_2'}{x_2''}. \tag{9.5-5}$$

Substitution of Eqs. (9.5-2), (9.5-3), and (9.5-4) into (9.5-5) gives

$$\ln K_2 = \alpha''(1 - x_2'')^2 - \alpha'(1 - x_2')^2. \tag{9.5-6}$$

This result shows that, in general, the distribution coefficient is not a constant but depends on composition; it is a function of either x_2' or x_2''.

Only in the limiting case where the solute is very dilute in both solvents ($x_2' \ll 1$ and $x_2'' \ll 1$) is the distribution coefficient a constant:

$$\lim_{\substack{x_2' \to 0 \\ x_2'' \to 0}} K_2 = \exp(\alpha'' - \alpha'). \qquad (9.5\text{-}7)$$

The interaction constant α' can be found from the solubility of the solute in solvent ('). Let this solubility be designated by $x_{2,\text{sat}}'$, where the subscript "sat" stands for saturation. When the pure solid is in equilibrium with its saturated solution the equation of equilibrium is:

$$f_2^{\diamond} = x_{2,\text{sat}}' \gamma_{2,\text{sat}}' f_2^{L}. \qquad (9.5\text{-}8)$$

Substituting Eq. (9.5-3) and solving for α' we obtain:

$$\alpha' = \frac{\ln\left(\dfrac{f_2^{\diamond}}{f_2^{L} x_{2,\text{sat}}'}\right)}{(1 - x_{2,\text{sat}}')^2}. \qquad (9.5\text{-}9)$$

Similarly, we can obtain α'' from the solubility of the solid in phase (''). For α'' we obtain:

$$\alpha'' = \frac{\ln\left(\dfrac{f_2^{\diamond}}{f_2^{L} x_{2,\text{sat}}''}\right)}{(1 - x_{2,\text{sat}}'')^2}. \qquad (9.5\text{-}10)$$

The fugacity ratio $f_2^{\diamond}/f_2^{L}$ can be obtained from the thermodynamic properties of the pure solute as shown earlier; the reciprocal of this ratio is given by Eq. (9.2-9).

Once α' and α'' are known, the relationship between x_2' and x_2'' is given by Eq. (9.5-6).

An application of the equations just developed is provided by the distribution of iodine between water and chloroform at 25°C. Let ' refer to water and let '' refer to chloroform; the experimentally determined solubilities of iodine in these solvents at 25°C are

$$x_{2,\text{sat}}' = 2.418 \times 10^{-5}$$
$$x_{2,\text{sat}}'' = 1.37 \times 10^{-2}.$$

From the thermodynamic properties of iodine [Eq. (9.2-9)] the fugacity ratio at 25°C is:

$$\frac{f_2^{\diamond}}{f_2^{L}} = 0.2464.$$

The interaction parameters can now be calculated from Eqs. (9.5-9) and (9.5-10). They are:

$$\alpha' = 9.23$$

$$\alpha'' = 2.97.$$

The distribution of iodine is then determined by Eq. (9.5-6).

Figure 9-8 shows a plot of the equilibrium mole fraction of iodine in chloroform as a function of the mole fraction of iodine in water. The

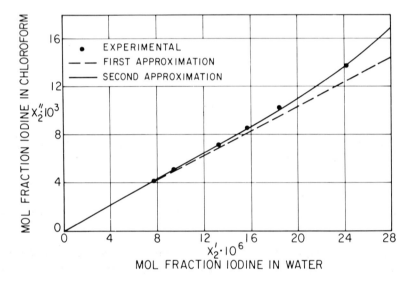

Fig. 9-8   Distribution of iodine between chloroform and water at 25°C.

experimental results[7] are compared with those of two calculations: The first approximation is based on the limiting distribution coefficient for very dilute solutions (Eq. 9.5-7) and the second approximation takes into account the variation of the distribution coefficient with composition as indicated by Eq. (9.5-6). In this particular ternary system, the mole fractions of iodine are quite small and therefore the first approximation already provides rather good agreement with the data. At larger solute concentrations, however, the second approximation provides significant improvement over the first approximation. A similar calculation by Mauser and Kortüm[8] for the distribution of iodine between water and carbon disulfide also gives excellent agreement with experiment.

9.6 Solid Solutions

In all of the previous sections we have assumed that whereas the solid has a finite solubility in the liquid solvent, there is no appreciable solubility of the solvent in the solid. As a result, we have been concerned only with the equation of equilibrium for component 2, the solid component. However, there are many situations where components 1 and 2 are miscible not only in the liquid phase but in the solid phase as well; in such cases we must write two equations of equilibrium, one for each component:

$$f_{1\,(\text{solid phase})} = f_{1\,(\text{liquid phase})} \qquad (9.6\text{-}1)$$

$$f_{2\,(\text{solid phase})} = f_{2\,(\text{liquid phase})}. \qquad (9.6\text{-}2)$$

Introducing activity coefficients we can rewrite these equations:

$$\gamma_1^\diamond x_1^\diamond f_{\text{pure 1}}^\diamond = \gamma_1^L x_1^L f_{\text{pure 1}}^L \qquad (9.6\text{-}3)$$

$$\gamma_2^\diamond x_2^\diamond f_{\text{pure 2}}^\diamond = \gamma_2^L x_2^L f_{\text{pure 2}}^L. \qquad (9.6\text{-}4)$$

If the system temperature T is above the triple-point temperature of component 1 but below that of component 2, then pure solid 1 and pure liquid 2 are both hypothetical. In Sec. 9.2 we showed how to compute the fugacity of a pure (hypothetical), subcooled liquid, and exactly the same principles may be used to calculate the fugacity of a pure, super-heated solid; the calculation is performed by using a thermodynamic cycle similar to that shown in Fig. 9-2 with the temperature T above, rather than below, the triple-point temperature T_t. When both components are miscible in both phases, Eqs. (9.6-3) and (9.6-4) indicate that information is required for the activity coefficients in both phases. Such information is rarely available. In some systems one may, to a good approximation, assume ideal behavior in both phases ($\gamma_1^\diamond = \gamma_2^\diamond = \gamma_1^L = \gamma_2^L = 1$) but in most systems this is a poor assumption.

When Eqs. (9.6-3) and (9.6-4) are applied over a range of temperatures, it is possible to calculate the freezing-point diagram for the binary system. If there is miscibility in both phases, such a diagram is qualitatively very different from that shown in Fig. 9-5. For example, Fig. 9-9 gives the experimentally determined phase diagram for the system naphthalene/β-naphthol, which exhibits approximately ideal behavior in both phases. On the other hand, the system mercuric bromide/mercuric iodide, shown in Fig. 9-10, exhibits appreciable nonideality resulting in the formation of a hylotrope (no change in composition upon melting). The thermodynamics of solid solutions is of considerable importance in metallurgy and many experimental studies have been made of binary and multicomponent metallic mixtures.

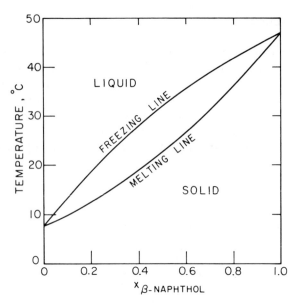

Fig. 9-9   Phase diagram for the system naphthalene/
β-naphthol.

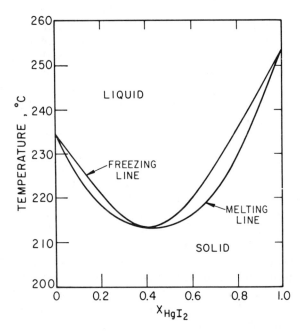

Fig. 9-10   Phase diagram for the system $HgBr_2/HgI_2$.

PROBLEMS

1. At 25°C, a solid A is in solution in a liquid solvent B. It is proposed to remove A from the solution by adsorption on a solid adsorbent S which is inert towards B. The adsorption equilibrium constant K is defined by a Langmuir-type expression

$$K = \frac{\theta}{(1 - \theta)\mathfrak{a}_A}$$

where θ is the fraction of surface sites on S which are occupied by A and $\mathfrak{a}_A$ is the activity of A ($\mathfrak{a}_A = 1$ when $x_A = 1$). $K = 130$ at 25°C.

What fraction of the surface sites on S is covered by A when x_A is one half of the saturated mole fraction of A in B at 25°C?

The following data are available for pure A:

$$\Delta h_{fusion} = 4.7 \text{ kcal/g-mol}$$
$$T_m = 412°K \text{ (melting point)}$$
$$c_{p(liquid)} = 8.0 \text{ cal/g-mol, } °K$$
$$c_{p(solid)} = 6.3 \text{ cal/g-mol, } °K$$
$$P^{sat}_{solid} = 2.1 \text{ mm Hg at } 25°C$$
$$x^{sat}_A = 0.050 \text{ at } 25°C \text{ (solubility in B)}$$

2. In a famous paper by Brown and Brady [*J. Am. Chem. Soc.*, **74**, 3570 (1952)] on the basicity of aromatics, the authors comment that they are puzzled by Klatt's measurements on the solubilities of aromatics in (acidic) liquid hydrogen fluoride. Klatt finds the solubilities are in the order *m*-xylene < toluene < benzene. However, there is ample evidence from other equilibrium measurements, from spectra and from kinetic data that the basicity of the aromatic ring *increases* with methyl substitution. Can you explain the apparent disagreement?

3. Estimate the solubility of naphthalene at 25°C in a mixed solvent consisting of 70 mol % isopentane and 30 mol % carbon tetrachloride. The following data are available for naphthalene:

Melting point = 80.2°C
Heat of fusion = 4540 cal/g-mol
$(c_{p(liquid)} - c_{p(solid)})$ is sufficiently small to be negligible in this problem.
Volume of "liquid" naphthalene at 25°C: 123 cc/g-mol (extrapolated).

4. A liquid mixture of benzene and *n*-heptane is to be cooled to as low a temperature as possible without precipitation of a solid phase. If the solution contains 10 mol % benzene, estimate what this lowest temperature is.

Data:	Benzene	Heptane
Melting Point	278.7° K	182.6° K
Heat of fusion	2351 $\dfrac{\text{cal}}{\text{g-mol}}$	3360 $\dfrac{\text{cal}}{\text{g-mol}}$

Data (Cont.):

Solubility parameter	$9.2 \left(\dfrac{cal}{cc}\right)^{1/2}$	$7.4 \left(\dfrac{cal}{cc}\right)^{1/2}$
Molar liquid volume	$89 \dfrac{cc}{g\text{-}mol}$	$148 \dfrac{cc}{g\text{-}mol}$

REFERENCES

1. McLaughlin, E., and H. A. Zainal, *J. Chem. Soc.*, 863 (March 1959).

2. Prigogine, I., and R. Defay, *Chemical Thermodynamics* (tr. D. H. Everett). London: Longmans Green & Co., Inc., 1954.

3. McLaughlin, E., and H. A. Zainal, *J. Chem. Soc.*, 2485 (June 1960).

4. Young, F. E., and J. H. Hildebrand, *J. Am. Chem. Soc.*, **64**, 839 (1942).

5. Myers, A. L., and J. M. Prausnitz, *I&EC Fundamentals*, **4**, 209 (1965).

6. Weimer, R. F., and J. M. Prausnitz, *J. Chem. Phys.*, **42**, 3643 (1965).

7. Wichterle, I., and B. Follprechtova, *Coll. Czech. Chem. Comm.*, **25**, 2492 (1960).

8. Mauser, H., and G. Kortüm, *Z. Naturforschung*, **10a**, 42 (1955).

High-Pressure Equilibria

10

Numerous chemical processes operate at high pressures and, primarily for economic reasons, many separation operations (distillation, absorption) are conducted at high pressures; further, phase equilibria at high pressures are of interest in geological exploration, such as in drilling for petroleum and natural gas. While the technical importance of high-pressure phase behavior has been recognized for a long time, quantitative application of thermodynamics toward understanding such behavior has only been attempted in recent years. In the early days of phase-equilibrium thermodynamics such an attempt could not be made because of computational complexity; realistic thermodynamic calculations for high-pressure equilibria are essentially impossible without electronic computers.

The adjective *high-pressure* is relative; in some areas of technology (e.g., outer-space research) one millimeter of mercury is a high pressure whereas in others (e.g., solid-state research) a pressure of a few hundred atmospheres is considered almost a vacuum. In this chapter we designate as "high-pressure" any pressure sufficiently large to have an appreciable effect on the thermodynamic properties of all the phases under consideration. In vapor-liquid equilibria, a high pressure may be anywhere between about 20 to 1000 atm, depending on the system and on the temperature; only in rare cases does the pressure exceed 1000 atm since in most cases of

common interest the vapor-liquid critical pressure of the system is below
1000 atm. In liquid-liquid equilibria or in gas-gas equilibria the pressure
may be considerably larger although experimental studies are rare for
fluid mixtures at pressures beyond 1000 atm.

10.1   Phase Behavior at High Pressure

In this chapter we discuss thermodynamic relations which govern the behavior of mixtures at high pressures; but before we do so, we shall take a
brief look at some typical results in order to obtain a qualitative picture
of how mixed fluids behave at pressures well above atmospheric.

For vapor-liquid behavior of a typical simple system, we consider mixtures of ethane and n-heptane; the critical temperature of ethane is 90.1°F
and that of n-heptane is 512.6°F. Figure 10-1 shows the relation between

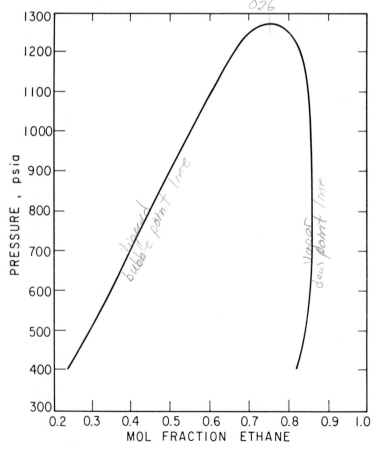

Fig. 10-1   Pressure-composition diagram for the system ethane/n-heptane at
300°F. (Mehta and Thodos)

pressure and composition at 300°F. The left-hand line gives the saturation pressure (bubble pressure) as a function of liquid composition and the right-hand line gives the saturation pressure (dew pressure) as a function of vapor composition. The two lines meet at the critical point where the two phases become identical. At 300°F, the critical composition is 76 mol % ethane and the critical pressure is 1275 psia. At this temperature, therefore, only one phase can exist at pressures higher than 1275 psia; further, regardless of pressure, it is not possible to have at 300°F a co-existing liquid phase containing more than 76 mol % ethane.

To characterize vapor-liquid equilibria for a binary system, measurements such as those shown in Fig. 10-1 must be repeated for other temperatures; for each temperature there is a critical composition and critical pressure. Figure 10-2 gives experimentally observed critical temperatures and pressures as a function of mole fraction for the ethane/n-heptane system.[1] While the critical temperature of this system is a monotonic function of composition, the critical pressure goes through a maximum; many, but by no means all, binary systems behave in this way.

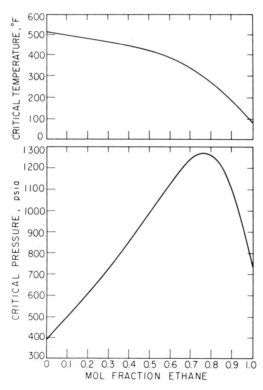

Fig. 10-2   Critical temperatures and pressures for the system ethane/n-heptane.   (Mehta and Thodos)

For typical technical calculations it is convenient to express phase-equilibrium relations in terms of K factors; by definition $K \equiv y/x$, where y is the mole fraction in the vapor phase and x is that in the liquid phase. The definition of K has no thermodynamic significance but K is commonly used in chemical engineering technology. Figure 10-3 shows K factors

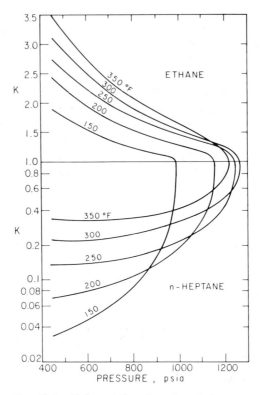

Fig. 10-3 K factors for the ethane/n-heptane system. (Mehta and Thodos)

for ethane and n-heptane in the ethane/n-heptane system at pressures above 400 psia.[1] As the pressure falls below 400 psia, K factors for ethane continue to increase while K factors for n-heptane go through a minimum and then increase.

Experimental K factors for another binary hydrocarbon system (methane/propane) are shown in Fig. 10-4. The lines for propane start at the saturation pressure for pure propane (where $K = 1$), decrease with rising pressure, and after going through minima, rise again to $K = 1$ at the critical point.[2]

In the vapor-liquid critical region, fluids may behave in a manner

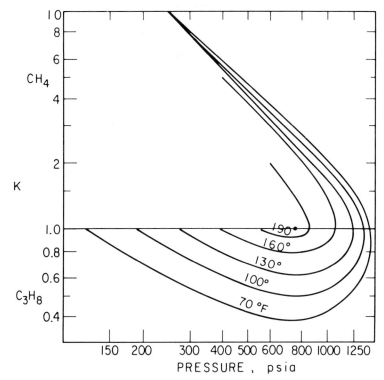

Fig. 10-4   *K* factors for the methane/propane system. (Sage and Lacey)

contrary to our usual experience. For a system at fixed composition, it may happen that isothermal compression produces vaporization rather than condensation; similarly, isobaric cooling may result in vaporization instead of condensation. These phenomena, called *retrograde condensation*, were first reported by Kuenen in 1892; they are technically important in oil and gas production and are more fully discussed elsewhere.[3,4,5]

Phase behavior of liquid mixtures containing only subcritical components is also affected by pressure, but in this case much higher pressures are usually required to produce significant changes. For example, at ordinary pressures, the system methyl ethyl ketone/water has an upper critical solution temperature and a lower critical solution temperature; in between these temperatures, the two liquids are only partially miscible. As observed by Timmermans,[6] increasing pressure lowers the upper critical solution temperature and raises the lower critical solution temperature as indicated in Fig. 10-5. As the pressure rises, the region of partial miscibility decreases and at pressures beyond 1100 bar† this region has disappeared. By contrast, the system triphenylmethane/sulfur[7] ex-

† 1 bar = 0.98692 atm.

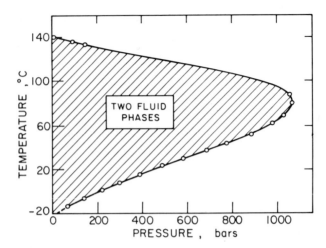

Fig. 10-5   Liquid-liquid phase behavior in the system methyl-
ethyl ketone/water. Lower and upper consolute
temperatures as functions of pressure.   (Timmermans)

hibits the opposite behavior; in this case, increasing pressure reduces the
region of complete miscibility as indicated by Fig. 10-6. Finally, Figs.
10-7 and 10-8 illustrate two additional types of phase behavior.[7] Figure
10-7 shows that for the system 2-methylpyridine/deuterium oxide, increasing pressure first induces miscibility and then, at still higher pres-

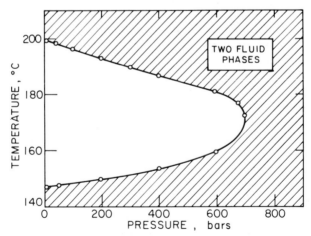

Fig. 10-6   Liquid-liquid phase behavior in the system tri-
phenyl methane/sulfur. Lower and upper consolute temperatures as functions of pressure.
(Schneider)

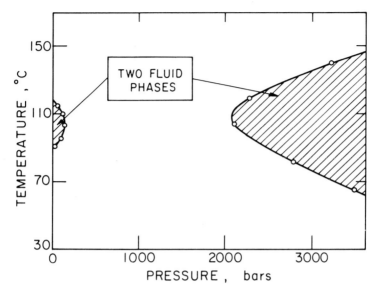

Fig. 10-7   Liquid-liquid phase behavior in the system 2-methyl pyridine/
deuterium oxide. Lower and upper consolute temperatures
as functions of pressure. (Schneider)

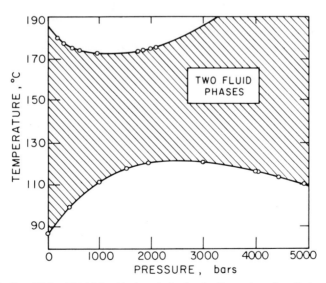

Fig. 10-8   Liquid-liquid phase behavior in the system 4-methyl
piperidine/water. Lower and upper consolute temperatures as functions of pressure. (Schneider)

sures, produces again a region of incomplete miscibility. On the other hand, Fig. 10-8 shows that for the system 4-methylpiperidine/water, increasing pressure does not eliminate the region of partial miscibility; very high pressures, after first causing this region to shrink, bring about an increase in the two-phase region. Behavior similar to that shown in Fig. 10-7 has been observed by Connolly[8] for some water/hydrocarbon systems.

Figures 10-1 to 10-8 illustrate only some of the types of phase behavior at high pressures; many other phase phenomena have been observed for binary systems and when ternary systems are considered, still different phase phenomena become possible as discussed elsewhere.[3,5,9,10,11,12] We do not here want to present a complete survey of high-pressure phase behavior but, having given a brief introduction to this vast subject, we want now to consider thermodynamic relations which determine phase equilibria at high pressures.

10.2   Thermodynamic Analysis

As discussed in each of the previous chapters, the fundamental equilibrium relation for multicomponent, multiphase systems is most conveniently expressed in terms of fugacities: For any component i, the fugacity of i must be the same in all equilibrated phases. However, this equilibrium relation is of no use until we can relate the fugacity of a component to directly measurable properties. The essential task of phase-equilibrium thermodynamics, therefore, is to describe the effects of temperature, pressure, and composition on the fugacity of each component in each phase of the system of interest. For any component i in a system containing m components, the total differential of the logarithm of the fugacity f_i is given by

$$d \ln f_i = \left(\frac{\partial \ln f_i}{\partial T}\right)_{P,x} dT + \left(\frac{\partial \ln f_i}{\partial P}\right)_{T,x} dP + \sum_{j=1}^{m-1} \left(\frac{\partial \ln f_i}{\partial x_j}\right)_{P,T,x_k} dx_j.$$

$$k = 1, \ldots, j-1, j+1, \ldots, m-1 \quad (10.2\text{-}1)$$

Thermodynamics gives limited information on each of the three coefficients which appear on the right-hand side of Eq. (10.2-1). The first term can be related to the partial molar enthalpy and the second to the partial molar volume; the third term cannot be expressed in terms of any fundamental thermodynamic property but it can be conveniently related to the excess Gibbs energy which, in turn, can be described by a solution model. For a complete description of phase behavior we must say something about each of these three coefficients, for each component, in every phase.

In high-pressure work, it is important to give particular attention to the second coefficient which tells us how phase behavior is affected by pressure.

When analyzing typical experimental data, it is often difficult to isolate the effect of pressure because, more often than not, a change in the pressure is accompanied by a simultaneous change in either the temperature or the composition, or both. A striking example of such simultaneous changes is given by experimental results in Fig. 10-9 which show the effect of pressure on the melting temperature of solid argon.[13] Line A gives the melting line for pure argon as reported by Clusius and Weigand;[14] the melting temperature rises with pressure, as predicted by the Clapeyron equation, since argon expands upon melting. Line B gives results of Mullins and Ziegler[13] for the melting temperature of argon in the presence

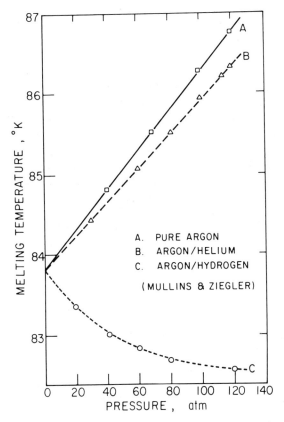

Fig. 10-9   Effect of pressure on melting temperature of argon. Qualitative difference between Lines B and C is due to effect of composition on liquid-phase fugacity of argon.

of high-pressure helium gas; these results are similar to those for pure argon, and again the melting temperature rises with increasing pressure. Finally, Line C gives the melting temperature for argon in the presence of high-pressure hydrogen gas; these data were also reported by Mullins and Ziegler. We are struck and perhaps puzzled by the fact that Line C shows completely different behavior: The melting temperature now falls with rising pressure.

The qualitative difference between Lines B and C can be clarified by thermodynamic analysis if we note that the three-phase equilibrium temperature is determined by two separate effects: First, the effect of pressure on the fugacities of solid and liquid argon is essentially the same for the three cases A, B, and C; and second, there is the effect of liquid composition on the fugacity of liquid argon. It is the effect of composition which for Case B is much different from that for Case C. At a given pressure, the solubility of hydrogen in liquid argon is much larger than that of helium; since pressure and composition simultaneously influence the fugacity of liquid argon, we find that the presence of hydrogen alters the equilibrium temperature in a manner qualitatively different from that found in the presence of helium. The fugacity of liquid argon is raised by pressure but lowered by solubility of another component. In the case of helium (low solubility) the pressure effect dominates, but for hydrogen (high solubility) the effect of composition is more important. We shall not here go into a more detailed analysis of this particular system. We merely wish to emphasize that the fugacity of a component is determined by the three variables temperature, pressure, and composition; that in a typical experimental situation these influences operate in concert; and that whatever success we may expect in explaining the behavior of a multicomponent, multiphase system is determined directly by the extent of our knowledge of the three coefficients given in Eq. (10.2-1).

At present we know least about the first coefficient, which is rigorously related to the enthalpy by

$$\left(\frac{\partial \ln f_i}{\partial T}\right)_{P,x} = \frac{h_i^+ - \bar{h}_i}{RT^2} \qquad (10.2\text{-}2)$$

where $\bar{h}_i$ is the partial molar enthalpy of i and h_i^+ is the molar enthalpy of i in the ideal-gas state at the same temperature. While we would like to use information on $\bar{h}_i$ to help us toward calculating f_i, we almost never can do so in high-pressure work because so little is known about enthalpies of fluid mixtures at high pressures. (In fact, in typical chemical engineering work, it is more common to reverse the procedure, viz., to differentiate fugacity data with respect to temperature, in order to estimate enthalpies.) As a result, the best we can do at present is always to analyze experimental phase-equilibrium data as a function of pressure and composition along an isotherm, and to allow any empirical parameters obtained from such analysis to vary with temperature as dictated by the experimental results.

We have a considerable body of knowledge concerning the third coefficient, the variation of fugacity with composition. Many empirical and semiempirical expressions (e.g., Margules, van Laar, Scatchard-Hildebrand) have been investigated toward that end. At present, most of our experience is limited to liquid mixtures at low pressures where we can safely neglect the second coefficient which relates the fugacity to the pressure according to

$$\left(\frac{\partial \ln f_i}{\partial P}\right)_{T,x} = \frac{\bar{v}_i}{RT} \tag{10.2-3}$$

where $\bar{v}_i$ is the partial molar volume. However, it is this coefficient which is of primary concern in high-pressure phase equilibria and no significant progress in high-pressure thermodynamics can be expected until we achieve improved quantitative understanding of partial molar volumes in liquid mixtures. We return to this key problem in Sec. 10.5 after a brief review of the relation between fugacities and volumetric properties and after some discussion on the normalization of liquid-phase activity coefficients at high pressures.

10.3 Fugacities from Volumetric Properties

In previous chapters (especially Chaps. 3 and 5) we discussed how fugacities can be calculated from volumetric properties which are frequently expressed by an equation of state. Equations (3.1-14) and (3.4-16) follow from thermodynamic considerations only and contain no physical assumptions. Calculations based on these equations, therefore, are not restricted to any phase nor to any range of pressure and, provided volumetric properties for a mixture are known as a function of composition and pressure over the pressure range zero to P, it is possible to calculate phase equilibria up to pressure P. As briefly discussed in Sec. 3.6, such calculations have been carried out with the Benedict-Webb-Rubin equation of state[15] for vapor-liquid equilibria in mixtures of light hydrocarbons at pressures up to several thousand pounds per square inch.

In Secs. 5.12 and 7.9 we described how volumetric properties, generalized by the theorem of corresponding states, may be used to calculate vapor-phase and liquid-phase fugacities in mixtures; in particular, in Figure 7-16 (p. 308) we illustrated how such calculations may be applied to phase-equilibrium problems. Such application, again for light hydrocarbon mixtures, has been discussed by Leland, Chappelear and Gamson.[16]

From a strictly thermodynamic point of view, there can be no objection to using volumetric properties for calculating *both* vapor-phase and liquid-phase fugacities. From a practical point of view however, such a procedure is not generally useful for liquid-phase fugacities; while vapor-phase volumetric properties (especially for mixtures of nonpolar fluids) can often be estimated with sufficient accuracy to yield good vapor-phase fugacities, volumetric data for condensed mixtures are usually not sufficiently well known to permit reliable calculations of fugacities in the liquid phase. Calculation of phase-equilibrium relations from volumetric data *alone* is useful only in those cases where the mixture's components are similar chemically.

Calculation of vapor-phase fugacities is discussed in detail in Chap. 5; in particular, Sec. 5.13 considers such calculations for components in mixtures at high pressures.

10.4   Liquid-Phase Activity Coefficients

For any component i in a liquid phase, the fugacity of i is most conveniently related to the mole fraction x_i through the activity coefficient, γ_i, according to

$$f_i = \gamma_i x_i f_i^0 \qquad (10.4\text{-}1)$$

where f_i^0 is the fugacity of i in its standard state. As discussed in Sec. 6.4, the choice of a standard state is arbitrary and is dictated by expedience; frequently it is convenient to choose a standard state for supercritical components which is different from that used for subcritical components. For a component in the liquid phase, the independent variables which determine the standard state are the temperature of the system, a fixed composition, and a specified pressure. The choice of standard state determines the normalization of the activity coefficient, i.e., the condition at which γ_i attains a fixed value, usually taken as unity.

For liquid mixtures at low pressures, it is not important to specify with care the pressure of the liquid standard state because at low pressures thermodynamic properties of liquids, pure or mixed, are not sensitive to pressure. However, at high pressures, liquid-phase properties are strong functions of pressure and we cannot be careless about the pressure-dependence of either the activity coefficient or the standard-state fugacity.

As discussed in Sec. 6.4, the most frequently used standard state is the pure liquid ($x_i = 1$) at the system temperature and pressure. When this standard state is used for all components in the mixture, the activity coefficients are said to be *symmetrically normalized* because in this case, for every component i,

$$\gamma_i \rightarrow 1 \quad \text{as} \quad x_i \rightarrow 1. \qquad (10.4\text{-}2)$$

However, whenever we are dealing with a supercritical component, there is a serious disadvantage in using a pure-liquid standard state. If the system temperature T is larger than T_{c_i} (the critical temperature of component i), pure liquid i cannot exist at T, and we are faced with the problem of finding the fugacity of a physically unattainable state. The fugacity of a hypothetical liquid state can only be found by some arbitrary method of extrapolation; while some extrapolations are more popular than others, there is no single, unambiguous way to compute the fugacity of any supercritical "liquid."

If we use hypothetical pure liquid i as our standard state, then f_i^0 must be chosen to satisfy the limiting relation $\gamma_i \rightarrow 1$ as $x_i \rightarrow 1$. However, experimental data for mixtures indicate that once we have calculated f_i^0 by a particular extrapolation, this limiting relation for γ_i may hold for a few solutions of i in various solvents, only to fail in others. Such failure is serious because unless we know how γ_i is normalized, we cannot write thermodynamically significant relations for the variation of γ_i with composition, nor can we meaningfully use binary data to estimate properties of multicomponent mixtures. Failure of the limiting relation $\gamma_i \rightarrow 1$ as $x_i \rightarrow 1$, regardless of the nature of the other component, is especially apparent whenever T is very much larger than T_{c_i} (e.g., solutions of hydrogen in common solvents near room temperature), since in that case the value of f_i^0 is extremely sensitive to the details of extrapolation. On the other hand, when T is only slightly larger than T_{c_i}, the use of hypothetical liquid i for the standard state often provides a useful thermodynamic procedure, since then f_i^0 is only slightly sensitive to the method of extrapolation.

The difficulties engendered by a hypothetical-liquid standard state can be eliminated by use of unsymmetrically normalized activity coefficients. In a binary liquid solution, we distinguish between subcritical component 1, called the solvent, (for which $T < T_{c_1}$) and supercritical component 2, called the solute ($T > T_{c_2}$). For component 1, we normalize γ_1 in the usual way:

$$\gamma_1 \rightarrow 1 \quad \text{as} \quad x_1 \rightarrow 1. \tag{10.4-3}$$

For component 2, however, we normalize γ_2 according to

$$\gamma_2^* \rightarrow 1 \quad \text{as} \quad x_2 \rightarrow 0. \tag{10.4-4}$$

As discussed in Sec. 6.4, Eqs. (10.4-3) and (10.4-4) constitute what is called the *unsymmetric convention of normalization*, because γ_1 and γ_2^* go to unity in different ways. The asterisk serves as a reminder that the activity coefficient so designated is normalized in a manner different from the customary one.

For the solvent, component 1, f_1^0 is the fugacity of pure saturated liquid 1 at the system temperature. However, the standard-state fugacity for the solute, component 2, is given by

$$f_2^0 = \underset{x_2 \to 0}{\text{limit}} \frac{f_2}{x_2} = H_{2,1} \qquad (10.4\text{-}5)\dagger$$

where $H_{2,1}$ stands for Henry's constant of solute 2 in solvent 1 at system temperature.

The standard state given by the unsymmetric convention for normalization avoids all arbitrariness about f_2^0 which, as given by Eq. (10.4-5), is an experimentally accessible quantity.

The unsymmetric convention of normalization can also be used for multicomponent solutions, but care must be taken to specify exactly the conditions that give $\gamma_2^* \to 1$. Whereas relation (10.4-3) is immediately applicable to solutions containing any number of components, relation (10.4-4) is not complete for a solution containing components in addition to 1 and 2. For solute 2, dissolved in a mixture of solvents 1 and 3, the normalization conditions are completely specified if we write

$$\gamma_1 \to 1 \quad \text{as} \quad x_1 \to 1 \qquad (10.4\text{-}6)$$

$$\gamma_3 \to 1 \quad \text{as} \quad x_3 \to 1 \qquad (10.4\text{-}7)$$

$$\gamma_2^* \to 1 \quad \text{as} \quad x_2 \to 0 \text{ for a fixed ratio } \frac{x_1}{x_1 + x_3}. \qquad (10.4\text{-}8)$$

It is sometimes convenient to use consistently one of the solvents (say, component 1) as the reference solvent; in that event, the fixed ratio is set equal to unity and $\gamma_2^* \to 1$ as $x_2 \to 0$ and $x_1 \to 1$. However, the fixed ratio $x_1/(x_1 + x_3)$ may be chosen to have any convenient value between zero and unity.

It is important to remind ourselves that thermodynamics neither provides nor demands any rigid convention for the choice of standard states. Conventions are determined strictly by their utility; while one particular convention may be useful for a particular class of solutions, it may be awkward for another class. Indeed, one of the main advantages of the activity-coefficient concept is its flexibility. The purpose of the activity coefficient is no more and no less than to relate the fugacity of a component at some condition of composition and pressure to what it is at some other (reference) condition of composition and pressure where we accurately know its numerical value. A standard state is like a point of reference on a map. If someone asks "Where is Berkeley?" a helpful reply is, "About ten miles East of San Francisco." The person asking the question knows where San Francisco is; it serves as his standard state. But if

---

†$H_{2,1}$ depends on pressure as well as temperature. See footnote to Eq. (6.3-19).

he had asked, "Where is Princeton?" then New York would be a more convenient standard state than San Francisco.

At constant temperature, the activity coefficient depends on both pressure and composition. One of the important goals of thermodynamic analysis is to consider separately the effect of each independent variable on the liquid-phase fugacity; it is therefore desirable to use constant-pressure activity coefficients which at constant temperature are independent of pressure and depend only on composition.[17] The definition of such activity coefficients follows directly from either of the exact thermodynamic relations:

$$\left(\frac{\partial \ln \gamma_i}{\partial P}\right)_{T,x} = \frac{\bar{v}_i}{RT} \qquad \text{(standard state is at a fixed pressure } P') \qquad (10.4\text{-}9)$$

$$\left(\frac{\partial \ln \gamma_i}{\partial P}\right)_{T,x} = \frac{\bar{v}_i - v_i^0}{RT} \qquad \text{(standard state is at the system total pressure } P) \qquad (10.4\text{-}10)$$

where v_i^0 is the molar volume of i in the standard state. Equations (10.4-9) and (10.4-10) are independent of the type of normalization used for γ_i; identical equations hold for γ_i^*.

For the fugacity of any component i in the liquid phase, at pressure P, temperature T and mole fraction x_i, we write

$$f_i = \gamma_i^{(P')} x_i f_i^{0(P')} \exp \int_{P'}^{P} \frac{\bar{v}_i \, dP}{RT} \qquad (10.4\text{-}11)$$

where $f_i^{0(P')}$ is the standard-state fugacity of i at system temperature T and arbitrary reference pressure P'. Equation (10.4-11) also holds for unsymmetrically normalized activity coefficient $\gamma_i^{*(P')}$. At a fixed temperature, $\gamma_i^{(P')}$ in Eq. (10.4-11) depends only on composition. Since it is our aim to relate activity coefficients to composition through some physical model, it is essential that we use constant-pressure activity coefficients $\gamma_i^{(P')}$ and $\gamma_i^{*(P')}$ when we attempt to test, interpret, and correlate high-pressure vapor-liquid equilibrium data.

The advantages of constant-pressure activity coefficients also become clear when we try to relate to one another the activity coefficients of all the components in a mixture through the Gibbs-Duhem equation (see Sec. 6.6 and App. IV). For variable-pressure activity coefficients at constant temperature the Gibbs-Duhem equation is:

$$\sum_i x_i \, d \ln \gamma_i = \frac{v \, dP}{RT} \qquad \text{(all standard states at fixed pressure } P') \qquad (10.4\text{-}12)$$

or

$$\sum_i x_i d \ln \gamma_i = \frac{\Delta v \, dP}{RT} \qquad \text{(all standard states at} \qquad (10.4\text{-}13)$$
$$\text{system total pressure } P)$$

where v is the molar volume of the liquid mixture and

$$\Delta v \equiv v - \sum_i x_i v_i^0. \qquad (10.4\text{-}14)$$

For constant-pressure activity coefficients, however, we obtain the much simpler relation

$$\sum_i x_i d \ln \gamma_i^{(P')} = 0 . \qquad (10.4\text{-}15)$$

Equations (10.4-12), (13), and (15) are all independent of the normalization used for activity coefficients.[17]

It is desirable to use activity coefficients which satisfy Eq. (10.4-15) rather than Eqs. (10.4-12) and (10.4-13) because all well-known mixture models (e.g., van Laar, Margules, Scatchard-Hildebrand) are particular mathematical solutions to Eq. (10.4-15); these models do not satisfy Eqs. (10.4-12) and (10.4-13) except in the limiting case where the right-hand sides of these equations vanish.

The two important features of this entire discussion are these: first, the activity coefficient must be carefully defined, and since the activity coefficient and standard state are inseparably linked (see footnote, p. 182), all standard states must be specified without ambiguity. Second, at constant temperature, the activity coefficient depends on both pressure and composition; the pressure effect is directly related to the partial molar volume. By defining constant-pressure activity coefficients, the effects of pressure and composition can be separated; under isothermal conditions, constant-pressure activity coefficients depend only on composition. We consider the composition dependence in Sec. 10.6.

10.5   Partial Molar Volumes

In Sec. 10.2 we indicated that significant progress in high-pressure thermodynamics of mixtures requires quantitative description of the variation of fugacity with pressure as given by Eq. (10.2-3). To obtain the effect of pressure on activity coefficient we substitute as follows:

$$\left(\frac{\partial \ln f_i}{\partial P} \right)_{T,x} = \left(\frac{\partial \ln \gamma_i x_i f_i^0}{\partial P} \right)_{T,x} = \frac{\bar{v}_i}{RT} \qquad (10.5\text{-}1)$$

which simplifies to

$$\left(\frac{\partial \ln \gamma_i}{\partial P} \right)_{T,x} + \left(\frac{\partial \ln f_i^0}{\partial P} \right)_{T,x} = \frac{\bar{v}_i}{RT} . \qquad (10.5\text{-}2)$$

If we define the standard-state fugacity f_i^0 at a fixed pressure, then the second term on the left side of Eq. (10.5-2) vanishes and we obtain Eq. (10.4-9). However, if we define f_i^0 at the total pressure of the system, we obtain Eq. (10.4-10).

Equations (10.4-9), (10.4-10), and (10.5-2) are independent of the convention used for normalization of activity coefficients; they apply equally to γ_i and γ_i^*.

For simplicity, it is tempting to assume $\bar{v}_i = v_i^0$, since we can then readily predict the effect of pressure on the activity coefficient.† For mixtures of liquids at pressures near 1 atm this is usually a valid assumption; for dilute solutions of gases in liquids at high pressures, it is also valid provided we use the unsymmetric convention for normalization. However, for concentrated solutions of gases in liquids, the assumption is poor, and at conditions approaching critical it is very poor. Partial molar volumes may be positive or negative, and near critical conditions they are usually strong functions of the composition. For example, in the saturated liquid phase of the system carbon dioxide/n-butane at 160°F, $\bar{v}$ for butane is $+1.8$ ft^3/lb-mol when the mole fraction of CO_2 is small, and it is -2.4 ft^3/lb-mol at the critical composition, $x_{CO_2} = 0.71$.

It is difficult to measure partial molar volumes and unfortunately many experimental studies of high-pressure vapor-liquid equilibria report no volumetric data at all; more often than not, experimental measurements are confined to total pressure, temperature, and phase compositions. Even in those cases where liquid densities are measured along the saturation curve, there is a fundamental difficulty in calculating partial molar volumes as indicated by the exact relations between partial molar volumes $\bar{v}_1$ and $\bar{v}_2$ and isothermal saturated molar volume v^s in a binary system:

$$\bar{v}_1 = v^s - x_2 \left[\left(\frac{\partial v^s}{\partial x_2} \right)_T + v^s \beta \left(\frac{\partial P}{\partial x_2} \right)_T \right] \qquad (10.5\text{-}3)‡$$

†v_i^0 is the molar volume of component i in the standard state, not necessarily that of the pure component.

‡To derive Eq. (10.5-3) we must note that at constant temperature v^s is a function of both x_2 and P:

$$dv^s = \left(\frac{\partial v^s}{\partial x_2} \right)_{T,P} dx_2 + \left(\frac{\partial v^s}{\partial P} \right)_{T,x_2} dP. \qquad (10.5\text{-}3a)$$

Therefore

$$\left(\frac{\partial v^s}{\partial x_2} \right)_T = \left(\frac{\partial v^s}{\partial x_2} \right)_{T,P} + \left(\frac{\partial v^s}{\partial P} \right)_{T,x_2} \left(\frac{\partial P}{\partial x_2} \right)_T. \qquad (10.5\text{-}3b)$$

The partial molar volume $\bar{v}_1$ is related to v^s by

$$\bar{v}_1 = v^s - x_2 \left(\frac{\partial v^s}{\partial x_2} \right)_{T,P}. \qquad (10.5\text{-}3c)$$

Simple substitution then gives Eq. (10.5-3).

$$\bar{v}_2 = v^s + x_1 \left[\left(\frac{\partial v^s}{\partial x_2} \right)_T + v^s \beta \left(\frac{\partial P}{\partial x_2} \right)_1 \right] \qquad (10.5\text{-}4)$$

where the compressibility β is defined by

$$\beta \equiv - \frac{1}{v^s} \left(\frac{\partial v^s}{\partial P} \right)_{T,x}. \qquad (10.5\text{-}5)$$

Experimental data for v^s are sometimes available, but experimental compressibilities for mixtures are rare.

For dilute solutions, the technical literature contains some direct (dilatometric) measurements of $\bar{v}_2$, the partial molar volume of the more volatile component, but the accuracy of these measurements is usually not high. A survey was made by Lyckman and Eckert,[18] who established the rough correlation shown in Fig. 10-10. On the ordinate, the partial molar

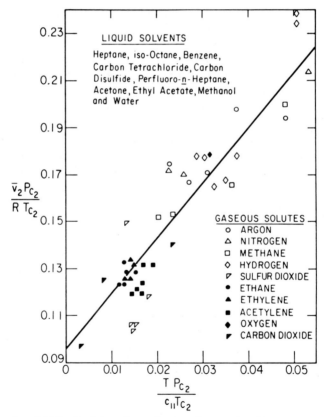

Fig. 10-10   Partial molar volumes of gases in dilute liquid solutions.

volume is nondimensionalized with the critical temperature and pressure of component 2; the abscissa is also dimensionless and includes c_{11}, the cohesive-energy density of the solvent, component 1 (see Sec. 7.2). Figure 10-10 is useful for rough approximations in systems remote from critical conditions. For expanded solvents, i.e., for solvents at temperatures approaching T_{c_i}, the partial molar volume of the solute tends to be much larger than that suggested by the correlation, as indicated in Fig. 10-11.

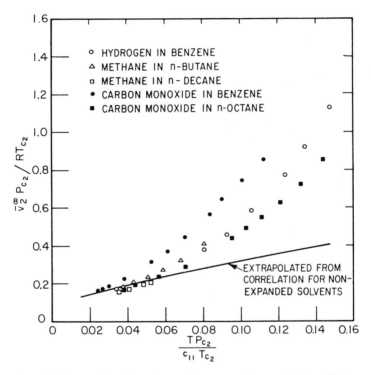

Fig. 10-11   Partial molar volumes of gaseous solutes at infinite dilution in expanded solvents.

If we can write an equation of state for liquid mixtures, we can then calculate partial molar volumes directly by differentiation. For a pressure-explicit equation, the most convenient procedure is to use the exact relation

$$\bar{v}_i \equiv - \frac{\left(\dfrac{\partial P}{\partial n_i}\right)_{T,V,n_j}}{\left(\dfrac{\partial P}{\partial V}\right)_{T,\,\text{all }n}} \tag{10.5-6}$$

where V is the total volume of the mixture containing n_1 moles of component 1, etc.

When a pressure-explicit equation of state for a liquid mixture is substituted into Eq. (10.5-6), we obtain an expression of the form

$$\bar{v}_i = \mathcal{f}(T, x_1, \ldots, v) \tag{10.5-7}$$

where v is the molar volume and $x_1, \ldots$ are the mole fractions of the liquid mixture. Given the pressure, temperature, and composition, we can, in principle, use the equation of state to calculate the molar volume of the mixture, but this is not a good procedure, since we rarely have an equation of state sufficiently accurate for that calculation. Instead, it is better to use in Eq. (10.5-7) an independent correlation for mixture volumes as shown by Chueh.[19] Confining his attention to saturated liquid mixtures, Chueh extended to mixtures the corresponding-states correlation for pure saturated liquids given by Lyckman and Eckert.[20] The correlation is of the form

$$\frac{v(T)}{v_c} = v_R^{(0)}\left(\frac{T}{T_c}\right) + \omega v_R^{(1)}\left(\frac{T}{T_c}\right) + \omega^2 v_R^{(2)}\left(\frac{T}{T_c}\right) \tag{10.5-8}$$

where v_c is the critical volume, T_c is the critical temperature, and ω is the acentric factor. The three generalized functions $v_R^{(0)}$, $v_R^{(1)}$ and $v_R^{(2)}$ depend only on the reduced temperature T/T_c and are represented by empirical algebraic equations. For mixtures, Chueh uses empirical mixing rules to calculate the composition-dependent parameters v_c, T_c, and ω.

To obtain an analytic function $\mathcal{f}$ in Eq. (10.5-7), Chueh uses the Redlich-Kwong equation; however, since application is directed at liquids, the two constants in that equation were not evaluated (as is usually done) from critical data alone, but rather from a fit of the pure-component, saturated-liquid volumes. The constants a and b in the equation of Redlich and Kwong are calculated from the relations

$$a = \frac{\Omega_a R^2 T_c^{2.5}}{P_c} \tag{10.5-9}$$

$$b = \frac{\Omega_b R T_c}{P_c} \tag{10.5-10}$$

where Ω_a and Ω_b are characteristic dimensionless constants determined from volumetric data for each saturated liquid. Table 10-1 gives Ω_a and Ω_b for 19 liquids. When a and b are evaluated from critical data alone, $\Omega_a = 0.4278$ and $\Omega_b = 0.0867$.

Table 10-1

DIMENSIONLESS CONSTANTS FOR SATURATED
LIQUIDS IN THE REDLICH-KWONG
EQUATION OF STATE

	Ω_a	Ω_b
Methane	0.4546	0.0872
Ethane	0.4347	0.0827
Ethylene	0.4290	0.0815
Acetylene	0.4230	0.0802
Propane	0.4138	0.0802
Propylene	0.4130	0.0803
n-Butane	0.4184	0.0794
iso-Butane	0.4100	0.0790
1-Butene	0.4000	0.0780
n-Pentane	0.3928	0.0767
iso-Pentane	0.3970	0.0758
n-Hexane	0.3910	0.0752
n-Heptane	0.3900	0.0740
n-Nonane	0.3910	0.0738
Cyclohexane	0.4060	0.0787
Benzene	0.4100	0.0787
Nitrogen	0.4540	0.0875
Hydrogen sulfide	0.4220	0.0823
Carbon dioxide	0.4184	0.0794

By adopting mixing rules similar to those given in Sec. 5.13, Chueh showed that Eq. (10.5-7) can be used for calculating partial molar volumes in multicomponent liquid mixtures.[19,29] Some results for binary systems are given in Figs. 10-12 and 10-13 which compare calculated partial molar volumes with those obtained from experimental data. Chueh's method gives consistently good results for mixtures of nonpolar (or slightly polar) components, except in the immediate vicinity of the critical ($T/T_{c_{mix}} >$ 0.93). For the critical region, his procedure was modified by using true critical constants rather than pseudocritical constants in Eq. (10.5-8). For this purpose a separate correlation was established for true critical volumes and temperatures (see App. VIII).

Chueh's method for calculating partial molar volumes is applicable to liquid mixtures containing any number of components. Required parameters are Ω_a and Ω_b (see Table 10-1), the acentric factor, the critical temperature, and critical pressure for each component, and a characteristic binary constant k_{ij} (see Sec. 5.13) for each possible unlike pair in the mixture. At present this method is restricted to saturated liquid solutions; for very precise work in high-pressure thermodynamics, it is also necessary to know how partial molar volumes vary with pressure at constant

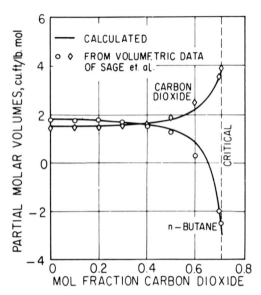

Fig. 10-12   Partial molar volumes in the saturated
             liquid phase of the *n*-butane/carbon di-
             oxide system at 160°F.

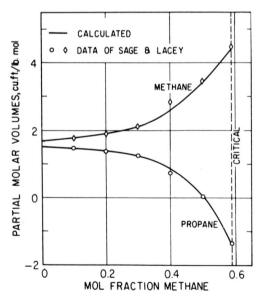

Fig. 10-13   Partial molar volumes in the saturated
             liquid phase of the propane/methane
             system at 100°F.

temperature and composition. An extension of Chueh's treatment may eventually provide estimates of partial compressibilities but in view of the many uncertainties in our present knowledge of high-pressure phase equilibria, such an extension is not likely to be of major importance for some time.

10.6   Excess Gibbs Energy: Dilated van Laar Model

We define the excess Gibbs energy per mole of binary solution by

$$\frac{g^{E*}}{RT} = x_1 \ln \gamma_1{}^{(P')} + x_2 \ln \gamma_2^{*(P')} \qquad (10.6\text{-}1)\dagger$$

where $\lambda_1{}^{(P')}$ and $\lambda_2^{*(P')}$ are constant-pressure activity coefficients. At constant temperature, once the reference pressure P' is chosen, g^{E*} is a function only of composition. In view of the unsymmetric normalization, g^{E*} vanishes at infinite dilution with respect to component 2 but not with respect to component 1; that is,

$$g^{E*} \to 0 \quad \text{as} \quad x_2 \to 0$$

but

$$g^{E*} \neq 0 \quad \text{as} \quad x_1 \to 0. \qquad (10.6\text{-}2)$$

As defined here, the ideal solution ($g^{E*} = 0$) is one where at constant temperature and pressure the fugacity of the light component is given by Henry's law and that of the heavy component by Raoult's law. In molecular terms this means that g^{E*} is zero whenever the concentration of component 2 in the liquid phase is sufficiently small to prevent molecules of component 2 from interacting with one another.

In a manner analogous to that used by Wohl,[21] the excess Gibbs energy can be represented by summing interactions of molecules (see Sec. 6.13):

$$\frac{g^{E*}}{RT(x_1 q_1 + x_2 q_2)} = -\alpha_{22(1)} z_2^2 - \dots \qquad (10.6\text{-}3)$$

where z_2 is the effective volume fraction of component 2:

$$z_2 = \frac{x_2 q_2}{x_1 q_1 + x_2 q_2} \qquad (10.6\text{-}4)$$

$$\dagger \ \gamma_1^{(P')} \equiv \frac{f_1^{(P)}}{x_1 f_{\text{pure }1}^{(P')}} \exp \int_P^{P'} \frac{\bar{v}_1 \, dP}{RT} \quad \text{and} \quad \gamma_2^{*(P')} \equiv \frac{f_2^{(P)}}{x_2 H_{2,1}^{(P')}} \exp \int_P^{P'} \frac{\bar{v}_2 \, dP}{RT}.$$

where q_i is the effective size of molecule i and where $\alpha_{22(1)}$ is the self-interaction constant of molecules 2 in the environment of molecules 1. In Eq. (10.6-3) only two-body interactions are considered; higher terms are neglected in order to keep the number of adjustable parameters to a minimum. The accuracy of most experimental data seldom warrants the use of more than two parameters.

The activity coefficients can be found from the exact relations (see Sec. 6.3):

$$\ln \gamma_1^{(P')} = \left(\frac{\partial n_T\, g^{E^*}/RT}{\partial n_1}\right)_{T,P,n_2} \qquad (10.6\text{-}5)$$

$$\ln \gamma_2^{*(P')} = \left(\frac{\partial n_T\, g^{E^*}/RT}{\partial n_2}\right)_{T,P,n_1} \qquad (10.6\text{-}6)$$

where n_i is the number of moles of i and n_T is the total number of moles.

Equations (10.6-3), (4), (5), and (6) yield the classical van Laar equations for unsymmetric normalization. Muirbrook[22] has shown that these equations, containing two adjustable parameters, are unsatisfactory for describing the properties of some systems which are at a temperature much above the critical temperature of the light component, or near the critical temperature of the heavy component. In addition, Muirbrook found that the three-suffix Margules equations were also unsatisfactory.

The probable reason for the failure of the classical van Laar treatment is due to van Laar's assumption that q_1 and q_2 are constants independent of composition. The q's are parameters which reflect the cross-sections, or sizes, or spheres of influence of the molecules; at conditions remote from critical, where the liquid molar volume is close to a linear function of the mole fraction, it is reasonable to assume that the q's are composition-independent. However, for a concentrated mixture of a noncondensable component 2 with a subcritical liquid 1, the molar volume of the mixture is a highly nonlinear function of the mole fraction, especially in the vicinity of the critical composition. The liquid solution dilates as x_2 rises, and van Laar's model must be modified to take this effect into account.

For practical reasons (since experimental data are usually not plentiful), it is desirable to derive equations for the pressure-independent activity coefficients which contain no more than two parameters. Because of this limitation, we assume for nonpolar (or slightly polar) systems that whereas q_1 and q_2 depend on composition, their ratio does not.[23] Since the van Laar treatment is a two-body (quadratic) theory, we assume that q_1 and q_2 are given by a quadratic function of the volume fraction Φ_2:

$$q_1 = v_{c_1}[1 + \eta_{2(1)}\, \Phi_2^2] \qquad (10.6\text{-}7)$$

$$q_2 = v_{c_2}[1 + \eta_{2(1)}\, \Phi_2^2] \qquad (10.6\text{-}8)$$

where

$$\Phi_2 = \frac{x_2 v_{c_2}}{x_1 v_{c_1} + x_2 v_{c_2}} .$$

In Eq. (10.6-7) and (10.6-8) we have arbitrarily used the pure-component critical volumes v_{c_i} as our measure of the molecular cross-sections at infinite dilution, when $\Phi_2 = 0$. Some other constant (for example, van der Waals' b or Lennard-Jones' σ^3) could just as easily be used. The dilation constant $\eta_{2(1)}$ is a measure of how effectively the light component dilates (swells) the liquid solution.

When Eqs. (10.6-7) and (10.6-8) are substituted into Eq. (10.6-3), the constant-pressure activity coefficients are

$$\ln \gamma_1^{(P^r)} = A' \Phi_2^2 + B' \Phi_2^4 \qquad (10.6\text{-}9)$$

$$\ln \gamma_2^{*(P^r)} = A' \left(\frac{v_{c_2}}{v_{c_1}}\right)(\Phi_2^2 - 2\Phi_2) + B'\left(\frac{v_{c_2}}{v_{c_1}}\right)(\Phi_2^4 - \tfrac{4}{3}\Phi_2^3) \qquad (10.6\text{-}10)$$

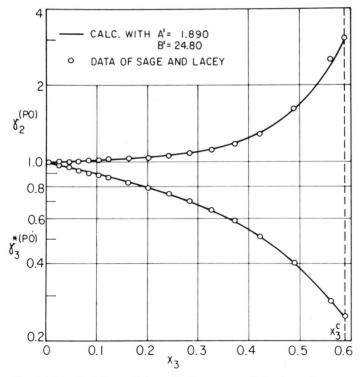

Fig. 10-14   Activity coefficients for the propane (2)/methane (3) system at 100°F. The superscript ($P0$) designates zero pressure.

where the dimensionless constants A' and B' are defined by

$$A' \equiv \alpha_{22(1)} v_{c_1} \tag{10.6-11}$$

$$B' \equiv 3\eta_{2(1)} \alpha_{22(1)} v_{c_1} . \tag{10.6-12}$$

Equations (10.6-9) and (10.6-10) are the desired two-parameter equations. These equations provide accurate representation of the constant-pressure activity coefficients of nonpolar binary mixtures from the dilute region up to the critical composition. To illustrate, Figs. 10-14 and 10-15

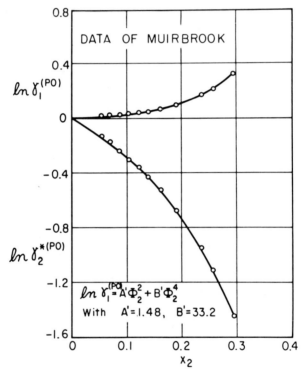

Fig. 10-15   Activity coefficients for the carbon dioxide (1)/ nitrogen (2) system at 32° F.

present typical results of data reduction for two binary systems, propane/methane[24] and carbon dioxide/nitrogen.[25]

Self-interaction constants, dilation constants, and Henry's constants for some binary systems are given in Tables 10-2, 10-3 and 10-4.[26] The magnitude of the dilation constant shows a consistent and meaningful variation with respect to the temperature and the properties of the constituent components; dilation constants are larger at temperatures approaching the critical temperature of the heavy component (component

Table 10-2

SELF-INTERACTION CONSTANTS FOR SOME BINARY SYSTEMS

System	T (°R)	$\alpha_{22(1)}$ (lb-mol/cu ft)	System	T (°R)	$\alpha_{22(1)}$ (lb-mol/cu ft)
Methane (2)/	259.7	0.425	Ethylene (2)/	424.7	0.305
Ethane (1)	309.7	0.305	Acetylene (1)	459.7	0.270
	359.7	0.182		499.7	0.244
	409.7	0.210			
	459.7	0.333	Ethane (2)/	424.7	0.538
	509.7	0.680	Acetylene (1)	459.7	0.490
Methane (2)/	259.7	0.342		499.7	0.365
Propane (1)	309.7	0.322		519.7	0.277
	359.7	0.322			
	409.7	0.355	Ethane (2)/	259.7	0.066
	459.7	0.415	Propane (1)	309.7	0.059
	491.7	0.462		359.7	0.051
	509.7	0.498		409.7	0.043
	559.7	0.593		459.7	0.034
	619.7	0.936		509.7	0.026
				559.7	0.025
				579.7	0.029
Methane (2)/	559.7	0.548		599.7	0.038
n-Pentane (1)	619.7	0.706		619.7	0.053
	679.7	0.939		639.7	0.099
	739.7	1.230			
			Propane (2)/	559.7	0.023
Ethylene (2)/	359.7	0.075	n-Pentane (1)	619.7	0.032
Ethane (1)	419.7	0.058		679.7	0.049
	459.7	0.039		739.7	0.080
	499.7	0.053		799.7	0.141
	519.7	0.069			

Table 10-3

DILATION CONSTANTS FOR SOME BINARY SYSTEMS

System	T (°R)	$\eta_{2(1)}$	System	T (°R)	$\eta_{2(1)}$
Methane (2)/	407.7	0.31	Methane (2)/	359.7	0.90
Propane (1)	499.7	1.46	Ethane (1)	409.7	1.29
	559.7	4.12		459.7	3.06
	619.7	28.35		509.7	27.10
Methane (2)/	559.7	1.19	Ethane (2)/	559.7	1.30
n-Pentane (1)	619.7	1.62	Propane (1)	599.7	4.41
	679.7	2.25		619.7	12.28
	739.7	8.39		639.7	43.54
Propane (2)/	679.7	0.27			
n-Pentane (1)	739.7	1.23			
	799.7	26.24			

Table 10-4

HENRY'S CONSTANTS FOR SOME SOLUTES IN SOLVENTS†

System	$T\,(^\circ R)$	$H_{2,1}^{(P0)}$ (psia)	System	$T\,(^\circ R)$	$H_{2,1}^{(P0)}$ (psia)
Methane (2)/	359.7	690	Ethane (2)/	559.7	449
Ethane (1)	409.7	1029	Propane (1)	579.7	503
	459.7	1330		599.7	573
	509.7	1500		619.7	616
				639.7	631
Methane (2)/	359.7	870			
Propane (1)	409.7	1360	Propane (2)/	619.7	289
	459.7	1800	n-Pentane (1)	679.7	447
	491.7	2044		739.7	610
	509.7	2130		799.7	750
	559.7	2141			
	619.7	1844			
Methane (2)/	559.7	2821			
n-Pentane (1)	619.7	3185			
	679.7	3256			
	739.7	2943			

†Henry's constants are here corrected to a reference pressure $P^r = 0$ as designated by superscript ($P0$).

1); also, they are larger for those systems in which the light component is highly supercritical. This behavior of dilation constants is in agreement with their physical significance in the dilated van Laar model, i.e., the liquid phase is swelled or dilated most when the subcritical, heavy component itself is near its critical temperature, or when the light component is far above its critical temperature. Under these conditions the liquid molar volume increases sharply with dissolved gas.

Plots of $\ln (\eta^{1/2})$ vs. $1/T$ show a similar shape for several nonpolar systems. It has been possible to unify the curves into a single reduced plot, as shown in Fig. 10-16. The curve can be represented by

$$\ln \left(\frac{\eta}{\eta\ddagger}\right)^{1/2} = -30.2925 + 39.1396\left(\frac{T\ddagger}{T}\right) - 17.2182\left(\frac{T\ddagger}{T}\right)^2$$
$$+ 2.81464\left(\frac{T\ddagger}{T}\right)^3 - 2.78571/\left(\frac{T\ddagger}{T}\right) - 5.26736\ln\left(\frac{T\ddagger}{T} - 0.9\right)$$

$$(10.6\text{-}13)$$

where $\eta\ddagger$ is a constant characteristic of the light component and $T\ddagger$ is a constant characteristic of the binary system. Some values of $\eta\ddagger$ and $T\ddagger$ are given in Fig. 10-16.

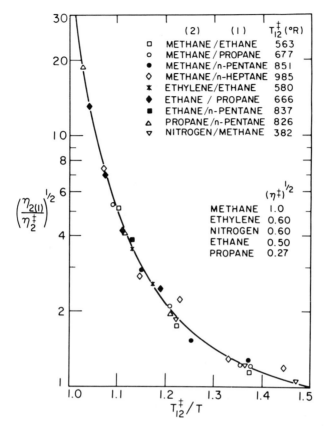

Fig. 10-16   Correlation of dilation constants $\eta_{2(1)}$ for some binary systems.

At temperatures sufficiently lower than the critical temperature of the light component (component 2), the dilation constant η obtained from data reduction for nonpolar systems becomes so small that it can be effectively equated to zero. Under these conditions, the constant-pressure activity coefficients of both components can be correlated with only one parameter, α. Chueh[27] found empirically that this occurs for a reduced temperature T_{R_2} less than 0.93. Therefore, components with a reduced temperature smaller than 0.93 are treated as heavy components (solvent), and those with T_R larger than 0.93 are treated as light components (solute). Nonpolar (or slightly polar) systems for which both T_{R_1} and T_{R_2} are smaller than 0.93 can be correlated with $\eta = 0$ and only one parameter, α. Systems for which the critical temperatures of the two components are very close (such as acetylene/ethane) are also analyzed with only one parameter, α, even though T_{R_2} is larger than 0.93; the terms "heavy" and "light" component lose their conventional meaning for such systems. In fact, it sometimes happens that the "heavy" component with the higher critical temperature may actually have a higher vapor pressure and critical pressure than the "light" component with the lower critical temperature.

For those nonpolar (or slightly polar) systems where both components can exist in the pure-liquid state, it is not necessary to use the unsymmetric convention for normalization of activity coefficients. Instead, such a system can often be analyzed with a one-parameter, symmetric-convention expression for the excess Gibbs energy:

$$\frac{g^E}{RT(x_1 v_{c_1} + x_2 v_{c_2})} = \alpha_{12}\, \Phi_1\, \Phi_2 . \tag{10.6-14}$$

From Eq. (10.6-5) we obtain

$$\ln \gamma_1^{(P')} = v_{c_1} \alpha_{12}\, \Phi_2^2 \tag{10.6-15}$$

$$\ln \gamma_2^{(P')} = v_{c_2} \alpha_{12}\, \Phi_1^2 \tag{10.6-16}$$

where $\gamma_2^{(P')}$ is given by

$$\gamma_2^{(P')} = \frac{f_2^{(P)}}{x_2 f_{\text{pure 2}}^{(P')}} \exp \int_P^{P'} \frac{v_2^L}{RT} \, dP . \tag{10.6-17}$$

It can be shown[27,28] that when both components are considered subcritical and when the excess Gibbs energy can be represented by Eq. (10.6-14), or by Eq. (10.6-3) with $q_1 = v_{c_1}$ and $q_2 = v_{c_2}$ (i.e., $\eta = 0$), there exist rigorous relations between the constants in the two conventions, viz.,

$$\alpha_{22(1)} = \alpha_{12} \tag{10.6-18}$$

$$\ln H_{2,1}^{(P')} = \ln f_{\text{pure 2}}^{(P')} + v_{c_2} \alpha_{12} . \tag{10.6-19}$$

The dilated van Laar model is readily extended to mixtures containing more than two components as shown elsewhere.[26,29] The resulting expressions for activity coefficients are tedious and to use them for practical work, an electronic computer is required. Since the dilated van Laar model is a quadratic (two-body) theory, all parameters which appear in these expressions can, with few assumptions, be obtained from binary data. As a result it is possible, at least for relatively simple systems, to reduce binary, high-pressure, vapor-liquid equilibrium data along lines suggested in the preceding paragraphs, and then to use the parameters obtained from such data reduction to predict vapor-liquid equilibria in systems containing three or more components.

10.7   Calculation of Multicomponent Vapor-Liquid Equilibria

Vapor-liquid equilibrium data at high pressures are plentiful for binary systems containing hydrocarbons, nitrogen, hydrogen, carbon dioxide, and hydrogen sulfide. Equilibrium data for ternary systems are much less plentiful and data for quaternary and higher systems are rare. Since industrial separations in the petroleum and related industries usually apply to multicomponent (i.e., ternary or higher) systems, it is important to estimate multicomponent equilibria from binary data.

Before data reduction, it is advisable whenever possible to test binary high-pressure data for thermodynamic consistency. A procedure for performing such tests is given in App. IX.

For vapor-liquid equilibria of an m-component system, the variables of interest are the temperature, total pressure, $m - 1$ independent liquid-phase mole fractions and $m - 1$ independent vapor-phase mole fractions. Since there are $2m$ variables but only m variables may be independently specified, the other m variables must be determined by solving m simultaneous equations, which are m equations of equilibrium,

$$f_i^V = f_i^L \qquad \text{for} \quad i = 1, 2, \ldots, m. \qquad (10.7\text{-}1)$$

In the m equations of equilibrium, the vapor-phase fugacities are given by

$$f_i^V = \varphi_i y_i P \qquad (10.7\text{-}2)$$

and the liquid-phase fugacities are given by

$$f_i^L = \gamma_i^{(P^r)} x_i f_i^{0\,(P^r)} \exp \frac{(P - P')\bar{v}_i^L}{RT}. \qquad (10.7\text{-}3)\dagger$$

Solution of the m equations of equilibrium must satisfy two stoichiometric relations:

$$\sum_{i=1}^{m} x_i = 1 \qquad (10.7\text{-}4)$$

and

$$\sum_{i=1}^{m} y_i = 1. \qquad (10.7\text{-}5)$$

---

†We have here assumed that $\bar{v}_i^L$ is a function of temperature and liquid-phase composition but not of pressure.

The phase-equilibrium calculations most often encountered in the design of separation processes are bubble-point and dew-point calculations. In the first case, pressure (or temperature) and all of the liquid-phase mole fractions are given; the temperature (or pressure) and the vapor-phase mole fractions are to be calculated. In the second case, pressure (or temperature) and all of the vapor-phase mole fractions are given, and the temperature (or pressure) and the liquid-phase mole fractions are desired. The calculations require simultaneous solution of the m equations given by Eq. (10.7-1) in addition to Eqs. (10.7-4) and (10.7-5). These calculations are most conveniently performed by iteration schemes.

The solution to each of these four problems involves the same thermodynamic quantities and relations; only the order of calculation and the convergence technique are different. It is therefore convenient to calculate the required thermodynamic quantities in separate subroutines which can then be used with any one of the main programs. The fugacity coefficients are calculated in subroutine PHIMIX, the activity coefficients $\gamma_i^{(P0)}$ in subroutine ACTCO,† the reference fugacities in subroutine RSTATE, and the partial molar volumes, $\bar{v}_i^L$ in subroutine VOLPAR. Details of computer programs are given elsewhere.[29]

To illustrate, we consider a bubble-pressure calculation which we solve with program BUBL P. Figure 10-17 shows a schematic diagram of the method of solution used. Data are read in by subroutine INPUT. Initial guesses for pressure and fugacity coefficients are then made for the first iteration. It is important that the solution always be approached from the ideal-gas side, i.e., for the first iteration we use a sufficiently low pressure and $\varphi_i = 1$. Use of an unreasonably high pressure for the first guess may cause divergence due to the large effect of a large Poynting correction. We arbitrarily set the first guess of pressure at 100 psia. The speed of convergence is essentially independent of this value; however, neither a value of zero nor a very high value should be used.

Next, the main program reads in the known quantities T and all x_i's. Subroutines RSTATE, ACTCO, and VOLPAR are then called to calculate those thermodynamic quantities which depend only on the known temperature and liquid-phase compositions.

Calculation of liquid-phase fugacities begins the loop of iteration. Vapor-phase compositions are calculated, for the first time, by

$$y_i = \frac{f_i^L}{P\varphi_i} \tag{10.7-6}$$

and pressure is calculated by

$$P = \sum_{i=1}^{m} \frac{f_i^L}{\varphi_i} \tag{10.7-7}$$

---

†The most convenient reference pressure is zero. Whenever we set $P^r = 0$, we designate the constant-pressure activity coefficient by $\gamma^{(P0)}$.

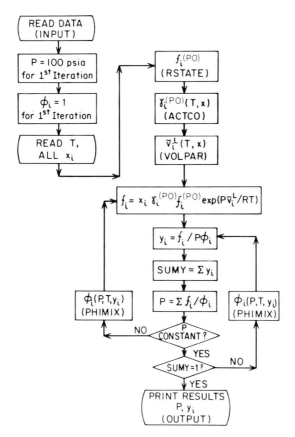

Fig. 10-17   Schematic diagram of bubble-pressure program.

The new pressure is then compared to the former value and, if it has changed, the vapor-phase fugacity coefficients (subroutine PHIMIX) are recalculated using the new pressure and vapor-phase compositions (after normalizing by SUMY). The loop is then reentered by recalculation of liquid-phase fugacity using Eq. (10.7-3).

When an unchanging value of pressure is achieved (within some small tolerance), the stoichiometry $\Sigma\, y_i = 1$ is tested. Usually this is satisfied when the pressure has attained an unchanging value. If $\Sigma\, y_i = 1$ is not satisfied, vapor-phase fugacity coefficients are recalculated and the loop is reentered at the calculation of vapor-phase compositions. When the two conditions of unchanging pressure and $\Sigma\, y_i = 1$ are met, the equations are all satisfied and the equilibrium results are printed out. Usually, convergence is attained after a few iterations, taking a total time of about half a second on an IBM 7094 computer (for a ternary system).

Figures 10-18, 10-19, and 10-20 and Tables 10-5 and 10-6 give ex-

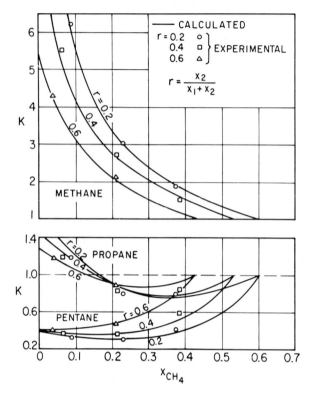

Fig. 10-18   Vapor-liquid equilibrium constants for the
n-pentane (1)/propane (2)/methane (3) system at
220° F.

amples of predicted results for some ternary systems using binary data only. In the calculation for the ternary system acetylene/ethane/ethylene, all three components are treated as solvents, using only one parameter, α, per binary, as explained in Sec. 10.6.

Very near the (true) critical point of the mixture, for the region $0.97 \leq T_R \leq 1.0$ for the liquid mixture, calculated results are extremely sensitive to small errors in any of the thermodynamic quantities involved. Results for this hypersensitive region can be best obtained by interpolating between calculations at a lower T_R and the known condition at the critical point: $K_i = 1$ for each component and the pressure is equal to the critical pressure of the mixture. Correlations of (true) critical properties of mixtures are given in App. VIII.

A detailed discussion of this and other computer programs is given in the monograph *Computer Calculations for High-Pressure Vapor-Liquid Equilibria.*[29]

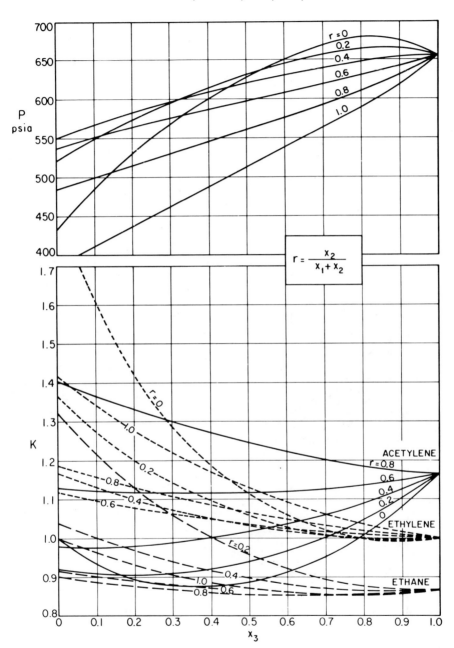

Fig. 10-19   Calculated vapor-liquid equilibrium constants and bubble pressures for the
acetylene (1)/ethane (2)/ethylene (3) system at 40°F.

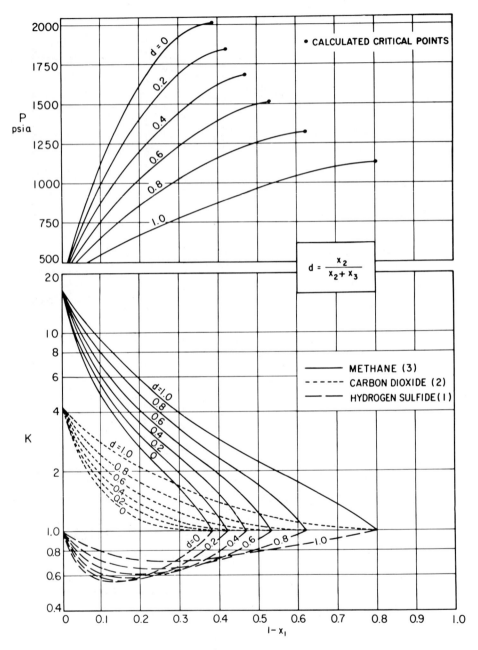

Fig. 10-20   Calculated vapor-liquid equilibrium constants and bubble pressures for the hydrogen sulfide (1)/carbon dioxide (2)/methane (3) system at 100°F.

Table 10-5

COMPARISON OF CALCULATED AND EXPERIMENTAL RESULTS FROM
PROGRAM BUBL P.
n-PENTANE (1)/PROPANE (2)/METHANE (3) SYSTEM

	Given				Calculated				Experimental[†]		
$T(°R)$	x_1	x_2	x_3	P (psia)	y_1	y_2	y_3	P (psia)	y_1	y_2	y_3
559.7	0.678	0.170	0.152	495	0.046	0.086	0.868	500	0.049	0.085	0.866
559.7	0.555	0.139	0.306	987	0.040	0.056	0.903	1000	0.043	0.056	0.901
559.7	0.442	0.110	0.448	1448	0.049	0.049	0.902	1500	0.051	0.049	0.900
559.7	0.315	0.079	0.606	1973	0.068	0.047	0.885	2000	0.083	0.046	0.871
619.7	0.701	0.175	0.124	515	0.113	0.143	0.744	500	0.112	0.145	0.743
619.7	0.587	0.147	0.266	1019	0.093	0.086	0.821	1000	0.094	0.080	0.826
619.7	0.482	0.121	0.397	1499	0.103	0.069	0.828	1500	0.103	0.068	0.829
619.7	0.363	0.091	0.546	2070	0.130	0.057	0.813	2000	0.138	0.073	0.789
679.7	0.730	0.182	0.088	495	0.243	0.227	0.529	500	0.238	0.215	0.547
679.7	0.616	0.154	0.230	1000	0.186	0.129	0.685	1000	0.181	0.122	0.697
679.7	0.501	0.125	0.374	1443	0.185	0.093	0.722	1500	0.206	0.099	0.695

[†] R. T. Carter, B. H. Sage, W. N. Lacey, *Trans. Am. Inst. Min. & Met. Engrs.*, **142**, 170 (1941); and H. R. Dourson, B. H. Sage, W. N. Lacey, *Trans. Am. Inst. Min. & Met. Engrs.*, **151**, 206 (1943).

Table 10-6

COMPARISON OF CALCULATED AND EXPERIMENTAL RESULTS FROM
PROGRAM BUBL P.
ACETYLENE (1)/ETHANE (2)/ETHYLENE (3) SYSTEM

	Given				Calculated				Experimental[†]		
$T(°R)$	x_1	x_2	x_3	P (psia)	y_1	y_2	y_3	P (psia)	y_1	y_2	y_3
424.7	0.187	0.330	0.483	197.4	0.203	0.252	0.545	200.3	0.202	0.260	0.538
424.7	0.061	0.316	0.623	199.4	0.073	0.230	0.697	200.3	0.084	0.234	0.682
424.7	0.084	0.104	0.812	218.9	0.084	0.075	0.841	220.0	0.084	0.080	0.836
424.7	0.142	0.630	0.228	170.8	0.205	0.501	0.294	175.1	0.206	0.511	0.283
459.7	0.060	0.204	0.736	356.2	0.069	0.156	0.775	359.9	0.065	0.157	0.778
459.7	0.025	0.437	0.538	316.9	0.035	0.346	0.619	320.2	0.034	0.347	0.619
459.7	0.143	0.795	0.062	277.0	0.237	0.685	0.078	280.1	0.220	0.684	0.096
459.7	0.056	0.464	0.480	315.8	0.079	0.370	0.551	319.7	0.077	0.374	0.549
499.7	0.074	0.767	0.159	468.2	0.114	0.692	0.194	465.5	0.113	0.698	0.189
499.7	0.031	0.741	0.228	462.7	0.049	0.672	0.279	464.8	0.046	0.677	0.277
499.7	0.843	0.082	0.075	517.0	0.766	0.119	0.115	514.4	0.778	0.114	0.108
499.7	0.042	0.423	0.535	545.5	0.057	0.362	0.581	564.0	0.050	0.363	0.586
499.7	0.103	0.068	0.829	663.2	0.118	0.058	0.824	664.0	0.104	0.060	0.834
499.7	0.021	0.026	0.953	654.6	0.026	0.022	0.952	664.0	0.026	0.023	0.951

[†] R. J. Hogan, W. T. Nelson, G. H. Hanson, M. R. Cines, *Ind. Eng. Chem.*, **47**, 2210 (1955); and J. L. McCurdy, D. L. Katz, *Ind. Eng. Chem.*, **36**, 674 (1944).

10.8 Liquid-Liquid and Gas-Gas Equilibria

In the previous sections we have been primarily concerned with high-pressure equilibria in systems containing one liquid phase and one vapor phase. We now briefly consider the effect of pressure on equilibria between two liquid phases and later, between two gaseous phases. In particular, we are concerned with the question of how pressure may be used to induce miscibility or immiscibility in a binary system.

Two liquids are miscible in all proportions if Δg, the molar Gibbs energy of mixing at constant temperature and pressure, satisfies the relations

$$\Delta g < 0 \qquad (10.8\text{-}1)$$

and

$$\left(\frac{\partial^2 \Delta g}{\partial x^2}\right)_{T,\,P} > 0. \qquad (10.8\text{-}2)$$

Since Δg is a function of pressure, it follows that under certain conditions, a change in pressure may produce immiscibility in a completely miscible system, or conversely, such a change may produce complete miscibility in a partially immiscible system. The effect of pressure on miscibility in binary liquid mixtures is closely connected with the volume change on mixing as indicated by the exact relation

$$\left(\frac{\partial \Delta g}{\partial P}\right)_{T,\,x} = \Delta v \qquad (10.8\text{-}3)$$

where Δv is volume change on mixing at constant temperature and pressure.

To fix ideas, consider a binary liquid mixture which at normal pressure is completely miscible and whose isothermal Gibbs energy of mixing is given by Curve a in Fig. 10-21. Suppose that for this system Δv is positive; an increase in pressure raises Δg, and at some higher pressure the variation of Δg with x_1 may be given by Curve b. As indicated in Fig. 10-21, Δg at the high pressure no longer satisfies Eq. (10.8-2) and the liquid mixture now has a miscibility gap in the composition interval $x_1' < x_1 < x_1''$.

For contrast, consider also a binary liquid mixture which at normal pressures is incompletely miscible as shown by Curve a in Fig. 10-22. If Δv for this system is negative, then an increase in pressure lowers Δg and at some high pressure the variation of Δg with x_1 may be given by Curve b, indicating complete miscibility. It follows from these simple considerations that the qualitative effect of pressure on phase stability of

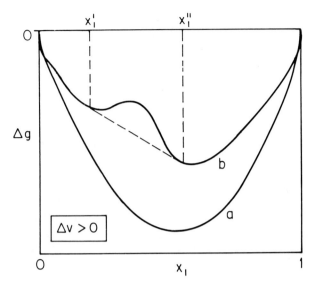

Fig. 10-21   Effect of pressure on miscibility: a–low pressure, no immiscibility; b–high pressure, immiscible for $x_1' < x_1 < x_1''$.

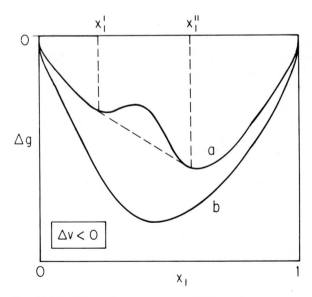

Fig. 10-22   Effect of pressure on miscibility: a–low pressure, immiscible for $x_1' < x_1 < x_1''$; b–high pressure, no immiscibility.

binary liquid mixtures depends on the magnitude and sign of the volume change of mixing. In order to carry out quantitative calculations at some fixed temperature, it is necessary to have information on the variation of Δv with x and P in addition to information on the variation of Δg with x at a single pressure.

To illustrate, we consider a simple, symmetric binary mixture at some fixed temperature and 1 atm pressure. For this liquid mixture, we assume that

$$\Delta g = RT(x_1 \ln x_1 + x_2 \ln x_2) + Ax_1 x_2 \qquad (10.8\text{-}4)$$

$$\Delta v = Bx_1 x_2 \qquad (10.8\text{-}5)$$

where A and B are experimentally determined constants. Further, we assume that the liquid mixture is incompressible for all values of x; i.e., we assume that B is independent of pressure. At any pressure P then, we have for Δg

$$\Delta g = RT(x_1 \ln x_1 + x_2 \ln x_2) + [A + B(P - 1)]x_1 x_2 \quad (10.8\text{-}6)$$

where P is in atmospheres. Substituting Eq. (10.8-6) into Eq. (10.8-2), we find that the mixture is partially immiscible when

$$\frac{A + B(P - 1)}{RT} > 2. \qquad (10.8\text{-}7)$$

Equation (10.8-7) tells us that if $A/RT < 2$ (complete miscibility at 1 atm) and if $B > 0$, then there is a certain pressure P (larger than 1 atm) at which immiscibility is induced. On the other hand, if $A/RT > 2$ (incomplete immiscibility at 1 atm) and if $B < 0$, then there is a certain pressure P (larger than 1 atm) at which complete miscibility is attained.

When two liquid phases exist, the compositions of the two phases α and β are governed by the equality of fugacities for each component:

$$f_1^\alpha = f_1^\beta \qquad (10.8\text{-}8)$$

$$f_2^\alpha = f_2^\beta \qquad (10.8\text{-}9)$$

or, equivalently, if the same standard-state fugacity is used for any component in both phases,

$$(\gamma_1 x_1)^\alpha = (\gamma_1 x_1)^\beta \qquad (10.8\text{-}10)$$

$$(\gamma_2 x_2)^\alpha = (\gamma_2 x_2)^\beta. \qquad (10.8\text{-}11)$$

For the simple mixture described by Eqs. (10.8-4) and (10.8-5), we can substitute into Eqs. (10.8-10) and (10.8-11) and we then obtain

$$x_1^\alpha \exp \frac{[A + B(P - 1)][1 - x_1^\alpha]^2}{RT} = x_1^\beta \exp \frac{[A + B(P - 1)][1 - x_1^\beta]^2}{RT}$$

$$(10.8\text{-}12)$$

$$x_2^\alpha \exp \frac{[A + B(P - 1)][1 - x_2^\alpha]^2}{RT} = x_2^\beta \exp \frac{[A + B(P - 1)][1 - x_2^\beta]^2}{RT}.$$

$$(10.8\text{-}13)$$

Simultaneous solution of these equilibrium relations (coupled with the conservation equations $x_1^\alpha + x_1^\alpha = 1$ and $x_1^\beta + x_2^\beta = 1$) gives the co-existence curve for the two-phase system as a function of pressure.†

Experimental studies of liquid-liquid equilibria at high pressures were reported many years ago by Roozeboom[30] and by Timmermans[6] and Poppe.[31] More recently, experimental work has been reported by Schneider[7] and by Winnick and Powers.[32] These latter authors made a detailed study of the acetone/carbon disulfide system at 0°C; at normal pressure this system is completely miscible and for an equimolar mixture the volume increase upon mixing is of the order of 1 cc/g-mol, which represents a fractional change of about 1.5 percent. Winnick and Powers measured the volume change on mixing as a function of both pressure and composition; these measurements, coupled with experimentally deter-mined activity coefficients at low pressure, were then used to calculate the phase diagram at high pressures, using a thermodynamic procedure similar to the one outlined above. The calculations showed that incom-plete miscibility should be observed at pressures larger than about 75,000 psia. Winnick and Powers also made experimental measurements of the phase behavior of this system at high pressures; they found, as shown in Fig. 10-23, that their observed results are in good agreement with those calculated from volumetric data. This is not surprising since the calculations follow from rigorous thermodynamic relations and are not based on any physical or mathematical approximations. Perfect agreement between predicted and observed results was not obtained because the calculations are sensitive to small inaccuracies in the data used.

Schneider[7] has presented a thorough review of the effect of pressure on liquid-liquid equilibria. Some typical results from that review have already been given in Sec. 10.1 in Figs. 10-5 to 10-8.

---

†Because of symmetry, Eq. (10.8-12) and Eq. (10.8-13) are always satisfied by the trivial solution $x_1^\alpha = x_2^\alpha = x_1^\beta = x_2^\beta = \frac{1}{2}$.

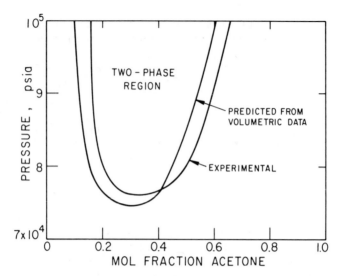

Fig. 10-23   Liquid-liquid equilibria for a system completely miscible at normal pressure. Calculated and observed behavior of the acetone/carbon disulfide system at 0°C. (Winnick and Powers)

Thermodynamic analysis of incomplete miscibility in liquid mixtures at high pressures [Eqs. (10.8-1), (2), and (3)] can also be applied to gaseous mixtures at high pressures. It has been known since the beginning of recorded history that not all liquids are completely miscible with one another, but only in recent times have we learned that gases may also, under suitable conditions, exhibit limited miscibility. The possible existence of two gaseous phases at equilibrium was predicted on theoretical grounds by van der Waals as early as 1894 and again by Onnes and Keesom in 1907.[33] Experimental verification, however, was not obtained until about forty years later.

Immiscibility of gases is observed only at high pressures where gases are at liquid-like densities. The term "gas-gas equilibria" is therefore somewhat misleading because it refers to fluids whose properties are similar to those of liquids, very different from those of gases under normal conditions. For our purposes here, we define two-component, gas-gas equilibria as the existence of two equilibrated, stable, fluid phases at a temperature in excess of the critical of either pure component, both phases being at the same pressure but having different compositions.

Some experimental results for the helium/xenon system[11] are shown in Fig. 10-24. (The critical temperature of xenon is 16.6°C.) At temperatures several degrees above the critical of xenon, the two phase-compositions are significantly different even at pressures as low as

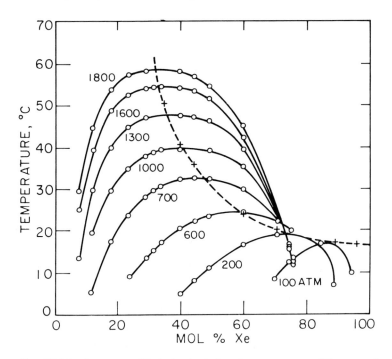

Fig. 10-24   Gas-gas equilibria in the helium/xenon system. (Diepen and De Swaan Arons)

200 atm. However, to obtain the same degree of separation at higher temperatures, much higher pressures are required.

A theoretical analysis of the helium-xenon system was reported by Zandbergen, Knaap and Beenakker[34] who based their calculations on the Prigogine-Scott theory of corresponding states for mixtures[35,36] (see Sec. 7.10). We cannot here go into the details of their analysis; we merely indicate the essential elements. Zandbergen, Knaap and Beenakker use the *three-liquid* theory to obtain an expression for the volumes of helium/xenon mixtures as a function of temperature, pressure, and composition. This expression is

$$v_{\text{mixture}} = y_1^2 v_1^0 + 2y_1 y_2 v_{12}^0 + y_2^2 v_2^0 \qquad (10.8\text{-}14)$$

where y is the mole fraction, v_1^0 and v_2^0 are the pure-component volumes at the temperature and pressure of interest and where v_{12}^0 is a function of pressure P_1' and temperature T_1' given by

$$v_{12}^0 = \left(\frac{\sigma_{12}}{\sigma_{11}}\right)^3 v_1'(P_1', T_1') \qquad (10.8\text{-}15)$$

where v_1' is the molar volume of pure 1 at pressure P_1' and temperature T_1' defined by

$$P_1' = P \frac{\epsilon_{11}}{\epsilon_{12}} \left(\frac{\sigma_{12}}{\sigma_{11}}\right)^3 \tag{10.8-16}$$

$$T_1' = T \frac{\epsilon_{11}}{\epsilon_{12}} . \tag{10.8-17}$$

In these equations, σ and ϵ are parameters in the Lennard-Jones potential function; for interactions between unlike molecules the customary mixing rules were used:

$$\sigma_{12} = \tfrac{1}{2}(\sigma_{11} + \sigma_{22}) \tag{10.8-18}$$

$$\epsilon_{12} = (\epsilon_{11}\epsilon_{22})^{1/2} . \tag{10.8-19}$$

For helium, $\sigma = 2.56$ Å and $\epsilon/k = 10.22°$K, where k is Boltzmann's constant. For xenon, $\epsilon/k = 221°$K and $\sigma = 4.10$ Å. Because of symmetry, subscript 1 may refer to either helium or xenon; from the assumption of corresponding-states behavior, v_{12}^0 should be independent of the component designated by subscript 1. Equation (10.8-14) is an equation of state for the binary mixture and from it the phase behavior can be calculated without further assumptions.

Some results reported by Zandbergen and co-workers are shown in Fig. 10-25. Considering the severe simplifying assumptions made, the

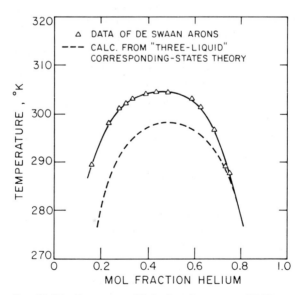

Fig. 10-25   Gas-gas equilibria for the system H3/Xe at 686 atm. (Zandbergen, Knaap and Beenakker)

calculated phase boundary is in gratifying agreement with that found experimentally. Because of symmetry with respect to mole fraction in the three-liquid model, the calculated T-y curve shown is necessarily a parabola whose maximum is at the composition midpoint. However, the vertical position and the width of the calculated parabola are subject to adjustment upon making small changes in the semiempirical mixing rules for Lennard-Jones parameters, Eqs. (10.8-18) and (10.8-19).

In the previous paragraphs we indicated how, under certain conditions, pressure may be used to induce immiscibility in liquid and gaseous binary mixtures which at normal pressures are completely miscible. Finally, we want to consider briefly how the introduction of a third component can bring about immiscibility in a binary mixture that is completely miscible in the absence of the third component. In particular, we are concerned with a binary liquid mixture where the added (third) component is a gas; in this case, elevated pressures are required in order to dissolve an appreciable amount of the added component in the binary liquid solvent. For the situation to be discussed, it should be clear that phase instability is not a consequence of the effect of pressure on the chemical potentials, as was the case in the previous sections, but results instead from the presence of an additional component which affects the chemical potentials of the components to be separated. High pressure enters into our discussion only indirectly, because we want to use a highly volatile substance for the additional component.

The situation under discussion is similar to the familiar salting-out effect in liquids, where a salt added to an aqueous solution serves to precipitate one or more organic solutes. Here, however, instead of a salt, we add a gas; and, in order to dissolve an appreciable quantity of gas, the system must be at an elevated pressure.

We consider a binary liquid mixture of components 1 and 3; to be consistent with our previous notation, we reserve the subscript 2 for the gaseous component. Components 1 and 3 are completely miscible at room temperature; the (upper) critical solution temperature T^c is far below room temperature, as indicated by the lower curve in Fig. 10-26. Suppose now that we dissolve a small amount of component 2 in the binary mixture; what happens to the critical solution temperature? This question was considered by Prigogine[37] who assumed that for any binary pair which can be formed from the three components 1, 2, and 3, the excess Gibbs energy (symmetric convention) is given by

$$g_{ij}^E = \alpha_{ij} x_i x_j \qquad (i,j = 1, 2; 1, 3 \text{ or } 2, 3) \qquad (10.8\text{-}20)$$

where α_{ij} is an empirical (Margules) coefficient determined by the properties of the ij binary. Prigogine has shown that the change in critical solution temperature which results upon adding a small amount of com-

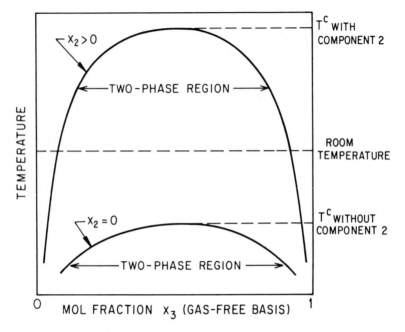

Fig. 10-26   Effect of gaseous component (2) on mutual solubility of liquids (1) and (3).

ponent 2 is given by

$$\frac{\partial T^c}{\partial x_2} = -\frac{1}{2R}\frac{(\alpha_{13} - \alpha_{12} + \alpha_{23})(\alpha_{13} - \alpha_{23} + \alpha_{12})}{\alpha_{12}}. \quad (10.8\text{-}21)$$

In order to induce the desired immiscibility in the 1-3 binary at room temperature, we want $\partial T^c/\partial x_2$ to be positive and large.

Let us now focus attention on the common case where all three binaries exhibit positive deviations from Raoult's law, i.e., $\alpha_{ij} > 0$ for all ij pairs. If T^c for the 1-3 binary is far below room temperature, then that binary is only moderately nonideal and α_{13} is small. We must now choose a gas which forms a highly nonideal solution with one of the liquid components (say component 3) while it forms with the other component (component 1) a solution which is only modestly nonideal. In that event,

$$\alpha_{23} \gg \alpha_{12} \qquad\qquad\qquad (10.8\text{-}22)$$

and also

$$\alpha_{23} \gg \alpha_{13}. \qquad\qquad\qquad (10.8\text{-}23)$$

Equations (10.8-21), (22), and (23) indicate that under the conditions just described, $\partial T^c/\partial x_2$ is both large and positive, as desired; i.e., dissolution of a small amount of component 2 in the 1-3 mixture raises the critical solution temperature as shown in the upper curve of Fig. 10-26. From Prigogine's analysis we conclude that if component 2 is properly chosen, it can induce binary miscible mixtures of components 1 and 3 to split at room temperature into two liquid phases having different compositions.

As shown by Balder,[10] thermodynamic considerations may be used to establish the phase diagram of a ternary system consisting of two miscible liquid components and a supercritical gas at high pressures. Such diagrams have been obtained experimentally[38,39,40] for a variety of ternary systems and have led to suggestions for separations using a high-pressure gas as the selective solvent.

PROBLEMS

1. Consider a high-pressure ternary liquid mixture containing one heavy (subcritical) component 1 and two light (supercritical) components 2 and 3. Isothermal P-x-y and v^L data are available for the two binaries 1-2 and 1-3. We want to predict the behavior of the ternary system at the same temperature using only the binary data. Describe how this may be done using the concept of excess Gibbs energy for unsymmetrically normalized activity coefficients. Use a three-suffix Margules expansion for the Gibbs energy and devise physically reasonable mixing rules to characterize those interactions which cannot be obtained from the binary data.

2. A binary liquid mixture follows the equation

$$g^E = A x_1 x_2$$

where A is a constant. At atmospheric pressure and 0°C it is observed that $A/RT = 1.90$, indicating that liquids 1 and 2 are completely miscible, although not far from splitting into two liquid phases. In order to separate an equimolar mixture of components 1 and 2 it is proposed to compress it isothermally at 0°C in the hope that two liquid phases will be formed. Is this likely? What simple experimental information obtainable at 1 atm can be called upon to answer this question? Assuming that phase separation can be obtained by isothermal compression and that 1 and 2 are typical liquids, what order of magnitude pressure would be required to produce two liquid phases?

REFERENCES

1. Mehta, V. S., and G. Thodos, *J. Chem. Eng. Data*, **10**, 211 (1965).

2. Sage, B. H., and W. N. Lacey, *Ind. Eng. Chem.*, **30**, 1299 (1938).

3. Dodge, B. F., *Chemical Engineering Thermodynamics*. New York: McGraw-Hill Book Company, 1944.

4. Katz, D. L., and F. Kurata, *Ind. Eng. Chem.*, **32,** 817 (1940).

5. Rowlinson, J. S., *Liquids and Liquid Mixtures*. London: Butterworths Scientific Publications., 1959.

6. Timmermans, J., *J. Chim. Phys.*, **20,** 491 (1923).

7. Scheneider, G., *Ber. der Bunsengesellschaft*, **70,** 497 (1966).

8. Connolly, J. F., *J. Chem. Eng. Data*, **11,** 13 (1966).

9. Zernike, J., *Chemical Phase Theory*. Antwerp: Kluwer, 1956.

10. Balder, J. R., and J. M. Prausnitz, *I&EC Fundamentals*, **5,** 449 (1966).

11. De Swaan Arons, J., and G. A. M. Diepen, *J. Chem. Phys.*, **44,** 2323 (1966).

12. Gonikberg, M. G., *Chemical Equilibria and Reaction Rates at High Pressures* (trans. from Russian). Washington, D. C.: Office of Technical Services, U.S. Dept. of Commerce, 1963.

13. Mullins, J. C., and W. T. Ziegler, *Intern. Advan. Cryogenic Eng.*, **10,** 171 (1964).

14. Clusius, K., and K. Weigand, *Z. Phys. Chem.*, **B46,** 1 (1940).

15. Benedict, M., G. B. Webb and L. C. Rubin, *J. Chem. Phys.*, **8,** 334 (1940); **10,** 747 (1942); *Chem. Eng. Progr.*, **47,** 419 (1951).

16. Leland, T. W., P. S. Chappelear, and B. W. Gamson, *A. I. Ch. E. Journal*, **8,** 482 (1962).

17. Prausnitz, J. M., *Chem. Eng. Sci.*, **18,** 613 (1963).

18. Lyckman, E. W., C. A. Eckert, and J. M. Prausnitz, *Chem. Eng. Sci.*, **20,** 685 (1965).

19. Chueh, P. L., and J. M. Prausnitz, *A. I. Ch. E. Journal*, **13,** 1099 (1967).

20. Lyckman, E. W., C. A. Eckert, and J. M. Prausnitz, *Chem. Eng. Sci.*, **20,** 703 (1965).

21. Wohl, K., *Trans. A. I. Ch. E.*, **42,** 215 (1946).

22. Muirbrook, N. K., Dissertation, Univ. of California, Berkeley (1964).

23. Chueh, P. L., N. K. Muirbrook, and J. M. Prausnitz, *A. I. Ch. E. Journal*, **11,** 1097 (1965).

24. Reamer, H. H., B. H. Sage, and W. N. Lacey, *Ind. Eng. Chem.*, **42,** 534 (1950).

25. Muirbrook, N. K., and J. M. Prausnitz, *A. I. Ch. E. Journal*, **11,** 1092 (1965).

26. Chueh, P. L., and J. M. Prausnitz, *Ind. Eng. Chem.*, **60,** 34 (March 1968).

27. Chueh, P. L., Dissertation, Univ. of California, Berkeley (1967).

28. O'Connell, J. P., and J. M. Prausnitz, *I&EC Fundamentals*, **3,** 347 (1964).

29. Chueh, P. L., and J. M. Prausnitz, *Computer Calculations for High-Pressure Vapor-Liquid Equilibria*. Englewood Cliffs, N. J.: Prentice-Hall, Inc., 1968.

30. Roozeboom, H. W., *Die Heterogenen Gleichgewichte.* Braunschweig, 1918.

31. Poppe, G., *Bull Soc. Chim. Belg.*, **44**, 640 (1935).

32. Winnick, J., Dissertation, Univ. of Oklahoma (1963).

33. Onnes, H. K., and W. H. Keesom, *Commun. Phys. Lab. Univ. Leiden,* Suppl. No. 16 (1907).

34. Zandbergen, P., H. F. P. Knaap and J. J. M. Beenakker, *Physica,* **33**, 379 (1967).

35. Prigogine, I., *The Molecular Theory of Solutions.* Amsterdam: North Holland Publishing Co., 1957.

36. Scott, R. L., *J. Chem. Phys.*, **25**, 193 (1956).

37. Prigogine, I., *Bull. Soc. Chim. Belg.*, **52**, 115 (1943).

38. Elgin, J. C., and J. J. Weinstock, *J. Chem. Eng. Data*, **4**, 3 (1959).

39. Francis, A. W., *Liquid-Liquid Equilibria.* New York: Interscience Publishers Inc., 1963.

40. Chappelear, D. C., and J. C. Elgin, *J. Chem. Eng. Data*, **6**, 415 (1961).

APPENDIX I

Outline of a Proof of the Uniformity of Intensive
Potentials as a Criterion of Phase Equilibrium

As discussed in Sec. 2.3, we use the function U for the purpose of showing that the temperature and pressure and the chemical potential of each species must be uniform throughout a closed, heterogeneous system at equilibrium internally with respect to heat transfer, boundary displacement, and mass transfer across phase boundaries. Since we identify equilibrium processes (variations) with reversible processes, the criterion for equilibrium in a closed system is that U be a minimum, and that any variation in U at constant total entropy and total volume vanishes; that is,

$$dU_{S, V} = 0. \tag{1}$$

An expression for the total differential dU can be written by summing over all the phases, the extension of Eq. (2.2-5a) to a multiphase system:

$$dU = \sum_{\alpha} T^{(\alpha)} dS^{(\alpha)} - \sum_{\alpha} P^{(\alpha)} dV^{(\alpha)} + \sum_{\alpha} \sum_{i} \mu_i^{(\alpha)} dn_i^{(\alpha)}, \tag{2}$$

where α is a phase index, taking values 1 to π, and i is a species index, taking values 1 to m.

On expansion, Eq. (2) becomes

$$
\begin{aligned}
dU = \ & T^{(1)} dS^{(1)} & - P^{(1)} dV^{(1)} + \mu_1^{(1)} dn_1^{(1)} + \cdots + \mu_m^{(1)} dn_m^{(1)} \\
& + T^{(2)} dS^{(2)} & - P^{(2)} dV^{(2)} + \mu_1^{(2)} dn_1^{(2)} + \cdots + \mu_m^{(2)} dn_m^{(2)} \\
& \ \ \vdots & \vdots \qquad \vdots \qquad\qquad \vdots \\
& + T^{(\pi)} dS^{(\pi)} & - P^{(\pi)} dV^{(\pi)} + \mu_1^{(\pi)} dn_1^{(\pi)} + \cdots + \mu_m^{(\pi)} dn_m^{(\pi)}.
\end{aligned} \tag{3}
$$

The individual variations $dS^{(1)}$, etc., are subject to the constraints of constant *total* entropy, constant *total* volume and constant *total* moles of each species (chemical reaction excluded). These may be written as:

$$dS = dS^{(1)} + \cdots + dS^{(\pi)} = 0, \tag{4}$$

$$dV = dV^{(1)} + \cdots + dV^{(\pi)} = 0, \tag{5}$$

$$\sum_\alpha dn_i^{(\alpha)} = dn_i^{(1)} + \cdots + dn_i^{(\pi)} = 0, \qquad i = 1, \ldots, m. \tag{6}$$

There are thus $\pi(m + 2)$ independent variables in Eq. (3), and there are $m + 2$ constraints. The expression for dU may be written in terms of $m + 2$ fewer independent variables by using the constraining equations to eliminate, for example, $dS^{(1)}$, $dV^{(1)}$, and the m $dn_i^{(1)}$. The result is an expression for dU in terms of $(\pi - 1)(m + 2)$ truly independent variables; that is, all the variations expressed as $dS^{(\alpha)}$, etc., are then truly independent, since the constraints have been used to eliminate certain variables. The resulting expression, if we eliminate $dS^{(1)}$, $dV^{(1)}$, and all $dn_i^{(1)}$, as indicated above, is

$$
\begin{aligned}
dU = &(T^{(2)} - T^{(1)})dS^{(2)} - (P^{(2)} - P^{(1)})dV^{(2)} + (\mu_1^{(2)} - \mu_1^{(1)})dn_1^{(2)} \\
&+ \cdots + (\mu_m^{(2)} - \mu_m^{(1)})dn_m^{(2)} \\
&+ (T^{(3)} - T^{(1)})dS^{(3)} - (P^{(3)} - P^{(1)})dV^{(3)} + (\mu_1^{(3)} - \mu_1^{(1)})dn_1^{(3)} \\
&+ \cdots + (\mu_m^{(3)} - \mu_m^{(1)})dn_m^{(3)} \\
&\quad \vdots \qquad\qquad\qquad \vdots \qquad\qquad\qquad \vdots \\
&+ (T^{(\pi)} - T^{(1)})dS^{(\pi)} - (P^{(\pi)} - P^{(1)})dV^{(\pi)} + (\mu_1^{(\pi)} - \mu_1^{(1)})dn_1^{(\pi)} \\
&+ \cdots + (\mu_m^{(\pi)} - \mu_m^{(1)})dn_m^{(\pi)}.
\end{aligned} \tag{7}
$$

All variations $dS^{(2)}$, $dV^{(2)}$, $dn_1^{(2)}$, $dn_2^{(2)}$, etc., are truly independent. Therefore, at equilibrium in the closed system where $dU = 0$, it follows that:[†]

$$\frac{\partial U}{\partial S^{(2)}} = 0, \quad \frac{\partial U}{\partial V^{(2)}} = 0, \quad \frac{\partial U}{\partial n_1^{(2)}} = 0, \quad \frac{\partial U}{\partial n_2^{(2)}} = 0, \text{ etc.} \tag{8}$$

Hence $T^{(2)} - T^{(1)} = 0$, or:

$$T^{(2)} = T^{(1)}, \quad T^{(3)} = T^{(1)}, \text{ etc.} \tag{9}$$

Similarly,

$$P^{(2)} = P^{(1)}, \quad P^{(3)} = P^{(1)}, \text{ etc.} \tag{10}$$

and

$$\mu_1^{(2)} = \mu_1^{(1)}, \quad \mu_1^{(3)} = \mu_1^{(1)}, \quad \mu_2^{(2)} = \mu_2^{(1)}, \mu_2^{(3)} = \mu_2^{(1)}, \text{ etc.} \tag{11}$$

[†] F. B. Hildebrand, *Methods of Applied Mathematics* (Englewood Cliffs, N. J.: Prentice-Hall, Inc., 1952).

Equations (9), (10), and (11) tell us that at internal equilibrium with respect to the three processes (heat transfer, boundary displacement, and mass transfer), temperature, pressure, and chemical potential of each species are uniform throughout the entire heterogeneous, closed system. This uniformity is expressed by Eqs. (2.3-1) to (2.3-3).

Although chemical reactions have been excluded from consideration in this section, it can be shown that Eq. (2.3-3) is not altered by the presence of such reactions. For any species i at equilibrium, the chemical potential of i is the same in all phases, regardless of whether or not species i can participate in a chemical reaction in any (or all) of these phases. This is true provided only that all such chemical reactions are also at equilibrium.

However, the existence of chemical reactions does affect the phase rule Eq. (2.5-1). In that equation, m is the number of distinct chemical species only in the absence of chemical reactions. If chemical reactions are considered, then m is the number of independent components, i.e., the number of chemically distinct species minus the number of chemical equilibria interrelating these species.

APPENDIX II

Virial Equation as a Power Series in
Density or Pressure

The compressibility factor of a gas can be expressed by an expansion using either the density or the pressure as the independent variable:

$$z = \frac{Pv}{RT} = 1 + B\rho + C\rho^2 + D\rho^3 + \cdots \tag{1}$$

or

$$z = \frac{Pv}{RT} = 1 + B'P + C'P^2 + D'P^3 + \cdots \tag{2}$$

where for a pure gas the virial coefficients $B, C, D, \ldots$ and $B', C', D', \ldots$ are functions only of the temperature. Equation (1) is frequently called the Leiden form and Eq. (2) is often referred to as the Berlin form of the virial equation in recognition of early workers in these cities who first used these equations. We now show how the coefficients of one series are related to those of the other.

First, we multiply both equations by $RT\rho$ and obtain:

$$P = RT\rho + BRT\rho^2 + CRT\rho^3 + DRT\rho^4 + \cdots \tag{1a}$$

and

$$P = RT\rho + B'\rho RTP + C'RT\rho P^2 + D'RT\rho P^3 + \cdots \tag{2a}$$

Next, we substitute Eq. (1a) into Eq. (2a):

$$\begin{aligned}
P = {} & RT\rho + B'RT\rho[RT\rho + BRT\rho^2 + CRT\rho^3 + DRT\rho^4 + \cdots] \\
& + C'RT\rho[RT\rho + BRT\rho^2 + CRT\rho^3 + DRT\rho^4 + \cdots]^2 \\
& + D'RT\rho[RT\rho + BRT\rho^2 + CRT\rho^3 + DRT\rho^4 + \cdots]^3 + \cdots.
\end{aligned} \tag{3}$$

Equations (3) and (1a) are both power series in ρ. If we compare like terms in the two equations, we obtain from terms in ρ^2:

$$BRT\rho^2 = B'(RT)^2\rho^2 \tag{4}$$

459

or

$$\frac{B}{RT} = B'. \tag{5}$$

From terms in ρ^3:

$$CRT\rho^3 = B'B(RT)^2\rho^3 + C'(RT)^3\rho^3 \tag{6}$$

or

$$\frac{C - B^2}{(RT)^2} = C'. \tag{7}$$

Similarly, from terms in ρ^4 we obtain:

$$D' = \frac{D - 3BC + 2B^3}{(RT)^3}. \tag{8}$$

Equations (5), (7), and (8) are exact provided we compare the two *infinite* series given by Eqs. (1) and (2). In other words, these relations are correct only if we evaluate the coefficients in both series from isothermal experimental z data according to

$$B = \lim_{\rho \to 0}\left(\frac{\partial z}{\partial \rho}\right)_T \quad \text{and} \quad B' = \lim_{P \to 0}\left(\frac{\partial z}{\partial P}\right)_T \tag{9}$$

$$C = \tfrac{1}{2}\lim_{\rho \to 0}\left(\frac{\partial^2 z}{\partial \rho^2}\right)_T \quad \text{and} \quad C' = \tfrac{1}{2}\lim_{P \to 0}\left(\frac{\partial^2 z}{\partial P^2}\right)_T \tag{10}$$

$$D = \tfrac{1}{3}\lim_{\rho \to 0}\left(\frac{\partial^3 z}{\partial \rho^3}\right)_T \quad \text{and} \quad D' = \tfrac{1}{3}\lim_{P \to 0}\left(\frac{\partial^3 z}{\partial P^3}\right)_T. \tag{11}$$

In practice it is not possible to evaluate virial coefficients from experimental data very near zero density or zero pressure because experimental measurements at these conditions are insufficiently accurate, especially for the second and third derivatives. As a result, Eqs. (5), (7), and (8) are only approximations when they are used to convert actual experimental virial coefficients from one series to those of the other. To illustrate the approximate nature of these equations we consider two methods of data reduction discussed by Scott and Dunlap.†

Suppose we want to evaluate B from low-pressure volumetric measurements of some gas at constant temperature. We can evaluate B either by fitting our data to the truncated virial equation

$$\frac{Pv}{RT} = 1 + B\rho, \tag{12}$$

or else by fitting it to the truncated virial equation

$$\frac{Pv}{RT} = 1 + \frac{B}{RT}P. \tag{13}$$

In either case, we obtain B by fitting the experimental data over a *finite* region of density or pressure. If we use Eq. (12) we are in effect assuming that over the density range used, $C = 0$. If we use Eq. (13) we are in effect assuming that over the pressure range used, $C' = 0$. Since B is generally not zero, we can see from Eq. (7) that if one of these assumptions is valid, then the other one is not. As a result it is not surprising that the value of B which we obtain by reduction of actual data depends on the method used and we find that Eq. (5) cannot, in practice, be satisfied exactly.

Scott and Dunlap made accurate volumetric measurements at low pressures for n-butane at 29.88°C. When they used Eq. (12) for data reduction they found:

$$B = -715 \pm 5 \text{ cm}^3/\text{g-mol}.$$

But when Eq. (13) was used, they obtained:

$$B = -745 \pm 6 \text{ cm}^3/\text{g-mol}.$$

We can see from these results that even if experimental uncertainties are considered, Eq. (5) is only an approximation.

A preferable procedure for determining B from experimental data is to use for data reduction either one of the virial equations truncated after the third virial coefficient. When this was done, Scott and Dunlap found for the same data, using the density series,

$$B = -695 \pm 25 \text{ cm}^3/\text{g-mol}$$

$$C = (-4 \pm 4)10^5 \ (\text{cm}^3/\text{g-mol})^2$$

†R. L. Scott and R. D. Dunlap, *J. Phys. Chem.*, **66**, 639 (1962).

and, using the pressure series,

$$B = -691 \pm 26 \, \text{cm}^3/\text{g-mol}$$
$$C = (-10 \pm 5)10^5 \, (\text{cm}^3/\text{g-mol})^2.$$

With this method of data reduction the uncertainties in B are larger than before but agreement between the two methods is now very much improved. The results obtained for C are of little value but for obtaining B, it is preferable to include even a rough estimate for C rather than to include none at all.

APPENDIX III

Virial Coefficients for Hydrogen, Helium and Neon

Because of their small masses, the properties of hydrogen, helium and neon cannot be described by classical statistical mechanics. As discussed elsewhere† it is possible to write an expression for the second virial coefficient of light gases based on quantum mechanics and it is found that

$$\frac{B}{b_0} = \oint \left(\frac{kT}{\epsilon}, \Lambda^* \right) \tag{1}$$

where

$$b_0 = \tfrac{2}{3} \pi N_A \sigma^3$$

and

$$\Lambda^* = \frac{h}{\sigma \sqrt{m\epsilon}} \; .$$

Λ^* is called the *reduced de Broglie wavelength* and it depends on Planck's constant h, the molecular mass m and the intermolecular potential parameters ϵ and σ.

Figure III-1 gives reduced experimental second virial coefficients of hydrogen and helium at very low temperatures. The parameters used for data reduction are:

	ϵ/k (°K)	b_0 (cm³/g-mol)	Λ^*
Hydrogen	37.0	31.67	1.73
Helium	10.22	21.07	2.67

For comparison, Fig. III-1 also shows reduced second virial coefficients as calculated from the Lennard-Jones potential using classical statistical mechanics.

†J. O. Hirschfelder, C. F. Curtiss, and R. B. Bird, *Molecular Theory of Gases and Liquids* (New York: John Wiley & Sons, Inc., 1954).

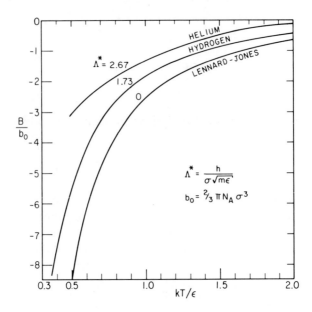

Fig. III-1   Reduced second virial coefficients for helium and hydrogen at low temperatures.

The results shown in Fig. III-1 are useful for estimating second virial cross-coefficients B_{12} whenever component 1 or 2 (or both) is a light (quantum) gas. To estimate B_{12} we use the customary mixing rules

$$\epsilon_{12} = (\epsilon_1 \epsilon_2)^{1/2} \tag{2}$$

and

$$\sigma_{12} = \tfrac{1}{2}(\sigma_1 + \sigma_2), \tag{3}$$

and in addition we have

$$\Lambda_{12}^* = \frac{h}{\sigma_{12}\sqrt{m_{12}\epsilon_{12}}}, \tag{4}$$

where

$$\frac{1}{m_{12}} = \frac{1}{2}\left(\frac{1}{m_1} + \frac{1}{m_2}\right). \tag{5}$$

For further details, see Hirschfelder, et al.,† and the work of Myers.‡

† *Molecular Theory of Gases and Liquids* (New York: John Wiley & Sons, Inc., 1954).
‡ J. M. Prausnitz and A. L. Myers, *A. I. Ch. E. Journal*, **9**, 5 (1963).

Experimental results

Goodwin and co-workers† have established empirical correlations for second and third virial coefficients of hydrogen over a wide temperature range.

For the second virial coefficient, their results are given in the form

$$B = \sum_{i=1}^{4} b_i x^{(2i-1)/4},$$ (6)

where

$$x = \frac{109.83}{T},$$

$T = $ Temperature (°K),

$b_1 = +42.464$ $b_3 = -2.2982$

$b_2 = -37.1172$ $b_4 = -3.0484$ (cm³/g-mol).

Equation (6) covers the temperature range 15 to 423°K and the mean deviation from experimental results is ±0.066 cm³/g-mol.

The third virial coefficient (in cm⁶/g-mol²) is given by the expression

$$C = 1310.5 x^{1/2}[1 + 2.1486 x^3][1 - \exp(1 - x^{-3})].$$ (7)

Equation (7) covers the temperature range 24 to 423°K and the mean deviation from experimental results is 17.4 cm⁶/g-mol².

Virial coefficients of helium have not been fitted to any empirical expressions although experimental results have been reported over a wide temperature range by several authors. Agreement among results from different laboratories is not always very good but it appears that the best consistent results now available are those reported by Keesom.‡ His values at very low temperature (below about 4°K) are subject to considerable uncertainty but those at temperatures close to and above the critical of helium (5.26°K) are probably reliable. A summary of Keesom's results is given in Table III-1.

Virial coefficient data for neon are not as plentiful as those for hydrogen and helium. Very good data of Michels§ have been reported for the temperature range 0–150°C, and some old data are available for temperatures down to 55.64°K; the critical temperature of neon is 44.5°K.

---

† R. D. Goodwin, D. E. Diller, H. M. Roder, and L. A. Weber, *J. Res. Natl. Bur. Std.*, **68A**, No. 1, 121 (1964).

‡ W. H. Keesom, *Helium* (Amsterdam: Elsevier Publishing Co., 1942).

§ A. Michels, T. Wassenaar, and P. Louwerse, *Physica*, **26**, 539 (1960).

Table III-1

SECOND AND THIRD VIRIAL COEFFICIENTS
FOR HELIUM†

T (°K)‡	B (cm³/g-mol)	C (cm⁶/g-mol²)§
4.5	−70.4	151
5	−62.1	322
6	−49.1	458
8	−33.0	513
10	−23.3	518
12	−16.9	508
14	−12.3	498
16	− 8.86	488
20	− 4.04	468
30	+ 2.42	417
40	5.63	382
60	8.86	327
80	10.4	292
173.2	11.7	201
273.2	11.5	156
373.2	11.1	126
573.2	10.1	101

†W. H. Keesom, *Helium* (Amsterdam: Elsevier Publishing Co., 1942).

‡For high temperatures (300–10,000°K) virial coefficients for helium have been calculated using results from molecular-beam experiments. The second and third virial coefficients, as reported by E. F. Harrison [*A. I. A. A. Journal*, **2**, 1854 (1964)], are given by

$$B = 1.3436 \times 10^{-2} [15.8922 - \ln T]^3$$
$$-4.39 \exp [-(2.4177 \times 10^{-3})T]$$

$$C = 9.0263 \times 10^{-5} [15.8922 - \ln T]^6$$

where the units are the same as those in Table III-1.

§The accuracy of these third virial coefficients is low.

A compilation of second and third virial coefficients for neon is given in Table III-2. The accuracy of the third virial coefficients is not high and for temperatures below 0°C the accuracy of the second virial coefficients is only fair. For very low temperatures, estimates of B can be made using Fig. III-1 with the parameters $\epsilon/k = 34.9°K$, $b_0 = 27.1$ cm³/g-mol, and $\Lambda^* = 0.593$.

Table III-2

SECOND AND THIRD VIRIAL COEFFICIENTS FOR NEON†

$T\,(°K)$	$B\,(cm^3/g\text{-}mol)$	$C\,(cm^6/g\text{-}mol^2)$‡
55.6	−32.30	900
60.1	−25.13	−
65.3	−20.98	563
90.7	− 8.18	442
123.2	+ 0.10	309
173.2	6.45	−
223.2	9.12	228
273.2	10.77	246
323.2	11.82	234
373.2	12.48	238
423.2	13.08	208
473.2	13.42	−
573.2	13.69	−

†G. A. Nicholson and W. G. Schneider, *Can. J. Chem.*, **33**, 589 (1955); A. Michels, T. Wassenaar and P. Louwerse, *Physica*, **26**, 539 (1960); J. Otto, in Landolt-Börnstein, *Physikalisch-Chemische Tabellen* 5th Ed., 2nd Supplement, p. 45 (Berlin: Springer, 1931); C. A. Crommelin, J. P. Martinez, and H. Kamerlingh-Onnes, *Commun. Phys. Lab. Univ. Leiden*, **154a** (1919).

‡The accuracy of these third virial coefficients is low.

APPENDIX IV

The Gibbs-Duhem Equation

A brief discussion of the Gibbs-Duhem equation is given in Sec. 2.4 and some applications are given in Chap. 6. In this appendix we give a derivation of the fundamental equation and we present special forms of the equation as applied to activity coefficients.

Let M be an extensive property of a mixture. For a homogeneous phase, M is a function of temperature, pressure, and the mole numbers. The total differential of M is given by

$$dM = \left(\frac{\partial M}{\partial T}\right)_{P,\,\text{all}\,n} dT + \left(\frac{\partial M}{\partial P}\right)_{T,\,\text{all}\,n} dP + \sum_i \overline{m}_i\, dn_i, \qquad (1)$$

where

$$\overline{m}_i \equiv \left(\frac{\partial M}{\partial n_i}\right)_{T,P,n_j}. \qquad (2)$$

The extensive property M is related to the partial molar properties $\overline{m}_1, \overline{m}_2, \ldots$, etc. by Euler's theorem:

$$M = \sum_i \overline{m}_i n_i. \qquad (3)$$

Differentiation of Eq. (3) gives

$$dM = \sum_i \overline{m}_i\, dn_i + \sum_i n_i\, d\overline{m}_i. \qquad (4)$$

Equations (1) and (4) yield the general Gibbs-Duhem equation:

$$\boxed{\left(\frac{\partial M}{\partial T}\right)_{P,\,\text{all}\,n} dT + \left(\frac{\partial M}{\partial P}\right)_{T,\,\text{all}\,n} dP - \sum_i n_i\, d\overline{m}_i = 0.} \qquad (5)\dagger$$

†Notice that the general Gibbs-Duhem equation applies to any extensive property, not just the Gibbs energy as discussed in Sec. 2.4.

Suppose M is the Gibbs energy G. As indicated in Chap. 2,

$$\left(\frac{\partial G}{\partial T}\right)_{P,\,\text{all }n} = -S \tag{6}$$

$$\left(\frac{\partial G}{\partial P}\right)_{T,\,\text{all }n} = V \tag{7}$$

$$\left(\frac{\partial G}{\partial n_i}\right)_{T,P,n_j} = \mu_i \tag{8}$$

where μ is the chemical potential. Equation (5) then becomes

$$S\,dT - V\,dP + \sum_i n_i\,d\mu_i = 0. \tag{9}$$

In terms of excess functions for one mole of mixture, Eq. (9) is

$$s^E\,dT - v^E\,dP + \sum_i x_i\,d\mu_i^E = 0. \tag{10}$$

The excess chemical potential of component i is related to activity coefficient γ_i by

$$\mu_i^E = RT \ln \gamma_i. \tag{11}$$

At constant temperature and pressure, we then have

$$\sum_i x_i\,d\ln\gamma_i = 0. \tag{12}$$

The phase rule tells us that in a binary, two-phase system it is not possible to change the composition of either phase while holding both temperature and pressure constant. Therefore, in a binary system, experimental data used to compute activity coefficients may be either isothermal or isobaric but not both. As a result, Eq. (12) is not strictly applicable to activity coefficients for a binary system. To obtain an equation similar to, but less restrictive than, Eq. (12), we consider how Eq. (9) can be rewritten in terms of activity coefficients. First we treat the case of constant pressure and variable temperature and then the case of constant temperature and variable pressure.

The isobaric, nonisothermal case

Equation (9), on a molar basis, now is

$$\sum_i x_i \, d\mu_i = -s \, dT, \tag{13}$$

where s is the entropy per mole of mixture. Introducing the activity coefficient,

$$\frac{\mu_i}{T} - \frac{\mu_i^0}{T} = R \ln \gamma_i x_i, \tag{14}$$

where superscript 0 stands for the standard state where

$$a_i = \gamma_i x_i = 1.$$

Differentiating Eq. (14) and rearranging,

$$d\left(\frac{\mu_i}{T}\right) = d\left(\frac{\mu_i^0}{T}\right) + R \, d \ln \gamma_i + R \, d \ln x_i. \tag{15}$$

Next, we utilize the Gibbs-Helmholtz equation,

$$d\left(\frac{\mu_i^0}{T}\right) = -\frac{h_i^0 \, dT}{T^2} \tag{16}$$

and the mathematical identity

$$d\left(\frac{\mu_i}{T}\right) = \frac{d\mu_i - \frac{\mu_i}{T} dT}{T}. \tag{17}$$

Substituting Eqs. (16) and (17) into Equation (15) and solving for $d\mu_i$,

$$d\mu_i = -\frac{h_i^0 dT}{T} + RT \, d \ln \gamma_i + RT \, d \ln x_i + \frac{\mu_i}{T} \, dT. \tag{18}$$

To simplify this result we recall that

$$\frac{\mu_i}{T} = \frac{\bar{h}_i}{T} - \bar{s}_i. \tag{19}$$

Substituting Eqs. (18) and (19) into (13) and using the relations

$$s = \sum_i x_i \bar{s}_i \quad \text{and} \quad \sum_i x_i \, d \ln x_i = 0,$$

we finally obtain

$$\sum_i x_i \, d \ln \gamma_i = -\frac{h^E}{RT^2} \, dT, \qquad (20)$$

where $h^E = h - \sum_i x_i h_i^0$ and h is the molar enthalpy of the mixture.
Equation (20) is the desired result. It shows that the activity coefficients of
a multicomponent system at constant pressure are related to one another
through a differential equation which includes the enthalpy of mixing.

In many cases the standard state for component i is taken as pure
liquid i at the temperature and pressure of the system. In that case h^E is
the enthalpy change which results upon mixing the pure liquids isothermally and isobarically to form the solution. However, in some cases when
one of the components is a gaseous (or solid) solute, the standard-state
fugacity for the solute is often taken to be Henry's constant evaluated at
the system temperature and pressure. In that case, for the solute,
$h_i^0 = \bar{h}_i^\infty$, the partial molar enthalpy of i in an infinitely dilute solution
at system temperature and pressure.

The isothermal, nonisobaric case

Equation (9) on a molar basis now is:

$$\sum_i x_i \, d\mu_i = v \, dP, \qquad (21)$$

where v is the molar volume of the mixture.

Again, we introduce the activity coefficient

$$\mu_i - \mu_i^0 = RT \ln \gamma_i x_i. \qquad (22)$$

Differentiating Eq. (22) at constant temperature,

$$d\mu_i = d\mu_i^0 + RT \, d \ln \gamma_i + RT \, d \ln x_i. \qquad (23)$$

In order to say something about $d\mu_i^0$ we must now distinguish
between two cases which we call Case A and Case B. These cases correspond to our choice of pressure for the standard state.

Case A

Let the standard state for component i be at the system temperature, at a fixed composition, and at some constant pressure which does not vary with composition. Then $d\mu_i^0 = 0$ and Eqs. (21) and (23) combine to

$$\sum_i x_i \, d \ln \gamma_i = \frac{v \, dP}{RT}. \qquad (24)$$

Case B

Let the standard state for component i be at the system temperature, at a fixed composition, and at the total pressure P of the system, which is not constant.but varies with composition. In this case,

$$d\mu_i^0 = v_i^0 dP, \qquad (25)$$

where v_i^0 is the molar volume of component i in its standard state.

Substitution of Eqs. (23) and (25) into Eq. (21) now gives:

$$\sum_i x_i \, d \ln \gamma_i = \frac{v^E \, dP}{RT}, \qquad (26)$$

where

$$v^E = v - \sum_i x_i v_i^0.$$

Equations (24) and (26) are the desired result. They show that the activity coefficients of a multicomponent system at constant temperature are related to one another through a differential equation which contains the volume of the liquid mixture.

Frequently the standard state is chosen as the pure liquid at the temperature and pressure of the mixture. In that case, v^E is the change in volume which results when the pure liquids are mixed at constant temperature and constant (system) pressure. An alternate application of Eq. (26) might be for a solution of a gas (or solid) in a liquid where the standard-state fugacity of the solute may be set equal to Henry's constant evaluated at system temperature and total pressure. In that case, for the solute, $v_i^0 = \bar{v}_i^\infty$, the partial molar volume of i in infinitely dilute solution at system temperature and total pressure.

APPENDIX V

Liquid-Liquid Equilibria in Binary Systems

Many pairs of liquids are only partially miscible. In this appendix to Sec. 6.15 we briefly discuss the thermodynamics of partially miscible liquid systems with particular emphasis on the relation between excess Gibbs energy and mutual solubilities.

At a certain temperature and pressure, we consider a binary system containing two liquid phases at equilibrium. Let $'$ (prime) designate one phase and let $''$ (double prime) designate the other phase. For component 1 the equation of equilibrium is

$$f_1' = f_1''. \tag{1}$$

Since both phases are liquids, it is convenient to use the same standard-state fugacity for both phases. Equation (1) can then be rewritten

$$a_1' = a_1'' \tag{2}$$

or

$$\gamma_1' x_1' = \gamma_1'' x_1'' \tag{3}$$

where x_1' and x_1'' are equilibrium mole fractions of component 1 in the two phases. Similarly, for component 2:

$$\gamma_2' x_2' = \gamma_2'' x_2'' \tag{4}$$

where x_2' and x_2'' are equilibrium mole fractions of component 2 in the two phases.

For a given binary system at a fixed temperature and pressure, we can calculate mutual solubilities if we have information concerning activity coefficients. Suppose we have such information in the form

$$g^E = \mathcal{f}(x_1, A, B, \ldots.) \tag{5}$$

473

where g^E, the molar excess Gibbs energy, is a function of composition with parameters $A, B, \ldots$ depending only on temperature (and to a lesser extent, on pressure). From Eq. (6.3-8) we can readily obtain activity coefficients of both components. Upon substitution, Eqs. (3) and (4) are of the form

$$f_1(x_1')x_1' = f_1(x_1'')x_1'' \tag{6}$$

$$f_2(x_2')x_2' = f_2(x_2'')x_2'' \tag{7}$$

where the functions f_1 and f_2 are obtained upon differentiating Eq. (5) as indicated by Eq. (6.3-8). There are two unknowns: x_2' and x_1''.† These can be found from the two equations of equilibrium, Eqs. (6) and (7). For example, suppose we use a two-suffix Margules equation [Eq. (6.5-1)] for the excess Gibbs energy in Eq. (5). We then have for our two equations of equilibrium:

$$\left[\exp \frac{A(1 - x_1')^2}{RT}\right] x_1' = \left[\exp \frac{A(1 - x_1'')^2}{RT}\right] x_1'' \tag{8}$$

$$\left[\exp \frac{A x_1'^2}{RT}\right][1 - x_1'] = \left[\exp \frac{A x_1''^2}{RT}\right] [1 - x_1'']. \tag{9}$$

For a given value of A/RT, Eqs. (8) and (9) give a solution for x_2' and x_1''. (In order that x_2' and x_1'' fall into the interval between zero and one, it is necessary that $A/RT \geq 2$.)

We have just described how mutual solubilities may be found if the excess Gibbs energy is known. Frequently, however, it is desirable to reverse the procedure (i.e., to calculate excess Gibbs energy from known mutual solubilities) because it is often a relatively simple matter to obtain mutual solubilities experimentally.

To calculate excess Gibbs energy from measured values of x_2' and x_1'', we must first choose a particular function for the excess Gibbs energy [Eq. (5)] containing no more than two unknown parameters, A and B. We can then find A and B by simultaneous solution of Eqs. (6) and (7). Once A and B are known, we can then calculate activity coefficients for both components in the two miscible regions. As a result, mutual solubility data may be used to calculate vapor-liquid equilibria.‡

Calculation of excess Gibbs energies from mutual solubility data is tedious but straightforward. The accuracy of such calculation is sensitive

---

†By stoichiometry, $x_1' + x_2' = 1$ and $x_1'' + x_2'' = 1$. If phase $'$ is rich in component 1 and phase $''$ is rich in component 2, then x_2' and x_1'' are the two mutual solubilities.

‡However, A and B are functions of temperature.

to the accuracy of the mutual solubility data but even if these are highly accurate, the results obtained are likely to be only approximate since the excess Gibbs energy function may contain no more than two parameters. Further, the results obtained depend strongly on the arbitrary algebraic function chosen to represent the excess Gibbs energy.   This sensitivity is illustrated by calculations made by P. L. T. Brian for five partially miscible binary systems shown in Table V-1. Using experimental mutual solubility data, excess Gibbs energy parameters A and B were calculated once using the van Laar equation and once using the three-suffix Margules equation (see Sec. 6.13).   With these parameters, Brian calculated activity coefficients at infinite dilution for both components. Although the same experimental data were used, results obtained with the van Laar equation differ markedly from those obtained with the Margules equation. Brian found that the differences are especially large in strongly asymmetric systems, i.e., in systems where the ratio x_1''/x_2' is far removed from unity. In the last system (propylene oxide/water), where this ratio is approximately 2, results obtained with the van Laar equation are close to those obtained with the Margules equation; however, for the system phenol/water, where the ratio is approximately 35, the limiting activity coefficient for water as obtained with the van Laar equation is several orders of magnitude larger than that obtained with the Margules equation. Brian found that when predicted limiting activity coefficients were compared with experimental results, the Margules equation frequently gave poor results.   The van Laar equation gave reasonable results but quantitative agreement with experiment was at best fair.

Table V-1

LIMITING ACTIVITY COEFFICIENTS AS CALCULATED FROM
MUTUAL SOLUBILITIES IN FIVE BINARY AQUEOUS SYSTEMS†

Component	Temp.	Solubility Limits		$\log_{10} \gamma_1^\infty$		$\log_{10} \gamma_2^\infty$	
(1)	(°C)	x_1''	x_2'	van Laar	Margules	van Laar	Margules
Aniline	100	0.01475	0.372	1.8337	1.5996	0.6076	−0.4514
iso-Butyl alcohol	90	0.0213	0.5975	1.6531	0.6193	0.4020	−3.0478
1-Butanol	90	0.0207	0.636	1.6477	0.2446	0.3672	−4.1104
Phenol	43.4	0.02105	0.7325	1.6028	−0.1408	0.2872	−8.2901
Propylene oxide	36.3	0.166	0.375	1.1103	1.0743	0.7763	0.7046

†P. L. T. Brian, *I&EC Fundamentals*, **4**, 101 (1965).

As briefly indicated in Sec. 6.14, Renon found that his NRTL equation may be useful for calculating excess Gibbs energies from mutual solubility data. The NRTL equation [Eq. (6.14-6)] contains three parameters

but one of them, the nonrandomness parameter α_{12}, can often be estimated for a given binary mixture from experimental results for other
mixtures of the same class. Once a value of α_{12} has been chosen†, the two
remaining parameters can be obtained from simultaneous solution of
Eqs. (6) and (7). The computational effort required is tedious and therefore Renon has computer-programmed the equations of equilibrium and
has presented the calculated results in graphical form as shown in Figs.
V-1, V-2 and V-3. In these figures, S and D are related to the NRTL
parameters $(g_{12} - g_{11})$ and $(g_{12} - g_{22})$ by

$$S = \frac{1}{2RT} [(g_{12} - g_{11}) + (g_{12} - g_{22})] \tag{10}$$

$$D = \frac{1}{2RT} [(g_{12} - g_{11}) - (g_{12} - g_{22})]. \tag{11}$$

Further, x_1^a is the mole fraction of component 1 in phase a and x_2^b is the
mole fraction of component 2 in phase b such that $x_1^a / x_2^b \leq 1$. To use
Figs. V-1, V-2, and V-3, one must first decide on a value for α_{12}; using
the known quantities $(x_1^a x_2^b)$ and (x_1^a / x_2^b), one may then find S
and D, which yield the parameters $(g_{12} - g_{22})$ and $(g_{12} - g_{11})$ through
Eqs. (10) and (11).

---

†For $\alpha_{12} > 0.426$ there can be no phase separation.

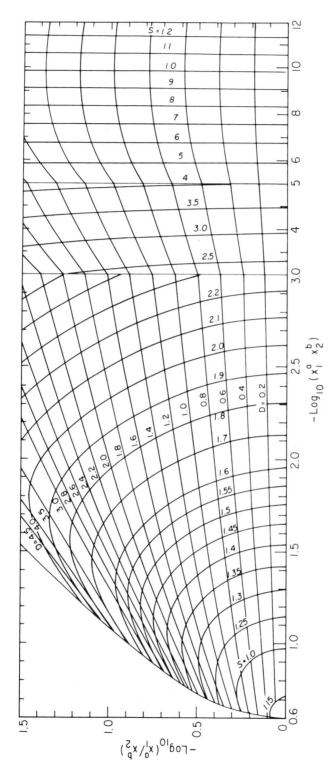

Fig. V-1   Parameters in the NRTL equation from mutual solubilities for $\alpha_{12} = 0.2$.

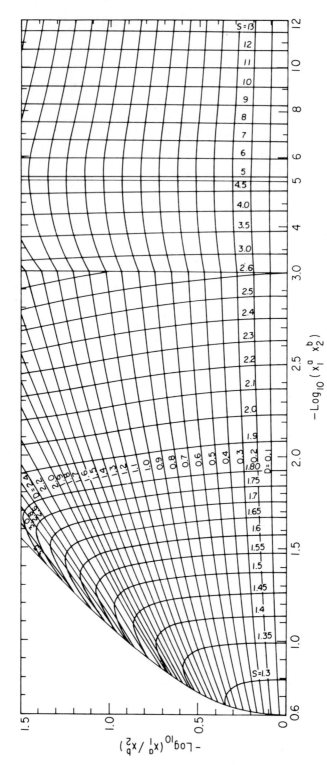

Fig. V-2   Parameters in the NRTL equation from mutual solubilities for $\alpha_{12} = 0.3$.

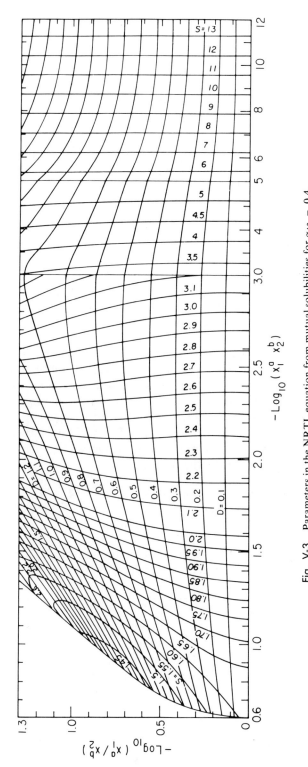

Fig. V-3   Parameters in the NRTL equation from mutual solubilities for $\alpha_{12} = 0.4$.

APPENDIX VI

Derivation of the Regular-Solution Equation
Using the Radial Distribution Function

In Sec. 7.2 we derived the equation of Scatchard and Hildebrand using a procedure similar to that used by van Laar. It was shown by Hildebrand and Wood[†] that the same equation can be derived using a more fundamental molecular procedure based on the concept of a radial distribution function. We present here the essential steps of this derivation as given by Hildebrand and Scott.[‡]

First, we must define the radial distribution function. We consider a pure liquid which consists of spherical molecules and we use a radially symmetric coordinate system whose origin is at the center of one of the molecules. From this origin we proceed in a straight line to some distance r away and we imagine a sphere of radius r drawn around the center of the central molecule where $r = 0$. On the surface of this sphere we imagine a thin spherical shell of thickness dr and we now define the radial distribution function $g(r)$ by the statement

$$\begin{matrix} \text{No. of molecular centers located in} \\ \text{the spherical shell of thickness } dr \\ \text{on the surface of a sphere of radius } r \end{matrix} = \frac{4\pi r^2 g(r)\,dr}{(V/N)} \qquad (1)$$

where V is the volume of the liquid containing N molecules. If we let $N = N_A$ (Avogadro's number), we then have:

$$\frac{V}{N} = \frac{v}{N_A}, \qquad (2)$$

where v is the molar volume of the liquid.

The function $g(r)$ has the following properties: When r is less than d, where d is the hard-core diameter of one molecule, the left side of Eq.

[†]J. H. Hildebrand and S. E. Wood, *J. Chem. Phys.*, **1**, 817 (1933).

[‡]J. H. Hildebrand and R. L. Scott, *Regular Solutions* (Englewood Cliffs, N.J.: Prentice-Hall, Inc., 1962), p. 91.

(1) must be zero, and therefore

$$\text{for } r < d, \qquad g(r) = 0. \tag{3}$$

When r is very much larger than d, the positions of the molecules in the spherical shell are no longer affected by the presence of the central molecule and the left side of Eq. (1) must be equal to the volume of the shell divided by the liquid volume per molecule. Therefore,

$$\text{for } r \gg d, \qquad g(r) = 1. \tag{4}$$

Radial distribution functions of liquids depend on temperature and density. They can be obtained from x-ray or neutron diffraction measurements. Such measurements, however, are difficult to perform and only a few reliable results have been obtained. The radial distribution function for liquid argon is shown in Fig. VI-1.

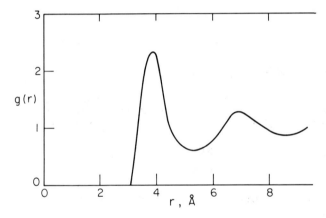

Fig. VI-1   Radial distribution function for liquid argon at 84°K from neutron diffraction data of Henshaw.

The potential energy of the liquid, relative to the ideal gas at the same temperature, is directly related to the radial distribution function and to the pair potential energy function $\Gamma(r)$. Assuming pairwise additivity of intermolecular potentials, the potential energy of one mole of pure liquid is:

$$u = \frac{2\pi N_A^2}{v} \int_0^\infty \Gamma(r) g(r) r^2 \, dr. \tag{5}$$

For a binary mixture of components 1 and 2 we have to consider three potential functions: Γ_{11} for a 1-1 pair, Γ_{22} for a 2-2 pair, and Γ_{12} for a 1-2 pair.† Similarly, we have to consider three radial distribution functions g_{11}, g_{22} and g_{12}.† If we assume that at constant temperature the two components mix without volume change, then the volume of one mole of mixture is given by

$$v = x_1 v_1 + x_2 v_2, \qquad (6)$$

where x is the mole fraction. The potential energy of the mixture is

$$u = \frac{2\pi N_A^2}{x_1 v_1 + x_2 v_2} \left[x_1 \left(x_1 \int_0^\infty \Gamma_{11} g_{11} r^2 \, dr + x_2 \int_0^\infty \Gamma_{12} g_{12} r^2 \, dr \right) \right.$$
$$\left. + x_2 \left(x_1 \int_0^\infty \Gamma_{12} g_{12} r^2 \, dr + x_2 \int_0^\infty \Gamma_{22} g_{22} r^2 \, dr \right) \right]. \qquad (7)$$

Equation (7) can be simplified by introducing the volume fractions Φ_1 and Φ_2, defined by

$$\Phi_1 \equiv \frac{x_1 v_1}{x_1 v_1 + x_2 v_2} \quad \text{and} \quad \Phi_2 \equiv \frac{x_2 v_2}{x_1 v_1 + x_2 v_2}. \qquad (8)$$

Upon substitution, Eq. (7) becomes

$$\frac{u}{2\pi N_A^2 (x_1 v_1 + x_2 v_2)} = \frac{\Phi_1^2}{v_1^2} \int_0^\infty \Gamma_{11} g_{11} r^2 \, dr + \frac{2\Phi_1 \Phi_2}{v_1 v_2} \int_0^\infty \Gamma_{12} g_{12} r^2 \, dr$$
$$+ \frac{\Phi_2^2}{v_2^2} \int_0^\infty \Gamma_{22} g_{22} r^2 \, dr. \qquad (9)$$

We want now to obtain an expression for Δu, the energy of mixing, defined by

$$\Delta u \equiv u - x_1 u_1 + x_2 u_2. \qquad (10)$$

To do so, we make several simplifying assumptions. First, we assume that the molecular arrangement in the mixture is completely random; as a result, g_{11} and g_{22} in the mixture are independent of composition and are, respectively, assumed to be equal to g_{11} and g_{22} of the pure liquids. Next, we assume that the three radial distribution functions can be superimposed by reducing the independent variable with a characteristic size parameter σ:

$$g_{11} \left(\frac{r}{\sigma_{11}} \right) = g_{22} \left(\frac{r}{\sigma_{22}} \right) = g_{12} \left(\frac{r}{\sigma_{12}} \right) = g(y). \qquad (11)$$

† Because of symmetry $\Gamma_{12} = \Gamma_{21}$ and $g_{12} = g_{21}$.

Finally, we assume that the three potential functions follow a two-parameter law of corresponding states (see Sec. 4.6):

$$\frac{\Gamma_{11}}{\epsilon_{11}} = \mathcal{f}\left(\frac{r}{\sigma_{11}}\right), \qquad \frac{\Gamma_{22}}{\epsilon_{22}} = \mathcal{f}\left(\frac{r}{\sigma_{22}}\right), \qquad \frac{\Gamma_{12}}{\epsilon_{12}} = \mathcal{f}\left(\frac{r}{\sigma_{12}}\right). \tag{12}$$

From Eq. (9) we then obtain

$$\frac{\Delta u}{2\pi N_A^2 (x_1 v_1 + x_2 v_2)} = \Phi_1 \Phi_2 \int_0^\infty f(y)g(y)y^2\, dy \left[\frac{2\epsilon_{12}\,\sigma_{12}^3}{v_1 v_2}\right.$$

$$\left. - \frac{\epsilon_{11}\,\sigma_{11}^3}{v_1^2} - \frac{\epsilon_{22}\,\sigma_{22}^3}{v_2^2}\right], \tag{13}$$

where

$$y = \frac{r}{\sigma}.$$

The similarity between Eq. (13) and Eq. (7.2-8) becomes evident if we write for the three cohesive-energy densities:

$$c_{11} = 2\pi N_A^2 \frac{\epsilon_{11}\,\sigma_{11}^3}{v_1^2} \int_0^\infty f(y)g(y)y^2\, dy, \tag{14}$$

$$c_{22} = 2\pi N_A^2 \frac{\epsilon_{22}\,\sigma_{22}^3}{v_2^2} \int_0^\infty f(y)g(y)y^2\, dy, \tag{15}$$

and

$$c_{12} = 2\pi N_A^2 \frac{\epsilon_{12}\,\sigma_{12}^3}{v_1 v_2} \int_0^\infty f(y)g(y)y^2\, dy. \tag{16}$$

To obtain the final regular-solution equation in terms of solubility parameters, two further assumptions are required:

1. The excess entropy is zero:

$$s^E = 0. \tag{17}$$

2. The cohesive-energy density for the 1-2 interaction is given by the geometric mean of those for the 1-1 and 2-2 interactions:

$$c_{12} = (c_{11} c_{22})^{1/2}. \tag{18}$$

Using Eq. (6.3-8) we then obtain, finally:

$$RT \ln \gamma_1 = v_1 \Phi_2^2 (\delta_1 - \delta_2)^2, \tag{19}$$

$$RT \ln \gamma_2 = v_2 \Phi_1^2 (\delta_1 - \delta_2)^2, \tag{20}$$

where the solubility parameter δ is the square root of the cohesive-energy density.

The derivation given here is considerably longer and more detailed than that given in Sec. 7.2. The advantage of a detailed, molecular derivation lies in its ability to point out clearly the simplifying assumptions about molecular behavior which are required to obtain a useful result. As indicated above, many severe assumptions are needed in order to derive the final regular-solution equations. It is therefore surprising that these equations work as well as they do. It is likely that the success of the regular-solution treatment is due to a considerable amount of fortuitous cancellation of those errors which are introduced by the many simplifying assumptions.

APPENDIX VII

Solubility of Gases in Mixed Solvents

Solubility data for gases in liquid mixtures are rare and it is therefore important to consider how to utilize thermodynamic relations to calculate such solubilities from solubility data in single solvents. If we know at some temperature T the solubility of gaseous solute r in solvent i, and in solvent j, and in solvent k, etc., we would like to predict the solubility of r in a mixture of the solvents. We seek an expression for Henry's constant of solute r in the solvent mixture; we designate this by $H_{r,M}$. To obtain the desired expression, we outline a procedure given by Chueh[1] which in principle is similar to that of O'Connell[2] and of Kehiaian.[3]

For the liquid mixture containing m components (the gaseous solute is one of the components) we expand the molar excess Gibbs energy in a manner similar to that of Wohl (see Sec. 6.13):

$$\frac{g_{1\ldots m}^E}{RT\sum\limits_{i=1}^{m} x_i q_i} = \sum_{i=1}^{m} \sum_{j=1}^{m} A_{ij} z_i z_j + \sum_{i=1}^{m} \sum_{j=1}^{m} \sum_{k=1}^{m} A_{ijk} z_i z_j z_k + \cdots,$$

(1)

where x_i is the mole fraction of i and where z_i is the effective volume fraction of i defined by

$$z_i \equiv \frac{x_i q_i}{\sum\limits_{i=1}^{m} x_i q_i},$$

and where q_i is the effective molecular size (or cross-section) of molecule i.

The constant A_{ij} is a two-body coefficient characteristic of the i-j interaction; the constant A_{ijk} is a three-body coefficient characteristic of the i-j-k interaction and so on; these coefficients depend on the temperature and on the nature of the components. The expansion in Eq. (1) has been terminated after the three-body term. In Eq. (1) the excess Gibbs energy is based on the symmetric convention for normalization of activity coefficients (see Sec. 6.4). Since $g^E \to 0$ as $x_i \to 1$ for every i, it follows that

$$A_{ii} = A_{jj} = \ldots = A_{iii} = A_{jjj} = \ldots = 0.$$

For gaseous solute r in the liquid mixture, the activity coefficient $\gamma_{r,M}$ can be found from Eq. (1) by differentiation according to Eq. (6.3-8):

$$\ln \gamma_{r,M} = \frac{1}{RT} \left(\frac{\partial n_T g^E_{1 \ldots m}}{\partial n_r} \right)_{T,P,n_{i(i \neq r)}}, \tag{2}$$

where n_r is the number of moles of r and n_T is the total number of moles. The differentiation is straightforward but tedious and we need not repeat it here. We merely indicate the result in the form

$$\ln \gamma_{r,M} = q_r \mathcal{f}(A_{ij}\text{'s}, A_{ijk}\text{'s}, z\text{'s}). \tag{3}$$

From Eq. (3) we can readily obtain an expression for $\ln \gamma_{r,i}$ in a single solvent i, or for $\ln \gamma_{r,j}$ in a single solvent j, etc. Since we want to calculate Henry's constant $H_{r,M}$ in terms of Henry's constants in the single solvents $H_{r,i}$, $H_{r,j}$, etc., we use the relation

$$H_{r,M} \equiv \lim_{x_r \to 0} \frac{f_{r,M}}{x_r} = \lim_{x_r \to 0} (\gamma_{r,M})(f^L_{\text{pure } r}) \tag{4}$$

or

$$\ln H_{r,M} = \lim_{x_r \to 0} (\ln \gamma_{r,M}) + \ln f^L_{\text{pure } r}. \tag{5}$$

Similarly,

$$\ln H_{r,i} = \lim_{x_r \to 0} (\ln \gamma_{r,i}) + \ln f^L_{\text{pure } r} \tag{6}$$

and

$$\ln H_{r,j} = \lim_{x_r \to 0} (\ln \gamma_{r,j}) + \ln f^L_{\text{pure } r}, \tag{7}$$

etc., where $f^L_{\text{pure } r}$ is the fugacity of pure liquid r at the system temperature and at some reference pressure (see Sec. 10.4). At the system temperature, $f^L_{\text{pure } r}$ may be a hypothetical quantity but its numerical value need not concern us; as shown below, it cancels out in the final expression.

Let us now multiply Eq. (6) by z_i, Eq. (7) by z_j, etc. Adding all of these equations and substituting Eq. (5) we find that in the limit, as $x_r \to 0$,

$$\ln H_{r,M} = \sum_{\substack{i=1 \\ (i \neq r)}}^{m} z_i \ln H_{r,i} - \sum_{i=1}^{m} z_i \lim_{x_r \to 0} (\ln \gamma_{r,i})$$

$$+ \lim_{x_r \to 0} (\ln \gamma_{r,M}). \tag{8}$$

From Eq. (3) we can obtain the limiting activity coefficient of solute r in the solvent mixture as $x_r \to 0$. Equation (3) may also be used to find the limiting activity coefficient of solute r in a single solvent i or j, etc.

Substituting Eq. (3) into Eq. (8) we find after much algebraic rearrangement that

$$\ln H_{r,M} = \sum_{\substack{i=1 \\ (i \neq r)}}^{m} z_i \ln H_{r,i} - \frac{q_r}{2} \sum_{\substack{i=1 \\ (i,j \neq r)}}^{m} \sum_{j=1}^{m} \frac{g_{ij}^E}{RT(x_i q_i + x_j q_j)} - q_r \Delta, \qquad (9)$$

where g_{ij}^E is the excess Gibbs energy of the binary i-j solvent mixture at the temperature of interest and at a composition where the ratio of x_i to x_j is the same as that in the solute-free solvent mixture.† The quantity Δ depends on the three-body coefficients according to

$$\Delta = 2 \sum_{\substack{i=1 \\ (i,j,k \neq r)}}^{m} \sum_{j=1}^{m} \sum_{k=1}^{m} A_{ijk} z_i z_j z_k - 3 \sum_{\substack{i=1 \\ (i,j \neq r)}}^{m} \sum_{j=1}^{m} A_{ijr} z_i z_j$$

$$+ 3 \sum_{\substack{i=1 \\ (i \neq r)}}^{m} A_{iir} z_i . \qquad (10)$$

If we terminate Wohl's expansion [Eq. (1)] after the two-body terms, then $\Delta = 0$. In that event we can compute $H_{r,M}$ using only binary data. However, if we retain three-body terms, then unless we make further simplifying assumptions, we require data for ternary mixtures in order to calculate $H_{r,M}$.

For many simple, nonpolar mixtures a sufficiently good approximation may be provided by setting $\Delta = 0$. In such cases, it is often advantageous to reduce the number of adjustable parameters by setting $q_r = v_{c_r}, q_i = v_{c_i}$, etc., where v_c is the critical volume.

Further simplification may be achieved by letting $q_r = q_i \ldots = 1$ in addition to assuming $\Delta = 0$. In that case, Eq. (9) reduces to

$$\ln H_{r,M} = \sum_{\substack{i=1 \\ (i \neq r)}}^{m} x_i \ln H_{r,i} - \frac{1}{2} \sum_{\substack{i=1 \\ (i,j \neq r)}}^{m} \sum_{j=1}^{m} A_{ij} x_i x_j . \qquad (11)$$

Finally, if all the solvent pairs form ideal solutions, all A_{ij}'s for all solvent pairs vanish; in that event the logarithm of Henry's constant in the

† g_{ij}^E stands for quadratic terms in volume fraction only.

solvent mixture is a mole fraction average of the logarithms of Henry's constants in the individual solvents.

REFERENCES

1. Chueh, P. L., Dissertation, University of California, Berkeley, 1967.

2. O'Connell, J. P., and J. M. Prausnitz, *I&EC Fundamentals*, **3**, 347 (1964).

3. Kehiaian, H., *Bull. Acad. Polon. Sci., Ser. Sci. Chim.*, **12**, 323 (1964).

APPENDIX VIII

Vapor-Liquid Critical Properties of Mixtures

In a multicomponent mixture, the vapor-liquid critical conditions are those where, at equilibrium, the vapor phase and the liquid phase become identical. For a pure component, the critical point is coincident with the maximum temperature and maximum pressure at which the liquid phase can exist, but this coincidence does not hold for mixtures.

Experimental critical properties for mixtures have been correlated by Chueh.† Almost all mixture critical data are for binary mixtures and most of these are for mixtures of nonpolar (or slightly polar) components.

Critical temperatures and critical volumes

Chueh found that to a good approximation, the (true) critical temperature T_{c_M} and the (true) critical molar volume v_{c_M} are quadratic functions of the composition expressed in terms of the surface fraction θ. For a binary mixture

$$T_{c_M} = \theta_1 T_{c_1} + \theta_2 T_{c_2} + 2\theta_1 \theta_2 \tau_{12} \tag{1}$$

$$v_{c_M} = \theta_1 v_{c_1} + \theta_2 v_{c_2} + 2\theta_1 \theta_2 \nu_{12} \tag{2}$$

where the surface fraction is defined by

$$\theta_i \equiv \frac{x_i v_{c_i}^{2/3}}{\sum\limits_i x_i v_{c_i}^{2/3}} \tag{3}$$

where x_i is the mole fraction of component i. In Eq. (1), T_{c_1} and T_{c_2} are, respectively, the critical temperatures of pure components 1 and 2 and in Eqs. (2) and (3) v_{c_1} and v_{c_2} are, respectively, the critical molar volumes of pure components 1 and 2. The binary parameters τ_{12} and ν_{12} characterize the intermolecular forces between dissimilar molecules; they are found from binary data.

†P. L. Chueh and J. M. Prausnitz, *A. I. Ch. E. Journal*, **13**, 1107 (1967).

Table VIII-1

COMPARISON OF CALCULATED AND EXPERIMENTAL CRITICAL TEMPERATURES
OF BINARY SYSTEMS AND THEIR REDUCED CORRELATING PARAMETERS†

System	Avg. Dev. (%)	$2\tau_{12}/(T_{c_1} + T_{c_2})$
Methane/argon	0.05	0.0044
Methane/nitrogen	0.33	0.0198
Methane/oxygen	0.51	-0.0400
Methane/propane	1.39	0.1237
Methane/propane	0.28	0.1410
Methane/propane	0.39	0.1775
Methane/n-butane	0.81	0.1826
Methane/isobutane	0.45	0.1444
Methane/n-pentane	0.73	0.2378
Methane/isopentane	0.02	0.1953
Methane/n-heptane	3.39	0.2773
Acetylene/ethane	0.20	-0.0866
Acetylene/ethylene	0.84	-0.0545
Acetylene/propane	0.62	-0.0468
Acetylene/propylene	0.17	-0.0304
Ethane/propane	0.13	0.0211
Ethane/propylene	0.24	-0.0078
Ethane/n-butane	0.13	0.0267
Ethane/n-pentane	0.73	0.0438
Ethane/cyclohexane	0.47	0.0695
Ethane/n-heptane	0.61	0.0743
Ethylene/ethane	0.17	-0.0006
Ethylene/propylene	0.14	0.0241
Ethylene/n-heptane	0.69	0.0799
Propane/n-butane	0.12	0.0144
Propane/n-pentane	0.14	0.0092
Propane/isopentane	0.06	0.0088
n-Butane/nitrogen	1.80	0.3500
n-Butane/n-heptane	0.03	0.0192
n-Pentane/neopentane	0.02	0.0038
n-Pentane/n-hexane	0.06	0.0031
n-Pentane/cyclohexane	0.03	0.0201
n-Pentane/n-hexane	0.05	0.0076
Neopentane/n-hexane	0.09	0.0064
Neopentane/cyclohexane	0.05	0.0047
n-Hexane/cyclohexane	0.03	0.0013
Benzene/ethane	0.82	0.0526
Benzene/propane	1.16	0.0264
Benzene/n-pentane	0.71	-0.0066
Benzene/neopentane	0.44	-0.0258
Benzene/n-hexane	0.14	-0.0182
Benzene/cyclohexane	0.01	-0.0128
Benzene/toluene	0.03	0.0008
Toluene/n-pentane	0.14	-0.0302
Toluene/n-hexane	0.09	-0.0028

†P. L. Chueh and J. M. Prausnitz, *A. I. Ch. E. Journal*, **13**, 1107 (1967).

Table VIII-1—*Cont.*

System†	Avg. Dev. (%)	$2\tau_{12}/(T_{c_1} + T_{c_2})$
Toluene/cyclohexane	0.04	−0.0061
Carbon dioxide/methane	1.61	0.0472
Carbon dioxide/ethane	0.10	−0.0911
Carbon dioxide/propane	0.99	−0.0573
Carbon dioxide/propane	0.67	−0.0693
Carbon dioxide/n-butane	0.91	−0.0313
Carbon dioxide/n-butane	0.74	−0.0707
Carbon dioxide/n-pentane	2.42	0.0156
Carbon monoxide/argon	0.13	−0.0015
Carbon monoxide/oxygen	0.07	−0.0005
Carbon monoxide/nitrogen	0.06	−0.0054
Carbon monoxide/methane	0.16	0.0220
Carbon monoxide/propane	0.20	0.3560
Hydrogen sulfide/methane	0.84	0.0577
Hydrogen sulfide/ethane	0.36	−0.0683
Hydrogen sulfide/propane	0.04	−0.0748
Hydrogen sulfide/carbon dioxide	0.14	−0.0666
Nitrogen/argon	0.08	0.0098
Nitrogen/oxygen	0.05	0.0163
Argon/oxygen	0.03	−0.0090

†The hydrogen sulfide n-pentane system has been correlated with an average deviation of 1.75% and a reduced correlating parameter of 0.0168. This parameter, however, is in disagreement with the trend of other hydrogen sulfide/paraffin systems shown in Fig. VIII-2.

Table VIII-2

COMPARISON OF CALCULATED AND EXPERIMENTAL CRITICAL VOLUMES OF BINARY SYSTEMS AND THEIR REDUCED CORRELATING PARAMETERS†

System	Avg. Dev. (%)	$2\nu_{12}/(v_{c_1} + v_{c_2})$
Methane/propane	1.9	−0.3653
Methane/n-butane	1.4	−0.6975
Methane/isobutane	0.3	−0.6503
Methane/n-pentane	3.4	−0.7153
Methane/n-heptane	5.9	−0.9808
Ethane/propylene	0.4	−0.1057
Ethane/n-butane	0.8	−0.2753
Ethane/n-pentane	1.0	−0.5250
Ethane/cyclohexane	2.9	−0.5931
Ethane/n-heptane	3.9	−0.6826
Ethylene/n-heptane	4.8	−0.8327
Propane/n-butane	1.3	−0.0061
Propane/isopentane	1.1	−0.2991
n-Butane/n-heptane	1.9	−0.3042
Carbon dioxide/propane	3.0	−0.3418

†P. L. Chueh and J. M. Prausnitz, *A. I. Ch. E. Journal*, **13**, 1107 (1967).

Table VIII-2—*Cont.*

System	Avg. Dev. (%)	$2v_{12}/(v_{c_1} + v_{c_2})$
Carbon dioxide/n-butane	1.0	-0.4513
Carbon dioxide/hydrogen sulfide	0.8	-0.0760
Hydrogen sulfide/methane	2.7	-0.6063
Hydrogen sulfide/ethane	0.5	-0.1279
Hydrogen sulfide/propane	1.0	-0.1746
Hydrogen sulfide/n-pentane	2.1	-0.5030
Benzene/ethane	3.8	-0.5588
Toluene/n-hexane	0.0	-0.1141
Nitrogen/methane‡	–	-0.07
Nitrogen/n-butane‡	–	-0.95

‡No critical volumes of mixtures available for these systems. Values of v_{12} are back-calculated from critical pressures of the mixture.

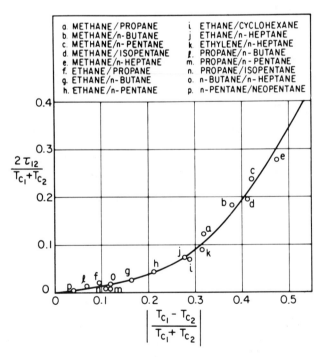

Fig. VIII-1   Correlating parameter τ_{12} for critical temperatures of some binary systems containing saturated hydrocarbons.

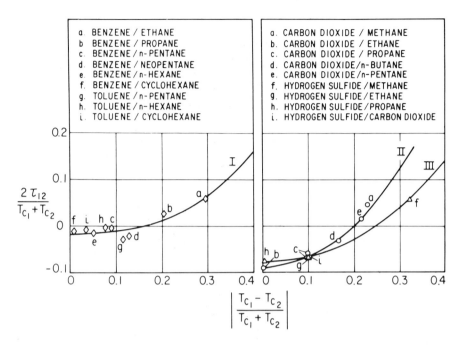

Fig. VIII-2   Correlating parameter τ_{12} for critical temperatures of some binary systems:
I-aromatic/paraffin;   II-carbon   dioxide/paraffin;   III-hydrogen   sulfide/
paraffin.

Table VIII-1 presents results for τ_{12} and Table VIII-2 presents results
for ν_{12}.† Also given are average deviations between experimental results
and those calculated with Eqs. (1) and (2) using the indicated values of
τ_{12} and ν_{12}. Within a chemical family, these parameters show systematic
trends as indicated by Figs. VIII-1, VIII-2, and VIII-3. As a result,
Chueh's correlation may be used to make estimates of critical temperatures and critical volumes for some systems that have not been studied
experimentally.

Critical pressures

Having correlated critical temperatures and critical volumes with
quadratic functions of the surface fraction, it is tempting to try a similar
correlation for the critical pressure. Such a correlation was not successful.
The dependence of the critical pressure on composition is much more

†Additional results are given by L. M. Schick and J. M. Prausnitz, *A. I. Ch. E. Journal*,
14, 673 (1968).

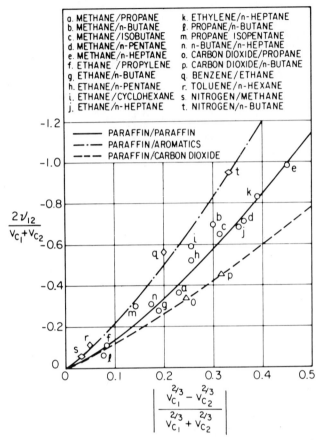

Fig. VIII-3   Correlating parameter ν_{12} for critical volumes of some binary systems.

strongly nonlinear than that of the critical temperature and the critical volume; in many systems, a plot of critical pressure versus mole fraction shows a sharp maximum and a point of inflection. The more complicated behavior of the critical pressure follows from its nonfundamental nature; subject to well-defined assumptions, both critical temperatures and critical volumes can be related directly to the intermolecular potential, but the critical pressure can be related to the intermolecular potential only indirectly through the critical temperature and critical volume.

To express the critical pressure as a function of composition, the correlations for critical temperature and critical volume can be coupled with an equation of state. For this purpose we use the Redlich-Kwong equation of state (see Sec. 5.13) with certain alterations. The Redlich-Kwong equation is

$$P = \frac{RT}{v - b} - \frac{a}{T^{1/2} v(v + b)} \cdot \tag{4}$$

For a pure component, the constants a and b are related to the critical temperature and pressure of that component by

$$a = \frac{\Omega_a R^2 T_c^{2.5}}{P_c} \tag{5}$$

$$b = \frac{\Omega_b R T_c}{P_c} \cdot \tag{6}$$

The dimensionless constants Ω_a and Ω_b may be found (as is commonly done) by equating to zero the first two isothermal derivatives of pressure with respect to volume at the critical point. This procedure gives $\Omega_a = 0.4278$ and $\Omega_b = 0.0867$. To do so, however, puts a severe strain on the equation of state, leading to a value of z_c (compressibility factor at the critical point) which is too large. Since any two-parameter equation of state is necessarily approximate when applied to a wide range of temperature and density, it is best to determine the dimensionless parameters Ω_a and Ω_b from experimental data available in the region of temperature and density where the equation of state is to be used. In Secs. 5.13 and 10.5 we evaluated the parameter Ω_b for a variety of fluids from pure-component volumetric data, once for saturated liquids and once for saturated vapors. For our present purpose, we use for Ω_b for each substance the arithmetic mean of the two values obtained from saturated liquid and saturated vapor volumes. For a variety of normal fluids, this averaged Ω_b may be represented by a function of the acentric factor, ω:

$$\Omega_b = 0.0867 - 0.0125\omega + 0.011\omega^2 \quad (0 \le \omega < 0.6) . \tag{7}$$

To force agreement for each pure component at the critical point, Ω_a is determined by the experimental critical temperature, pressure and volume of that component according to

$$\Omega_a = \left(\frac{RT_c}{v_c - b} - P_c \right) \frac{P_c v_c (v_c + b)}{(RT_c)^2} \tag{8}$$

where b is given by Eqs. (6) and (7).

To apply the Redlich-Kwong equation to mixtures, we require mixing rules for a and b. We propose, as in Secs. 5.13 and 10.5:

$$a = \sum_i \sum_j x_i x_j a_{ij} \quad (a_{ij} \ne \sqrt{a_{ii} a_{jj}}) \tag{9}$$

$$b = \sum_i x_i b_i \tag{10}$$

where

$$a_{ii} = \frac{\Omega_{a_i} R^2 T_{c_i}^{2.5}}{P_{c_i}} \tag{11}$$

$$b_i = \frac{\Omega_{b_i} R T_{c_i}}{P_{c_i}} \tag{12}$$

$$a_{ij} = \frac{\frac{1}{4}(\Omega_{a_i} + \Omega_{a_j}) R T_{c_{ij}}^{1.5} (v_{c_i} + v_{c_j})}{0.291 - 0.04(\omega_i + \omega_j)} \tag{13}$$

$$T_{c_{ij}} = \sqrt{T_{c_{ii}} T_{c_{jj}}} (1 - k_{ij}). \tag{14}$$

The constant k_{ij} is a small number (usually positive and of the order 10^{-2} or 10^{-1}) which is characteristic of the *i-j* interaction. To a good approximation, it is independent of temperature, density and composition; we use here the k_{ij}'s given in Table 5-10 (p. 158).

Using the equation of Redlich and Kwong together with the previously established correlations for critical temperatures and critical volumes, critical pressures were calculated and compared with experimental results

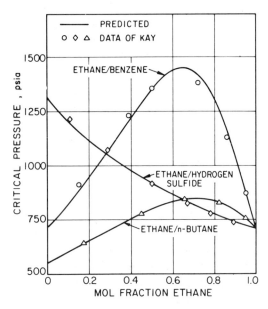

Fig. VIII-4 Critical pressures of three binary systems containing ethane.

for 36 systems; the mean of the average deviations was 3.6%.† In these calculations, critical temperatures and volumes of mixtures were obtained from Eqs. (1) and (2); experimental critical temperatures and volumes of mixtures were not used directly in the Redlich-Kwong equation.

Typical results are shown in Fig. VIII-4 for three binary systems containing ethane and in Fig. VIII-5 for three binary systems containing

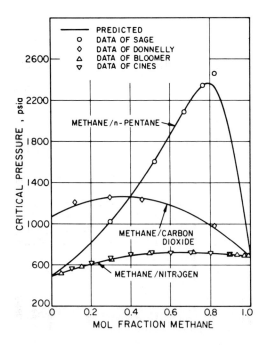

Fig. VIII-5   Critical pressures of three binary systems containing methane.

methane. The system ethane/hydrogen sulfide is unusual because unlike the behavior of most systems, the critical pressures fall below a straight line joining the pure-component critical pressures.

To illustrate the importance of k_{ij}, Fig. VIII-6 gives critical pressures for the n-butane/nitrogen system. Experimental results are compared with two sets of calculations; in one set, k_{ij} was zero and in the other it was 0.12 as found from second virial coefficient data. Figure VIII-6 shows that marked improvements can be obtained when corrections are applied to the (rough) rule that the temperature characteristic of the 1-2 interaction is given by the geometric mean of the pure-component critical temperatures.

† A. I. Ch. E. Journal, **13**, 1107 (1967).

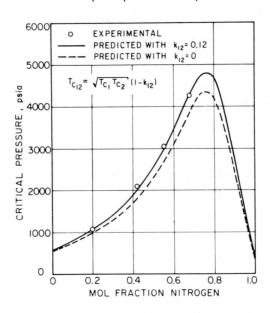

Fig. VIII-6   Effect of correction to geometric mean on predicted critical pressures of the nitrogen/n-butane system. ($k_{12} = 0.12$ obtained from second-virial-coefficient data.)

Multicomponent systems

Equations (1) and (2) are readily generalized to mixtures containing any number of components. The generalized equations are

$$T_{c_M} = \sum_i \theta_i T_{c_i} + \sum_i \sum_j \theta_i \theta_j \tau_{ij} \tag{15}$$

$$v_{c_M} = \sum_i \theta_i v_{c_i} + \sum_i \sum_j \theta_i \theta_j v_{ij} \tag{16}$$

where

$$\tau_{ii} = v_{ii} = 0.$$

The critical pressure of a multicomponent mixture is found from the equation of state, Eq. (4) with the mixing rules given by Eqs. (9) to (14).

For systems containing more than two components, directly measured critical temperatures and critical pressures are scarce, and directly measured critical volumes have not been reported at all. Critical constants obtained by extrapolation of vapor-liquid equilibrium (K factor) data are generally not reliable and in some cases may lead to considerable error. Using only directly measured experimental results, calculated and observed critical temperatures and critical pressures have been compared for six ternary systems, two quaternary systems, and two quinary systems. The average deviation for the critical temperature was 0.4% and that for the critical pressure 4.3%.[†] It appears, therefore, that the accuracy for calculating critical constants of multicomponent systems is close to that for calculating critical constants of binary mixtures.

[†] *A. I. Ch. E. Journal*, **13**, 1107 (1967).

APPENDIX IX

Thermodynamic Consistency Test for Binary
High-Pressure Vapor-Liquid Equilibrium Data

Vapor-liquid equilibrium data are said to be thermodynamically consistent when they satisfy the Gibbs-Duhem equation. When the data satisfy this equation, it is likely, but by no means guaranteed, that they are correct; however, if they do not satisfy this equation, it is certain that they are incorrect.

Thermodynamic consistency tests for binary vapor-liquid equilibria at low pressures have been described in Chap. 6. Extension of these tests to isothermal high-pressure equilibria presents two difficulties: First, it is necessary to have experimental data for the density of the liquid mixture along the saturation line; and second, since the ideal-gas law is not valid, it is necessary to calculate vapor-phase fugacity coefficients either from volumetric data for the vapor mixture or else from an equation of state.

A consistency test described by Chueh and Muirbrook[1] extends to isothermal high-pressure data the integral (area) test given by Redlich and Kister[2] and Herington[3] for isothermal low-pressure data. (A similar extension has been given by Thompson and Edmister.[4]) For a binary system at constant temperature, the Gibbs-Duhem equation is written

$$x_1 d \ln f_1 + x_2 d \ln f_2 = \frac{v^L dP}{RT}. \tag{1}$$

When one uses the identity $x_1 d \ln x_1 + x_2 d \ln x_2 = 0$, Eq. (1) can be rearranged to read

$$\ln \frac{f_2/x_2}{f_1/x_1} dx_2 + \frac{v^L dP}{RT} = d \left(\ln \frac{f_1}{x_1} + x_2 \ln \frac{f_2/x_2}{f_1/x_1} \right). \tag{2}$$

Introducing fugacity coefficients φ and K factors ($K_i = y_i/x_i$) into Eq. (2), one obtains

$$\left(\ln \frac{\varphi_2}{\varphi_1} + \ln \frac{K_2}{K_1} \right) dx_2 + \frac{v^L dP}{RT} = d \left[\ln K_1 + \ln \varphi_1 P + x_2 \left(\ln \frac{\varphi_2}{\varphi_1} + \ln \frac{K_2}{K_1} \right) \right]. \tag{3}$$

In Eqs. (1), (2), and (3), subscript 2 refers to the light component. Equation (3) is integrated from $x_2 = 0$ to some arbitrary upper limit x_2. The following boundary condition applies:

$$\text{when} \quad x_2 = 0: \quad \varphi_1 = \varphi_1^s; \quad P = P_1^s; \quad K_1 = 1$$

where superscript s denotes saturation.

The integrated form of Eq. (3) can be most conveniently written as

Area I + Area II + Area III =

$$\left[\ln K_1 + \ln \frac{\varphi_1 P}{\varphi_1^s P_1^s} + x_2 \left(\ln \frac{\varphi_2}{\varphi_1} + \ln \frac{K_2}{K_1} \right) \right] \quad \text{(at } x_2) \quad (4)$$

where

$$\text{Area I} = \int_{x_2=0}^{x_2} \ln \frac{K_2}{K_1} \, dx_2 \tag{5}$$

$$\text{Area II} = \int_{x_2=0}^{x_2} \ln \frac{\varphi_2}{\varphi_1} \, dx_2 \tag{6}$$

$$\text{Area III} = \frac{1}{RT} \int_{x_2=0}^{x_2} v^L \, dP. \tag{7}$$

The three areas are found by graphical integration. The thermodynamic consistency test consists of comparing the sum of the three areas [left-hand side of Eq. (4)] with the right-hand side of Eq. (4). The three areas depend upon equilibrium data for the composition range $x_2 = 0$ to $x_2 = x_2$. However, the right-hand side of Eq. (4) depends only on equilibrium data at the upper limit $x_2 = x_2$. The comparison indicated by Eq. (4) should be made for several values of x_2 up to and including the critical composition.

To illustrate this thermodynamic consistency test, Figs. IX-1, IX-2 and IX-3 show plots of the appropriate functions needed to calculate Areas I, II, and III for the nitrogen/carbon dioxide system at $0°C$; the data are taken from Muirbrook.[5] Fugacity coefficients were calculated with the modified Redlich-Kwong equation.[6]

A comparison of the left-hand side (LHS) and the right-hand side (RHS) of Eq. (4) is given in Table IX-1. Comparison is made at three different values of x_2 including the critical point. In order to assess their relative importance, values of all the individual terms in Eq. (4) are reported in Table IX-1. It is apparent that all the terms contribute significantly and that none may be neglected except that $\ln K_1$ must necessarily vanish at the critical mole fraction.

The final column in Table IX-1 reports the absolute value of the difference between LHS and RHS divided by their arithmetic mean. For the

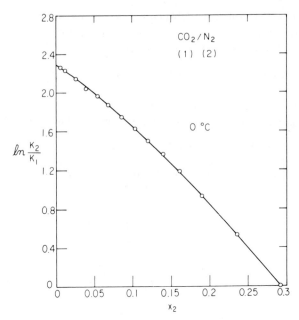

Fig. IX-1   First area in thermodynamic consistency test.

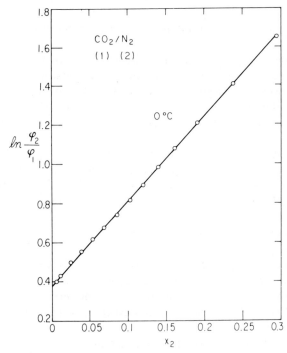

Fig. IX-2   Second area in thermodynamic consistency test.

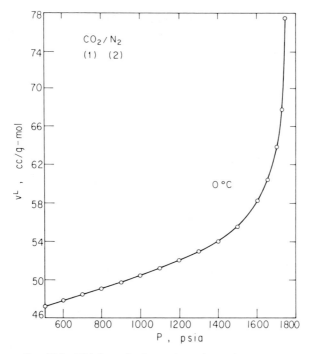

Fig. IX-3   Third area in thermodynamic consistency test.

Table IX-1

THERMODYNAMIC CONSISTENCY TEST.
CARBON DIOXIDE (1)/NITROGEN (2) AT $0°C$

x_2	Area I	Area II	Area III	LHS	$\ln K_1$	$\ln \dfrac{\varphi_1 P}{\varphi_1^s P_1^s}$	$x_2(\ln K_2/K_1 + \ln \varphi_2/\varphi_1)$	RHS	% Inconsistency
0.1030	0.2024	0.0618	0.1042	0.369	−0.3496	0.4855	0.2505	0.386	4.5
0.1902	0.3142	0.1500	0.1703	0.635	−0.2562	0.5044	0.404	0.652	2.6
0.2926^c	0.3618	0.2965	0.1968	0.855	0	0.3974	0.482	0.879	2.8

c = critical
LHS = left-hand side of Eq. (4)
RHS = right-hand side of Eq. (4)

system considered here, the percent inconsistency is always less than about 5%. Considering the uncertainties in the fugacity coefficients (resulting from a good but still approximate equation of state), the thermodynamic consistency of these data is fairly high.

The thermodynamic consistency test for binary systems described above can be extended to ternary (and higher) systems with techniques similar to those described by Herington.[7] The necessary calculations are tedious, and unless extensive multicomponent data are available, they are usually not worthwhile.

REFERENCES

1. Chueh, P. L., N. K. Muirbook, and J. M. Prausnitz, *A. I. Ch. E. Journal*, **11**, 1097 (1965).

2. Redlich, O., and A. T. Kister, *Ind. Eng. Chem.*, **40**, 345 (1948).

3. Herington, E. F. G., *Nature*, **160**, 610 (1947).

4. Thompson, R. E., and W. C. Edmister, *A. I. Ch. E. Journal*, **11**, 457 (1965).

5. Muirbrook, N. K., and J. M. Prausnitz, *A. I. Ch. E. Journal*, **11**, 1092 (1965).

6. Redlich, O., F. J. Ackerman, R. D. Gunn, M. Jacobson, and S. Lau, *I&EC Fundamentals*, **4**, 369 (1965).

7. Herington, E. F. G., *J. Appl. Chem.*, **2**, 11 (1952).

Formula Index

The arrangement of the systems is in the alphabetical order of the chemical formula. The elements in the inorganic compounds are written in the conventional order (e.g., HCl rather than ClH), while two identical atoms are taken as one (e.g., H_2 precedes HCl); the isotopes are given directly under the stable species (e.g., D_2 comes after H_2). The metal organic compounds are given under the symbol for the metal. The elements in the organic compounds are given in the order C, H, Br, Cl, F, I, O, N, S; all C_1 compounds precede C_2, etc., while C_n compounds with no hydrogen precede C_nH_1, etc. Entries with the generic name "hydrocarbons" appear at the end of the carbon compounds. The polymers are given at the end, with no chemical formula. Binary systems precede ternary systems.

505

Subject Index